T0215855

The Politics of Species

Reshaping our Relationships with Other Animals

The assumption that humans are cognitively and morally superior to other animals is fundamental to social democracies and legal systems worldwide. It legitimizes treating members of other animal species as inferior to humans. The last few decades have seen a growing awareness of this issue, as evidence continues to show that individuals of many other species have rich mental, emotional, and social lives.

Bringing together leading experts from a range of disciplines, this volume identifies the key barriers to a definition of moral respect that includes non-human animals. It sets out to increase concern, empathy, and inclusiveness by developing strategies that can be used to protect other animals from exploitation in the wild and from suffering in captivity. The chapters link scientific data with normative and philosophical reflections, offering unique insight into controversial issues around the ethical, political, and legal status of other species.

Raymond Corbey, a philosopher and anthropologist, is an associate professor at Tilburg University and holds an endowed chair at Leiden University, both in the Netherlands. He has a keen interest in animal cognition and human–animal relations in various settings, ranging from hominin evolution and extant foraging peoples to the globalized economy. He is the author of *The Metaphysics of Apes*, also published by Cambridge University Press (2005).

Annette Lanjouw is Vice-president for Strategic Initiatives and the Great Ape Program at the Arcus Foundation, the largest private funder of great ape conservation and sanctuaries in the world. She has studied bonobos, chimpanzees, and gorillas in the wild, and currently brings her experience in the areas of behavioral ecology, conservation strategy, organizational management, institutional development, and policy to her work across Africa and Southeast Asia.

This book is published in association with the Arcus Foundation (www.arcusfoundation. org), a leading global foundation advancing social justice and conservation issues. Specifically, Arcus works to conserve and protect the great apes, in addition to advancing lesbian, gay, bisexual, and transgender (LGBT) equality.

The Politics of Species

Reshaping our Relationships with Other Animals

Edited by

RAYMOND CORBEY

Tilburg University and Leiden University, the Netherlands

and

ANNETTE LANJOUW

Arcus Foundation, New York

CAMBRIDGE
UNIVERSITY PRESS

CAMBRIDGE
UNIVERSITY PRESS

University Printing House, Cambridge CB2 8BS, United Kingdom

Cambridge University Press is part of the University of Cambridge.

It furthers the University's mission by disseminating knowledge in the pursuit of education, learning and research at the highest international levels of excellence.

www.cambridge.org
Information on this title: www.cambridge.org/9781107434875

First published 2013
First paperback edition 2014

A catalogue record for this publication is available from the British Library

Library of Congress Cataloguing in Publication data
The politics of species : reshaping our relationships with other animals / edited by R. Corbey, Tilburg University and Leiden University, the Netherlands and A. Lanjouw, Arcus Foundation.
        pages   cm
ISBN 978-1-107-03260-6 (Hardback)
1. Speciesism.   2. Animal rights.   3. Human beings.   4. Human rights.   I. Corbey, Raymond, 1954–   II. Lanjouw, Annette.
HV4708.P675 2013
179'.3–dc23   2013010576

ISBN 978-1-107-03260-6 Hardback
ISBN 978-1-107-43487-5 Paperback

Contents

Contributors

Kristin Andrews is a philosophy professor and Director of the Cognitive Science Program at York University, in Toronto, Canada. Her interests in animal and child social cognition and communication have led her to work with dolphins in Hawaii (Kewalo Basin Marine Mammal Laboratory), children in Minnesota (Institute for Child Development), and, most recently, orangutans in Borneo (Samboja Lestari Reintroduction Project). In her recent book *Do Apes Read Minds? Toward a New Folk Psychology* (2012), Andrews brings together her empirical and theoretical work to argue that humans are not quite so fancy, and the other apes are not quite so simple, as some think.

Jet Bakels is an ethnologist specialized in human–animal relations, with a special interest in dangerous and categorially ambiguous animals, both in the wild and in captivity. She holds a PhD from Leiden University, the Netherlands. Bakels has conducted extensive fieldwork in several locations in Indonesia. As a museum curator – now at Artis, the Amsterdam zoo – and writer of children's books, she strives to bring issues around human–animal relations to a wider public.

Marc Bekoff is Professor Emeritus of Ecology and Evolutionary Biology at the University of Colorado and a former Guggenheim Fellow. In 2000 he was awarded the Exemplar Award from the Animal Behavior Society and in 2009 the Saint Francis of Assisi Award by the Auckland (New Zealand) Society for the Prevention of Cruelty to Animals (SPCA). Bekoff has published more than 500 scientific and popular essays and 23 books, including *Wild Justice: The Moral Lives of Animals* (2009) and *Ignoring Nature No More: the Case for Compassionate Conservation* (2013). In 2005 Bekoff was presented with the Bank One Faculty Community Service Award for his work with children, senior citizens, and prisoners as part of Jane Goodall's Roots & Shoots program.

Lucy Birkett is a PhD candidate at the School of Anthropology and Conservation at the University of Kent, Canterbury. She studied at Oxford and Kent Universities. The scientific work by Donald Griffin, Jane Goodall, and Bernard Rollin focused her main research interest on the intelligence and capabilities of non-human primates, with a special interest in suffering.

Chong Choe is a Faculty Fellow at the Kennedy Institute of Ethics at Georgetown University and former Senior Appellate Research Attorney with the California Court of Appeal. Her areas of specialization include the philosophy of law, political philosophy, applied ethics, and bioethics. Her current research focuses on human and non-human animal rights and, particularly, issues of international political legitimacy, procedural justice, gender and minority group equality, and animal justice. Choe defends that the minimum constraints of justice apply in every social context, whether comprising humans or non-human animals, even against other competing economic and institutional interests.

Raymond Corbey, a philosopher and anthropologist, is an associate professor at Tilburg University and holds an endowed chair at Leiden University, both in the Netherlands. He has a keen interest in animal behavior and in human–animal relations in various settings, ranging from hominin evolution and extant foraging peoples to the globalized economy. His book, *The Metaphysics of Apes: Negotiating the Animal–Human Boundary* (2005), deals with the reception and rebuttal of evolutionary approaches in twentieth-century and present-day continental-European philosophy and in the humanities.

Joan Dunayer, a vegan since 1989, is a writer whose work focuses on non-human rights. Her articles and essays have appeared in magazines, journals, college textbooks, and anthologies. She is the author of *Animal Equality: Language and Liberation* (2001) and *Speciesism* (2004). Dunayer defines speciesism as the failure, on the basis of species membership or species-typical characteristics, to accord any sentient being equal consideration and respect. She advocates legal rights to life, liberty, and property for every sentient being.

Hope Ferdowsian is a physician who specializes in internal medicine and preventive medicine and public health at the George Washington University in Washington, DC. Her clinical, research, and policy work has focused on the prevention and alleviation of suffering in vulnerable human and non-human populations. She has led novel and innovative projects in the United States, sub-Saharan Africa, and the Federated States of Micronesia on issues including torture, HIV/AIDS, animal research ethics, and chronic disease prevention and management in resource-limited settings. Her work has centered on the universality of basic needs, including freedom from exploitation and abuse and the importance of physical and psychological well-being.

Agustín Fuentes, trained in zoology and anthropology, is a Professor of Anthropology at the University of Notre Dame. Ranging from chasing monkeys in the jungles and cities of Asia, to exploring the lives of our evolutionary ancestors, to examining what people actually do across the globe, Fuentes is interested in both the big questions and the small details of what makes humans and our closest relatives tick. His current research includes cooperation and community in human evolution, ethnoprimatology and multi-species anthropology, evolutionary theory, and interdisciplinary approaches to human nature(s).

Lori Gruen is Professor of Philosophy, of Feminist, Gender, and Sexuality Studies, and of Environmental Studies at Wesleyan University where she also coordinates Wesleyan Animal Studies. Her work lies at the intersection of ethical theory and practice, with a particular focus on issues that impact those overlooked in traditional ethical investigations, e.g. women, people of color, and non-human animals. Her most recent book is *Ethics and Animals: An Introduction* (Cambridge, 2011). Her current work explores the ethical issues raised by captivity, which emerges from lessons learned from the lives of some of the chimpanzees she has come to know, respect, and love. She has documented the history of the first 100 chimpanzees in research in the US (http://first100chimps.wesleyan.edu/).

Daniel Hutto is Professor of Philosophical Psychology at the University of Hertfordshire. He is the author of several books, including *Folk Psychological Narratives* (2008) and *Radicalizing Enactivism: Basic Minds without Content* (2013). A special yearbook issue of *Consciousness and Emotion*, entitled *Radical Enactivism*, which focuses on his philosophy of intentionality, phenomenology, and narrative, was published in 2006. Hutto is a chief co-investigator for the Australian Research Council "Embodied Virtues and Expertise" project (2010–2013) and collaborator in the Marie Curie Action "Towards an Embodied Science of Intersubjectivity" initial training network (2011–2015), and the "Agency, Normativity, and Identity" project (2012–2015) funded by the Spanish Ministry of Innovation and Research.

Barbara J. King is Chancellor Professor of Anthropology at the College of William and Mary, Virginia. She is a biological anthropologist who specialized for many years in the behavior of monkeys and apes. Recently, via her love of bison, cats, frogs, and diverse other creatures, she has broadened her focus to animal cognition, emotion, and welfare. With her husband, she rescues homeless cats in southeastern Virginia. Her books include *Being with Animals* (2010) and *How Animals Grieve* (2013). She writes weekly at NPR's *13.7 Cosmos and Culture* blog and regularly for the *Times Literary Supplement*.

Eben Kirksey, a cultural anthropologist at the University of New South Wales, Australia, studies the political dimensions of imagination as well as the interplay of natural and cultural history. His first book, *Freedom in Entangled Worlds* (2012), is about an indigenous political movement in West Papua, the half of New Guinea under Indonesian control. As a guest co-editor of *Cultural Anthropology* he assembled a collection of original research articles from the emerging field of 'multispecies ethnography' (2010). His second book, an edited collection called *The Multispecies Salon: Gleanings from a Para-site*, is forthcoming from Duke University Press.

Annette Lanjouw is the Vice-president, Strategic Initiatives and the Great Ape Program for the Arcus Foundation, the largest private funder of great ape conservation and sanctuaries in the world. She holds a BA in zoology and psychology from Victoria

University (Wellington, New Zealand) and an MA in behavioural ecology from Utrecht University (the Netherlands), and has studied bonobos, chimpanzees, and gorillas in the wild. Before Lanjouw joined the Arcus Foundation in 2007 she worked for various non-governmental organizations over two decades to further develop a supportive institutional and policy environment for integrated conservation in central Africa. She presently brings her experience in the areas of behavioral ecology, conservation strategy, organizational management, institutional development, and policy to her work across Africa and South East Asia.

David Livingstone Smith is Professor of Philosophy and Director of the Human Nature Project at the University of New England. His research interests are mainly in the area of moral psychology, broadly construed, and he has particular interests in self-deception, ideology, war, dehumanization, and Freud. He is the author of seven books, the most recent of which, *Less than Human: Why we Demean, Enslave, and Exterminate Others* (2011), was awarded the 2012 Anisfield–Wolf prize for non-fiction. Livingstone Smith strongly believes that philosophers have a moral duty to address themselves to the general public and to make use of philosophy to try to leave the world a better place than they found it. Consequently, his work has received considerable attention in the national and international mass media. He lives in Portland, Maine, with his spouse Subrena, and their dog Zadie.

Edouard Machery is Associate Professor in the Department of History and Philosophy of Science at the University of Pittsburgh, a Fellow of the Center for Philosophy of Science at the University of Pittsburgh, and a member of the Center for the Neural Basis of Cognition (Pittsburgh-CMU). His research focuses on philosophical issues raised by psychology and cognitive neuroscience. He is the author of *Doing without Concepts* (2009) as well as the editor of *The Oxford Handbook of Compositionality* (2012) and of *Thinking about Human Nature* (2013). He has been an associate editor of the *European Journal for Philosophy of Science* since 2009 and the editor of the Naturalistic Philosophy section of *Philosophy Compass* since 2012. Machery is also involved in the development of experimental philosophy.

Lori Marino is a behavioral neuroscientist in the Department of Psychology and affiliated to the Center for Ethics at Emory University (Georgia). She specializes in cetacean and primate intelligence and brain evolution, including brain–behavior relationships, the evolution of intelligence, and self-awareness in other species. She is also interested in human–non-human relationships, non-invasive models of science, animal welfare, advocacy, and ethics. In 2001 she and her colleague Diana Reiss published the first evidence for mirror self-recognition in bottlenose dolphins.

William McGrew is Emeritus Professor of Evolutionary Primatology at the University of Cambridge. He has studied wild chimpanzees for 40 years, across Africa, from Senegal to Tanzania. He first met these fascinating creatures at Gombe, at the invitation of Jane Goodall, and was hooked for life. McGrew has worked less often with captive chimpanzees, but has served on the Executive of the Royal Zoological Society of Scotland and on the Board of Directors of Chimp Haven, Inc., Louisiana. He has degrees in anthropology, psychology, and zoology, and all have proven to be useful in tackling chimpanzee behavior.

Molly Mullin is a visiting scholar at Duke University (North Carolina) in the Department of Cultural Anthropology. Previously, she was Professor of Anthropology at Albion College, where she began teaching courses on the anthropology of animals in 1997 while researching connections between art patronage and animal breeding, as discussed in the epilogue to her book, *Culture in the Marketplace* (2001). Her subsequent research has focused on animals as commodities and in relation to consumerism and anti-consumerism as well as on the cultural politics of domestication. To the volume *Where the Wild Things are Now: Domestication Reconsidered* (2007), which she co-edited with Rebecca Cassidy, she contributed a paper on the pet food industry. Mullin is currently writing a book about urban and backyard chickens, and a memoir on animals and anthropology.

Erin P. Riley is an Associate Professor of Anthropology at San Diego State University, California. Drawing from primatology, conservation biology, and environmental anthropology, her research primarily focuses on primate behavioral and ecological flexibility in the face of anthropogenic change and the conservation implications of the ecological and cultural interconnections between human and non-human primates. With notable publications in *American Anthropologist*, *Evolutionary Anthropology*, *American Journal of Primatology*, and *Oryx*, her work has spearheaded the emerging field of "ethnoprimatology." She currently has two on-going field research projects: the behavioral ecology and ethnoprimatology of the macaque monkeys on Sulawesi, Indonesia, where she has worked for the past 13 years; and, the human–macaque interface along the Silver River in north central Florida.

Jon Stryker is the founder and President of the Arcus Foundation, a private, global grantmaking organization with offices in New York City, Kalamazoo, Michigan, and Cambridge, UK. Arcus supports the advancement of lesbian, gay, bisexual and transgender (LGBT) human rights, and conservation of the world's great apes. Stryker is a founding board member of the Ol Pejeta Wildlife Conservancy in Northern Kenya, Save the Chimps in Ft. Pierce, Florida, and Greenleaf Trust, a trust bank in Kalamazoo. He earned a bachelor's degree in biology from Kalamazoo College as well as a master's degree in architecture from the University of California, Berkeley, and is a registered architect in the State of Michigan.

Richard Twine is a Lord Kelvin Adam Smith Research Fellow in the Social Sciences at the University of Glasgow, Scotland. Before this he worked at Lancaster University for ten years. He researches at the intersection of critical animal studies, environmental studies, gender studies, and science and technology studies. He is the author of the book *Animals as Biotechnology: Ethics, Sustainability and Critical Animal Studies* (2010), as well as several articles and book chapters on ecofeminism, bioethics, and critical animal studies. A few years ago he took up running and can occasionally be seen representing the Vegan Runners club at various mass participation events.

Steven Wise holds a BSc in Chemistry from the College of William and Mary and a JD from Boston University School of Law. He currently teaches Animal Rights Jurisprudence at the Vermont Law School, Lewis and Clark Law School, and St. Thomas Law School, and has taught this class at the Harvard Law School, University of Miami Law School, and John Marshall Law School. He is the author of several books on animal rights and numerous law review articles. Wise directs the Nonhuman Rights Project (www.nonhumanrights.org), the purpose of which is to persuade courts that at least some non-human animals are legal persons with certain fundamental legal rights. Its first cases will be brought in 2013.

Preface

An interdisciplinary dialogue that took place in New York in August 2011 brought together experts from numerous fields to explore how humans define "others" and position themselves in relation to those others. In some cases, "otherness" has been defined by race, ethnicity, gender, culture, or religion. In the context of this debate, it is defined by species and the criteria used for assessing the value of other species, such as sentience, moral agency, or ability to suffer or feel empathy. The New York roundtable examined human relations with other animals, and in particular the conflicts with and discrimination against non-human animals. It explored how motivation and action can be harnessed to protect non-human animals in the wild and in captivity from harmful exploitation and suffering.

This volume brings together the contributors at the New York roundtable, from the fields of philosophy, ethnology, primatology, as well as ethology, neuroscience, law, journalism, conservation, sociology, and medical science. The positions represented varied from emphatic animal activism to a more anthropocentric and utilitarian economic pragmatism. Yet each presented an additional perspective on the reality that billions of non-human animals are exploited and/or killed each year for human use and enjoyment, with little regard for the impact of this behavior on the well-being of the individual animal and the natural world.

To unravel the complex historical and psychological underpinnings of a largely Western, or Western-influenced, perspective on non-human animals, and to strive for a more just and humane attitude to the numerous species on this planet, concepts such as entangled empathy, multispecies ethnography, compassionate conservation, and respectful coexistence were explored and discussed. Although the emphasis was on understanding the attitudes of primarily modernized Western societies, and the influence of the Judeo–Christian tradition, the discussion did examine the contrast with some traditional societies in various parts of the world.

Despite the varied backgrounds, perspectives, and motivations represented in this collection of papers, the shared objective was to improve the welfare and survival of all species on this planet, and to strive for a reshaping of our attitudes, tolerance, and ability to coexist respectfully. Although deep-rooted politics underlie our current behavior, there is a growing realization that our own survival and welfare is tied to a mutually dependent existence with other species.

We thankfully acknowledge input and feedback, in various stages of this project, from Gay Bradshaw, Hans Lindahl, Susanne Morrell, Joachim Nieuwland, Adam Phillipson, Wiktor Stoczkowski, Tom van den Berge and Rob van de Ven. A special word of thanks for the use of ten photographs to Canadian photojournalist Jo-Anne McArthur and her award-winning project We Animals (www.weanimals.org), which photographically documents animals in the human environment.

Introduction: between exploitation and respectful coexistence

Raymond Corbey and Annette Lanjouw

This book deals with a variety of aspects of the relentless exploitation of billions of non-human individuals in the profit- and consumption-driven global economy, and the marginalization and forced extinction of countless other (than human) species in the wild. This is a profoundly political process. Politics can be characterized as the activities that people engage in to define and exercise power, status, or authority, either among states or among groups within a state. This definition allows for the application of a political lens over the attitudes and behaviors of humans with regard to other species. It is all about control and power.

The destructive aspects of human domination of other species in the ecosystems of this planet at large stand in stark contrast with the treatment of companion species such as dogs and cats. Ecological domination is usually attributed to the species that has the most influence on other species in the same environment. The human presence and impact on this planet is substantial and ever-growing in scale: through population growth, large-scale and expansive agriculture, industrial-scale farming and fisheries, urbanization and infrastructure development, extractive industries (mining, oil, and gas), commercial forestry, global trade and commerce, trade and transport of tens of thousands of species, extraction and use of freshwater, and the altering and polluting of ecosystems.

Control and destruction of individual non-human lives for selfish human purposes is legitimized by worldviews such as humanism, several monotheistic religions, and by the legal systems and political stances rooted in these traditions. The legitimization of discrimination by *Homo sapiens* against other sentient beings, however, is even more complex than that. Not only does it occur on cultural and discursive levels, but equally on psychological and evolutionary ones. A current definition of speciesism is discrimination against individuals on the basis of their species membership, taken to be morally relevant. The concept was coined in the early 1970s by analogy to racism, sexism, and classism. In racism, vilified non-human animals (apes, dogs, pigs, etc.) are often called upon to vilify other humans. Recent research suggests a role for species-wide evolved taxonomies in racism and speciesism (see Livingstone Smith, this volume), next to and entwined with culturally and historically situated "folk taxonomies."

The Politics of Species: Reshaping our Relationships with Other Animals, eds R. Corbey and A. Lanjouw. Published by Cambridge University Press. © Cambridge University Press 2013.

People the world over, from Amerindian villages in the rainforest of Amazonia to traders in the stock exchange of Beijing, categorize everything around them. A substantial part of the messy complexities of human dealings with other species has to do with how we categorize on various levels. "Discrimination" can be seen in two different ways: making distinctions in a neutral sense – that is, categorizing; and discriminating in a prejudicial and value-laden way. How people categorize is both morally and politically relevant. The thoroughly political distinction between human and non-human has been probably the most pervasive dominator hierarchy in Western cultural discourse and practice. In processes and mechanisms of inclusion and exclusion, various categories of humans – the enslaved, women, peasants, the poor, criminals, the mentally sick, the colonized – at various times were construed as "less" than "human." Over time many of them have become, to some extent, emancipated. This is not the case for non-human individuals.

The institutionalized ecological dominance of humans can, ironically, itself again be analyzed – as has been done in detail by, among others, Jürgen Habermas and the Frankfurt School of cultural critique – in terms of a colonization of human personal lifeworlds by the blind systemic forces of technocracy and global markets. The latter tend to determine what humans should do and how they should live, instead of humans as individuals themselves, advancing their goals in a democratic political process. Thus when enslaving non-human animals, humans to some extent are enslaved themselves. However, the Frankfurt philosophers, in a humanist tradition, always had a sharper eye for the commodification and mistreatment within the human species than beyond it (but see Sanbonmatsu 2011; and Twine, this volume).

"Respectful coexistence" in the title of both this introduction and the third part of this volume refers to the ability to share resources and space, as well as to respect each other's needs and self. The expression implies an acknowledgment of moral and social relevance. It has connotations of concern, heedfulness, care, and regard for the other. It reminds us of shared lives on the commons, the common village meadows that provided for all before the "tragedy of the commons" (Hardin, 1968) happened: the depletion of shared resources by individuals acting according to their own self-interest and against the long-term best interests of all. Coexistence also suggests sharing space (Riley, this volume) and co-flourishing in environments thought of less as resources than as commons. It provides an angle on living with companion species, as well as on attitudes toward the rest of nature in many traditional, non-modern peoples.

"Respectful coexistence" may be seen by many as a rather Utopian phrase, not very analytical or pragmatic. It does, however, provide a provocative counter-image to the unbridled exploitation of animals in the globalized bioindustry and to the massive destruction of wildlife habitats. It resonates with the post-humanist idea of "multispecies ethnography" that puts numerous species and individuals center stage, focusing on how the "livelihoods of a multitude of organisms" both shape and are shaped by political, economic, and cultural forces (Kirksey and Helmreich, 2010: p. 545; Fuentes this volume; Kirksey, this volume; Riley, this volume). This fledgling paradigm provides a refreshing alternative to the two mainstream ways in which (other) animals are dealt with in traditional anthropology: animals as "good to eat" and animals as good

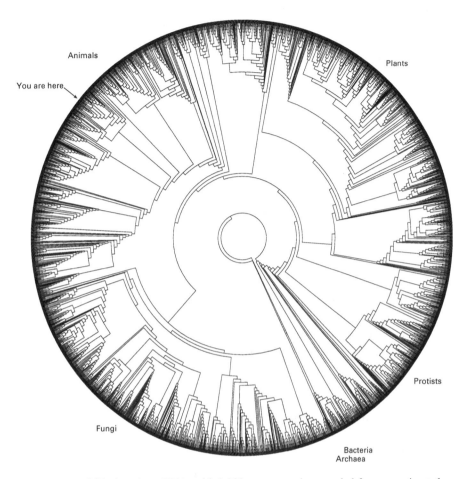

Figure 1 Tree of life, based on rRNA, with 3,000 extant species sampled from an estimated nine million living species on the circumference of the circle. Life and speciation events begin 3.8 billion years ago in the center of the circle. The human species (upper left corner) is one of 200 living primate species and about 1.3 million known living animal species. Yet "animals" excludes humans in most contexts – everyday discourse, law, religion, politics, etc., even much scientific literature. Diagram and research data David M. Hillis, Derrick Zwickl, and Robin Gutell, University of Texas.

to think (or symbolize) with. In both cases, animals figure as part of the backdrop of human social worlds, not as subjects in their own right.

The first part of this book deals with the way that speciesist views and practices draw sharp boundaries between human individuals and individuals of all other animal species (Figure 1). All contributions, but those in the third part in particular, suggest new views of spatial, conceptual, and moral boundaries – as not so much separations and divisions, but openings, or dynamic contact zones, in which animals of all species share spaces and are entwined. This suggests redefinitions of moral and legal communities and new approaches for the spaces in which species interface: laboratories, meat and dairy farms, homes, nature reserves, wild spaces, agricultural landscapes, urban spaces, veterinary hospitals, the entertainment industry, the wildlife trade, shelters, slaughterhouses, and zoos.

The chapters in the second part weigh various reasons for moving from speciesist exploitation to greater respectful coexistence. They look at what we have learned from research carried out in recent decades about the subjectivity and agency of individuals belonging to non-human species – the richness of their intelligence and emotions, the complexity of their interactions and communication. These new insights have played a key role in an immense increase in recent years of publications, academic disciplines, social movements, and political initiatives focusing on all sorts of aspects of the human interface with other animals.

Despite the continuing presence of Immanuel Kant (e.g., in Tom Regan's influential work, or in political and legal discourse at large), much of present-day thinking about animals can be labelled as broadly post-humanist. This means that, as the editors of *Humanimalia* put it when this journal was launched in 2009, the human species is no longer posited "as the sovereign agent of the Earth, privileged either as a metaphysical or evolutionary preference by its proven intelligence, rationality, technology, or whatever other explanation is provided for its demonstrable ability to shape and overpower other elements in the natural world" (Humanimalifesto, 2009). Along similar lines the present volume pleads for a non-anthropocentric, non-ethnocentric, and non-egocentric stance, in close affinity with the life sciences, which usually were, and still are, shunned by the humanist tradition and the humanities that issued from it.

Moving beyond speciesism (Part I)

Many categorizations, in particular those regarding human and other animals, are emotionally and value laden and are, or can quickly become, discriminatory. The drawing of lines between individuals belonging to different animal species is bad biology in view of the evidence for evolutionary continuity, argues ethologist **Marc Bekoff** in Chapter 1. It results in the establishment of false boundaries that have dire consequences for species deemed to be "lower" than others, such as ants, fish, birds, or rats. Most conservation efforts are directed at "higher" and charismatic animals. Speciesism, conscious and unconscious, is the main culprit in our interactions with other animals. It reinforces the property status of non-human animals and undermines our collective efforts to make the world a better place for all beings. Bekoff pleads for a "deep" ethology, studies of animals that take us not only into their minds but also into their hearts, as a beginning of expanding our "compassion footprint."

Joan Dunayer (Chapter 2) adds to Bekoff's challenge of the supremacy of "higher" species such as primates, cetaceans, elephants, and mammals in general, carrying the argument further: in her view, speciesism is the failure, on the basis of species membership or species-typical characteristics, to accord *any* sentient being equal consideration and respect. She criticizes old-speciesists, who limit rights to humans, but also new-speciesists, who advocate rights for relatively few non-humans, those who seem most human-like. Dunayer holds that it is fair, logical, and empirically justified to give all creatures with a nervous system the benefit of any doubt regarding sentience and accord them basic rights such as rights to life and liberty.

The next two chapters, both written by philosophers with a strong background in psychology, discuss the formidable psychological barriers that prevent us from including non-human animals in a moral community of some kind. **David Livingstone Smith** (Chapter 3) argues that hierarchical thinking with humans on top may reflect deeply rooted, innate intuitions about the structure of the cosmos, bound up with our tendency to carve the world into essentialized natural kinds. Humans have often demonstrated that their view of who is human is based not on membership of the taxon *Homo sapiens*, but on membership of a natural kind, a notion that allows exclusion on the basis of, for example, race, religion, or appearance. Livingstone Smith lays bare psychological and discursive ties between, on the one hand, the dehumanization of the enslaved, the colonized, non-citizens, and enemies in war, and, on the other hand, the treament of other than human animals.

Edouard Machery (Chapter 4) also compares the way we tend to think about races and about species. He examines whether three possible strategies that have been successful in addressing racism and some other forms of discrimination can be applied to our attitudes toward non-human animals, in particular other great apes. Machery suggests that of these three strategies individualization has the best chance of success, because moral emotions as well as capacities for empathy and sympathy are more likely to be engaged by individuals than by groups. This approach has been taken in several conservation spheres with the "adoption" concept.

Philosopher and anthropologist **Raymond Corbey** (Chapter 5) shows how, ironically, racism has often been combated on the basis of speciesist assumptions, in particular in the humanist post-World War II United Nations discourse on human rights. He traces those assumptions to various roots – in biological and anthropological thinking of the period, European metaphysics, middle-class cultural attitudes, and, ultimately, evolution. The subsequent Great Ape Project, which claimed moral respectability for all great apes, ran into a similar problem. Corbey also makes some observations on the ritual, performative character of various declarations of the rights of human and non-human beings.

In Chapter 6 sociologist **Richard Twine** approaches the mistreatment of livestock in the "animal–industrial complex" from the critical, left tradition in social science and political thought, in particular the emerging interdisciplinary field of critical animal studies. He shows how firmly speciesism is embedded in the global capitalist order with its routines of industrialized, commodified killing of contemptible, consumable, and enslaved non-human beings. Greater meat and dairy consumption is encouraged in the developing world and "superior" farming and breeding techniques and genetics are imposed on poor countries.

Sentience and agency (Part II)

Sentience is the condition or quality of being conscious and susceptible to sensation and emotion. Only sentient beings can suffer. There is no consensus among scientists how widespread sentience is in the animal kingdom. Agency is the ability or capacity to act in the world, to make choices. Traditionally and predominantly understood as *human* agency, the word carries connotations of personhood: of intentional, free, and therefore morally and legally relevant, action. Other animals, correspondingly, were understood not to be agents taking initiatives in the same, full sense as humans, but merely to be showing reactive behavior as a deterministic process. In Western moral philosophy, moral agency has usually been seen as a prerequisite for moral considerability.

A third relevant concept, next and connected to sentience and agency, is subjectivity, in the sense of having its, her, his own perspective, first person point of view; having thoughts, beliefs, feelings, desires that intend, mean, are about something. Subjectivity is what it is like to be something, what distinguishes a subject from a thing, in contrast to objectivity. Agency and subjectivity are central concepts in the hermeneutic or interpretive tradition in continental-European and anglophone analytic philosophy. They are associated with reasons rather than causes; actions rather than events; with the reasonableness and rationality of the subjects involved – usually, implicitly or explicitly, taken to be humans.

Lori Marino, a behavioral neuroscientist, deals with the sentience and agency of dolphins and other cetaceans (Chapter 7). They continue to be treated as non-sentient objects, commodities and resources, despite what is known about their intelligence, self-awareness, learning skills, social communication, and rich emotional lives. Marino discusses the effects of this mistreatment – such as aberrant behavior, stress, disease, and mortality – and the reasons it continues, pleading for the recognition of cetacean personhood and moral standing. But cetaceans, unlike other (than human) great apes, look and move differently, lack changes in clearly recognizable facial expressions, communicate in strange modalities, live in a very different physical environment, and seem to possess a level of social cohesion foreign even to us. This is an obstacle to moral inclusion. Cetaceans represent extremes of similarities and differences that challenge our ability to recognize them as moral equals. Therefore they are among the most vulnerable and the most bewildering non-human animals.

In Chapter 8 anthropologist **Barbara King** concurs with Marino's plea on the basis of an abundance of evidence for love and grief behaviors among non-human animals. A wide variety of animals – chimpanzees, elephants, ravens and geese, dogs and cats, maybe even turtles – respond differently to dead companions than to injured ones; express deep and prolonged sadness at the loss of a loved one; and in some cases participate in ritual behaviors around death. The occurrence and intensity of these behaviors varies between species and between individuals, and between wild and captive animals. This has profound implications for the humane treatment of apes in captivity, for example in allowing them time with a dying or dead member of their group.

The next two chapters on animal subjectivity, again by philosophers, both concern "theory of mind": the ability to attribute mental states of various sorts to others. Such complex cognitive abilities are often adduced as a reason for moral concern. **Kristin Andrews** (Chapter 9) holds that what appears to be a "higher" cognitive capacity in humans is in fact the result of many simpler heuristics, including capacities humans share with other animals. It is the need to explain behavior that drove the evolution of mindreading, and this, in turn, required a prior understanding of social norms. She suggests that researchers need to stop focusing on how complex animals are and first realize how simple humans may be. One consequence of this position is that mind-reading should not be a requirement for either moral or legal standing. **Daniel Hutto** (Chapter 10), in discussion with Andrews, also thinks that we are setting the bar too high for apes and mindreading. He too, from a slightly different angle, moves humans closer to other species who read minds. Both non-human and human apes, and human children too, engage with other minds in an emotionally charged, "enactive," and non-representational way, without "higher", representational cognition.

Primatologists **Lucy Birkett** and **William McGrew** (Chapter 11) compare ape behavior in nature with that in captivity. While they are spared predation and other threats, captive apes live in forced social environments; are deprived of the ability to roam, forage, engage in group fission/fusion; and may endure constant human exposure. Birkett and McGrew present plentiful evidence for emotional distress in captive chimpanzees, measured by abnormal behaviors such as eating feces, drinking urine, plucking out hair, and self-harming. This evidence shows how vulnerable, self-aware, emotive, and smart ape minds are to distortion and suffering in all captive environments, even in "good" zoos. They argue that the only defensible reason for keeping apes in captivity is to offer lifelong sanctuary to those who cannot be returned to nature.

A surprisingly different perspective on animal sentience and agency is present in many traditional, less modernized, small-scale societies like the Mentawai shifting cultivators and Kerinci farmers of Sumatra, as studied by **Jet Bakels**, a cultural anthropologist, in her fieldwork (Chapter 12). They conceive of animals such as the tiger and the crocodile as persons who are part of society as a moral order, and entertain respectful reciprocal relations with them. Forest spirits are owners of forest animals, and humans must behave as guests in the forest. Bakels sees the respect and generosity these animals are often treated with in such traditional settings worldwide as a source of inspiration for modernized societies in their struggle with moral inconsistencies with respect to animals. Although this set of beliefs has generally served these traditional societies and the natural world well, economic development is now threatening it and changing behavior.

Another exercise in multispecies ethnography and plea for entangled empathy in the sense of Lori Gruen (see below) is provided by anthropologist **Eben Kirksey**, in Chapter 13. He studied the complex agency, sociality, and multispecies entanglements of *Ectatomma* ant entrepreneurs in the shadow of humans in forested landscapes, agricultural fields, and parking lots of Central and South America. Kirksey analyzes these settings unorthodoxly as "cosmopolitical" worlds, where beings are involved in a complex web of mutual use and exploitation in which each has an interest in the survival of the other.

Toward respectful coexistence (Part III)

Although not explicitly following "multispecies ethnography" as a new paradigm, all contributors to Part III are close and sympathetic to it. Anthropologist–primatologist **Agustín Fuentes**, for one, explores an ethnoprimatological approach to the integration, engagement, and interface between ourselves and other primates (Chapter 14). Ethnoprimatology is the study of how a people ("ethnos") or society deals with other primates (alloprimates). Fuentes reappraises the way the human–alloprimate interface is shaped by shared histories, economies, cultures, and landscapes as well as each species' behavior and physiologies. Human relationships with other animals are complex, culturally contingent and contextual: no uniform or simple perspectives – ethical, ecological, ethnological, or literary – can effectively categorize them. Bonding and cooperation play a significant role in social niche construction among primates and in humans the emergence of regular altruism is likely related to an expansion of this core primate pattern. A distinctive evolutionary discontinuity is that humans can cast this physiological, social, and symbolic bonding "net" beyond biological kin and even beyond our species. This may point to an evolutionary pattern, a human adaptation, which is ready made to bring others into moral equivalencies and social kinship with us. A less positive evolutionary discontinuity is the distinctly human evolutionary trajectory of niche alteration for the benefit of human populations and the detriment of many other species.

Anthropologist–primatologist **Erin Riley**, concurring with Fuentes and adding to Bakels' analysis of non-modern views of nature, provides a case study from the highlands of Central Sulawesi, Indonesia (Chapter 15). She examines how native villagers and migrants perceive nature and how they share ecological space with Tonkean macaques in particular. Local people are more positive about implications of the Lore Lindu National Park, viewing it as a source of livelihood. Migrants are less positive, while locals view migrants as a threat to the forest. Both locals and migrants believe that monkeys are reincarnated or somehow recreated humans, who should be respected and never killed. But it is dangerous to rely on cultural traditions for conservation, as economic and development opportunities render them vulnerable. Riley discusses her ethnoprimatological findings in the context of change and the politically and ethically charged arena of biodiversity conservation.

Annette Lanjouw, a conservationist and primatologist, reflecting on her experiences in building institutional and individual capacity in wildlife conservation in Africa, takes a wider view again in Chapter 16. She elaborates on strategies for holistic approaches in conservation and engendering respect for non-human animals. Although the focus of her career has been on great apes, her argument for a focus on charismatic species can be applied to other species equally, and benefit broader ecosystems and landscapes.

The next chapter (Chapter 17), by **Molly Mullin**, a cultural anthropologist with a longstanding interest in human–animal relationships, is an essay in the cultural politics of poultry. Mullin reports on her two-year research project on the historical backgrounds and contradictions of backyard and urban chicken keeping in the United States. This practice raises questions about environments, animal welfare, political geography, relationships among species, and even neoliberal economics. It is a mistake, she argues, to dismiss anyone involved in exploiting animals for food as devoid of moral conscience. Ethnographic research on chickens as well as other animals often has revealed humans capable of caring for animals and exploiting them at the same time. There is no neat and tidy evolution from wild to domestic, from pre-modern to modern, or from not caring to caring about animal well-being.

Philosopher **Lori Gruen** (Chapter 18) identifies and criticizes discrimination against other animals as "humanormative" – presupposing that the human species provides the normative measure against which other species are to be judged deficient or deviant. "Entangled empathy" is a moral counterstrategy that helps to develop reflective responses to the lives and interests of others. Gruen discusses various empathetic failures: over- and under-empathetic responses are both a danger, particularly in relation to animals, who cannot easily correct such errors of perception. We can pay more attention to making better choices and promoting the well-being of those with whom we are entangled. Thinking harder not just about the nature of these relations, but also about how to be in ethical relations with a range of other beings, is an interesting under-explored project that becomes possible once we stop focusing exclusively on the properties that make us similar.

Hope Ferdowsian, physician, and **Chong Choe**, lawyer and philosopher, re-evaluate the ethics of using non-human animals in experimentation in Chapter 19. They do so from the perspective of human research protections in the United States, taking into account the potential for mental and physical suffering (cf. Chapter 11 by Birkett and McGrew, and Chapter 7 by Marino) of the species involved. The arbitrary nature of some current research legislation in the United States – which, for example, excludes birds, rats, and mice bred for research from being considered animals – makes this project all the more urgent. Animals grown for food are also excluded from the US Animal Welfare Act. Key concepts in their principled approach are autonomy, vulnerability, beneficence, and justice.

Finally **Steve Wise**, who has been practicing animal law for three decades, analyzes the modern legal system's treatment of animals and its historical background (Chapter 20). Humans are "legal persons," non-human animals, "legal things." A legal person has the capacity to possess legal rights; one who possesses a legal right is a legal person. "Legal things" lack the capacity to possess legal rights; they are invisible to civil law. Wise explains the Nonhuman Rights Project, which supports the filing of lawsuits intended to pave the way for an American state High Court to declare, for the first time, that a non-human animal is a legal person and has the capacity to possess a legal right. Many judges believe that common law structure requires them to bring a common law rule at odds with social morality, public policy, or human experience into harmony with modern understandings. Wise discusses qualities proposed for legal personhood, including autonomy; having preferences and being able to act to satisfy them; being able to cope with changed circumstances; and having desires and beliefs. Chimps and other apes and various other species clearly have this and so should be entitled to legal personhood and basic legal rights.

Overcoming resistance

The 20 chapters together offer an inventory of the determinants of our dealings with other animals: religious, moral, political, and everyday discourse; the global capitalist system and its ecological effects; evolved cognitive and motivational predispositions.

These determinants are in fact formidable obstacles to redefining moral and legal communities. To reflect this, perhaps a question mark could have been added to the optimistic title of the third part of this book: "Toward respectful coexistence."

As converging twentieth century developments in biology and philosophy have made abundantly clear, there are no fixed Aristotelian, scholastic, folk taxonomical or evolved essences to species, despite widespread reluctance to admit this (Livingstone Smith, this volume; Corbey, this volume). The concept of a "human nature" is the last bastion of essentialism, and a significant barrier to reshaping our relationships with other animals. The view that humans have a fixed essence, and, as self-conscious, free willing and morally responsible beings, a unique dignity and moral status, resists revision. It still permeates legal systems, political ideologies, religions, the humanities, and everyday discourse the world over, legitimizing our ongoing and massive exploitation of other species.

There is, however, now a broad scientific and philosophical consensus that discriminating on the basis of the underlying "essential nature" of living beings makes no sense. Living nature is to be explained in terms of variation, continuity, and change. It is variable and fuzzy. It underdetermines our value-laden pigeonholing and permits many classifications, in science and everyday life. How we classify depends on context and purpose; and is open to negotiation. As massive and pervasive the factors determining the human domination over nature may be, human discourse, including the values and discriminations it poses, is variable – historically situated and culturally specific. It changes over time. Post-Wittgensteinian philosophy, cultural anthropology, and cultural studies converge in stressing that vocabularies are but tools that are to be assessed in terms of the particular purposes they may serve, when ordering, describing, predicting, or providing a handle on things. They are man-made fabrics underdetermined by reality, leaving much latitude of choice. They are not true but useful.

As the philosopher Richard Rorty put it, in a pragmatist tradition: "No description of how things are from a God's-eye point of view, no skyhook provided by some contemporary or yet-to-be-developed science, is going to free us from the contingency of having been acculturated as we were. Our acculturation is what makes certain options live, or momentous, or forced, while leaving others dead, or trivial, or optional" (Rorty, 1991, p. 13). This even creates room for negotiating tendencies in our evolved human nature.

Perhaps this more optimistic observation can somewhat balance our pessimistic remarks in the foregoing pages. Rather than claiming that there is an absolute truth or objective reality, the needs and interests of all individuals who are concerned must be invoked in order to legitimize our inferential and other practices. This includes the changes and renewal that would be necessary in ethics, laws, laboratory routines, political programs, animal farming, conservation, and all other dealings with other sentient beings.

Chicken cauda

Because this book attempts, but does not entirely manage, to avoid a bias toward "higher" animals we would like to end this introduction with a word on chickens, who figure in the chapters by Bekoff and Dunayer on speciesism (Chapters 1 and 2),

Twine on global farmed animal production (Chapter 6), Bakels on animist Sumatran foragers (Chapter 12), and Mullin on urban backyards in the United States (Chapter 17).

The common chicken (*Gallus gallus domesticus* – a domesticated subspecies of Red Junglefowl) originates from omnivorous foragers roaming the Southern Asian jungle floor, living in stable social groups, and recognizing each other by their facial features. Being kept in small cages is extremely frustrating for them, despite the effects of domestication. The trimming of beaks interferes with facial recognition and social life. Far from being proverbial "bird brains," they are good at solving problems. They understand that an object, when taken away and hidden, nevertheless still exists, which is beyond the comprehension of young children. They use representational calls specific to kinds of food and specific predators to share information and express a range of emotions, including joy, fear, pain, and frustration (Evans and Evans, 2007).

The unceremonial exploitation and death of tens of billions of chickens every year, deprived of moral status in factory farms (see the back cover of this book), contrasts wryly with their respectful treatment as ontological equals in many small-scale traditional societies. An example of wonderfully complex human–chicken coexistence is provided by Roman Catholic speakers of Q'eqchi' Maya living in the cloud forests of highland Guatemala. Here chickens are cared for especially by women, who see them as strong-willed female affines, capable of prudent action based on experience. As reflexive selves, chickens contribute to the reflexive articulation of female identity (Kockelman, 2011).

Another example of non-modern ontology of poultry that subverts Western dichotomous views of moral humans and exploitable nature is that of the Sakkudei of Sumatra, as discussed by Bakels in Chapter 12 (see the front cover of this book). Chicken eggs are used for rituals and sometimes eaten, and chickens are slaughtered for feasts and rituals as gifts to the clan ancestors. In a wholly interconnected and interdependent cosmic society, which includes animal, ancestral, and other spirits, their souls serve as messengers who beg the ancestors for protection and favors. When the chickens are sent to them they are addressed solemnly, reminded of the good life they have enjoyed and the responsibilities ensuing after their death.

A third example is provided by *khokwa yabutama*, the chicken-who-is-about-to-lay-an-egg, in the cosmology of the Yaka of the Democratic Republic of the Congo (Devisch and Vervaecke, 2011). Moving between night and day, village and bush, earth and sky, the chicken is a much valued and sacred mediator between those domains. It symbolizes fertility, the regeneration of life, and even the coming into being of all things. When soothsaying during a trance, Yaka sorcerers do not just walk and cackle like them, but actually become chickens.

In such societies, humans and other beings are part of an overarching community with strong reciprocity and a single code of conduct for all. Arhem's characterization of the Makuna of Amazonia is much more widely relevant: "Rather than proclaiming the supremacy of humankind over other life forms, thus legitimizing human exploitation of nature, Makuna eco-cosmology emphasizes man's responsibility towards the environment and the interdependence of nature and society. Human life is geared to a single, fundamental and socially valued goal: to maintain and reproduce the interconnected totality of beings which constitute the living world; 'to maintain the world,' as the Makuna say" (Arhem, 1996: p. 201).

Such societies and their dealings with animals should not be romanticized, as they too often are. There are many examples of cruel treatment of animals here too. Often there is a manipulative and pragmatic aspect to respectful reciprocal relations with animals/ spirits, which serve to appease and guarantee the supply of meat. The Sakkudei chicken are treated politely, but slaughtered nevertheless. Rituals and prayers may also serve to relieve those who kill living beings from guilt. Yet the respectful treatment of chickens and other animals in many small-scale traditional societies can inspire eco- and bio-centric visions. It challenges the moral hubris and grotesqueness of large-scale battery farming, which commodifies these sentient animals into egg-laying and meat-providing machines.

Part I

Moving beyond speciesism

1 Who lives, who dies, and why?

How speciesism undermines compassionate conservation and social justice

Marc Bekoff

Man in his arrogance thinks of himself a great work, worthy the interposition of a deity.
[Yet it is] more humble and, I believe, true to consider him created from animals.

(Charles Darwin, 1871)

This book is very much about the question "Who lives, who dies, and why?" – why certain groups of non-human animals (hereafter: animals) are targeted as being lowly and unworthy of moral consideration, why others are seen as being more valuable and worthy of moral concern, and why we even care whose lives are trumped for our own benefit. Of course, we should care very much.

I've long been interested in the study of animal minds (cognitive ethology), especially the emotional and moral lives of these amazing and magnificent beings, and how what we know can be used to make their lives better in the wide variety of venues in which they are used, and often horrifically abused, and in conservation projects.

Despite some lingering and rapidly declining skepticism about whether or not other animals are conscious or experience deep and enduring emotions, it's time to stop ignoring who these animals really are and pretending that we don't know that they are indeed conscious and feeling beings. We must also make every effort to use this information on their behalf. Indeed, in July 2012 a group of renowned scientists met at Cambridge University and finally declared that animals are truly conscious. They produced a long overdue document they called the Cambridge Declaration on Consciousness (2012). Although we really have known for a much longer period of time that other animals are conscious, perhaps their highly publicized declaration will be helpful for radically improving animal well-being. We can only hope this declaration is not merely gratuitous hand waving.

Oxford University's animal welfarist Marian Dawkins (2012) has claimed that we really still do not know if other animals are conscious and that we should "remain skeptical and agnostic [about consciousness]... Militantly agnostic if necessary" (Dawkins, 2012: p. 177). However, there's now a wealth of scientific data that makes skepticism, and surely agnosticism, anti-science and harmful to animals. Now, at last, the Cambridge Declaration on Consciousness shows this to be so. Perhaps what I call

The Politics of Species: Reshaping our Relationships with Other Animals, eds R. Corbey and A. Lanjouw. Published by Cambridge University Press. © Cambridge University Press 2013.

"Dawkins' dangerous idea" (Bekoff, 2012) will now finally be shelved given the conclusions of the Cambridge Declaration. I don't see how anyone who has worked closely with any of a wide array of animals or who lives with a companion animal could remain uncertain and agnostic about whether they are conscious. It's time to move on with what we know.

We live in a troubled and wounded world that is in dire need of healing. We should all be troubled and terrified by what we have done and continue to do. Humans are a big-brained, big-footed, over-producing, over-consuming, invasive, and arrogant lot and we have made huge and horrific global messes that need to be repaired now. Losses of biodiversity are stinging and disheartening and many are irreversible. The overriding sense of turmoil is apparent to anyone who takes the time to pay attention. Researchers and non-researchers alike are extremely concerned about unprecedented global losses of biodiversity and how animals suffer because of our destructive ways. And, of course, despite wide-ranging destruction of Earth and its inhabitants there are numerous people around the world who are working hard to make Earth a better place, a safer, more peaceful, and more compassionate home for all animals, human and non-human alike (Bekoff, 2010, 2013a, b).

In this chapter I want to go beyond the confines of animals in captivity to consider how speciesism informs how we view and interact with other animals in broader contexts, in their homes and living rooms into which we freely intrude as we wantonly redecorate nature (Bekoff, 2006, 2013a). Speciesism is the main culprit in our inter-actions with other animals. Simply put, conscious and unconscious speciesism reinforces the property status of non-human animals and undermines our collective efforts to make the world a better place for all beings.

Against speciesism

"Naturalistic or Darwinist philosophy refuses to set humans apart from non-human primates and the rest of life as *the* unique species, and, instead, thinks of them as 'just' another unique species. Other philosophical positions maintain that features such as mind, rationality, humanness, and morality do set humans apart, but that these are not perceived by empirical science because it does not probe deep enough", writes Ray-mond Corbey (2005: p. 198). Our unique and ongoing contribution to the wanton decimation of the planet and its many life forms is an insult to other animals and also demeans us. In his extensive writings Charles Darwin (1859/1964, 1871) forcefully argued for evolutionary continuity, stressing that variations among species are differ-ences in degree rather than in kind. Simply stated, if we have something, they (other animals) have it too. Thus speciesist arguments ignore or violate well-accepted evolutionary theory and result in the establishment of false boundaries that have dire consequences for species deemed to be "lower" than others (Figure 1.1; Bekoff 1998a, 2007a, b, 2010, 2013). We are animals and we should be proud and aware of our membership in the animal kingdom. Scientific research tells us that we must stop ignoring nature (Bekoff and Bexell, 2010; Bekoff 2013) and that speciesism is "bad biology."

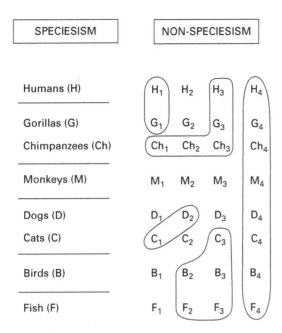

| SPECIESISM | NON-SPECIESISM |

Figure 1.1 Speciesist and non-speciesist categorization of animals (after Bekoff, 1998a). On the left a speciesist representation of eight species (chosen as examples). A linear hierarchy suggests, for example, that humans are "higher than" gorillas and chimpanzees and that monkeys are "higher than" dogs and cats. On the right a non-speciesist representation of four individuals in each of the same eight species. Lines encircling various individuals (H_1 and G_1; D_2 and C_1) indicate that individual characteristics "count." Individuals from different species may be "equivalent" with respect to various traits. Individuals of the same species may possess characteristics that are exclusively theirs. Individuals of species that are typically thought to be "lower" than others may be more skilled in certain areas or experience pain, anxiety, and suffering more than individuals of species that are thought to be "higher."

There aren't "lower" and "higher" species. We make that differentiation because it serves us well and makes life easier when deciding who lives and who dies. .

Humans freely construct barriers that include some species and exclude others. This speciesistic mentality fractures nature and constructs false boundaries among species. It puts us on a moving slippery slope and plays a significant role in misrepresenting who we are and who non-human animals are. When we probe deeply we discover Darwin was correct. The skeptics are simply wrong when they claim that probing deeply will reveal that humans are unique in mind, rationality, or morality, for example (for extensive discussion see Corbey, 2005; Bekoff and Pierce, 2009, 2012).

It somewhat surprises me that we are still considering the general topic of speciesism given what we know about animal cognition, emotions, and sentience. However, hierarchical speciesist thinking that suggests we're "higher, better, or more valuable" than other animals feeds into skeptical minds because it offers a psychological barrier justifying human superiority and exceptionalism. It's the same sort of denial that we encounter in discussions of global climate change and the role we play in unprecedented losses of biodiversity (Bekoff, 2010, 2013b; Norgaard, 2011).

Religious and cultural proclivities, economic contingencies, political alignments, and dogma all factor into misrepresenting animals and ignoring biology so that we can do anything we want to them. Mass media also are responsible for putting forth false views of animals (Freeman *et al.*, 2011). As Raymond Corbey (2005: p. 185) notes, "That there is an essential human nature and an essential difference between humans and other animals still informs much of what we do and how we view other animals – our thought, scientific research, laws, and practices." However, humans are not "better" or "higher" than individuals of other species. Detailed evolutionary and comparative studies of animal behavior and animal minds (cognitive ethology; Allen and Bekoff, 1997; Bekoff 1998b; Allen, n.d.) in a wide range of taxa readily inform discussions about speciesism and show how weak and misleading speciesistic thinking truly is. Simply put, animals have to do what they need to do to be card-carrying members of their species. We share many similarities with other animals but we're also rather different; *but different does not mean better.*

Speciesism also ignores individual differences within species that often are greater than the differences we observe between species (Bekoff and Gruen, 1993). Thus speciesism leads to dubious conclusions because while some of the comparisons that are made seem to make sense and are superficially pleasing, they actually lack real merit on close inspection.

Denialism also comes into the picture. We're extremely skilled at denying what is happening right in front of our eyes (Specter, 2009), including what we know about the amazing cognitive, emotional, and moral capacities of animals and the pain and suffering they endure (Wicks, 2011). Not only do many deny what science has well demonstrated, they also don't believe in it (Mooney, 2011). There's an amazing and self-serving disconnect between the way in which some people view animals and who they really are. As *Homo denialus* we readily "see no evil, hear no evil, or smell no evil," although if we open our senses even a little bit to our surroundings it's impossible to ignore or deny what's happening and the dire consequences of our actions, including the misclassification of other animals. We ignore and redecorate nature in incredibly self-serving ways, as if we're the only species that matters. It shouldn't be all about us but it often is.

"Surprises" in the study of animal sentience: how "low" can we go?

In recent years we've discovered a number of "surprises" in the cognitive and emotional capacities of animals (Bekoff, 2007a,b, 2010, 2013a,b; Benz-Schwarzburg and Knight, 2011). I put the word "surprises" in quotation marks because I don't think they're really all that surprising unless one believes that other animals can be placed on a hierarchical scale. For example, it's still an open question whether insects and other invertebrates feel pain. We do know bees are able to solve complex mathematical problems more rapidly than computers (Lihoreau *et al.*, 2010; for more on ants see also Kirksey, this

Figure 1.2 A female *Sitticus fasciger* jumping spider, 4 mm, somewhere in the United States. Jumping spiders are discerning predators with sophisticated cognition and communication. Photo Thomas Shahan.

volume). Bees solve what's called "the traveling salesman problem"[1] and rapidly learn to fly the shortest route between flowers using their tiny brains. Recent research on octopi shows clearly that they feel pain and perform complex patterns of behavior that readily yield to cognitive explanations invoking sentience (Bekoff, 2010; Montgomery, 2011). Indeed, octopi were protected from invasive research in the United Kingdom years before the great apes (Bekoff, 2010). Furthermore, the research in the field shows that what is observed in captivity is extremely misleading because captive octopi do not show nearly the range of behaviors that wild relatives show.

We also know spiders hunt with intention (see Figure 1.2; Wilcox and Jackson, 2002); that individuals of two unlikely species, chickens (Edgar *et al.*, 2011) and mice (Langford, 2006), display empathy; that birds manufacture and use sophisticated tools (more so than chimpanzees; Bluff *et al.*, 2010),[2] visually monitor group size and are aware of, and change their behavior, based on what other individuals are doing

(Bekoff, 1995, 1996), plan future meals, and display other unanticipated cognitive skills.[3] The results of these research projects are legitimized by being published in highly respected peer-reviewed professional journals.

Fish also have been victims of underhanded and shady speciesism. We know, for example, that fish display long-term memory and clearly are sentient beings (Braithwaite, 2010, 2011). Consider what Verheijen and Flight (1997: p. 362) write about fish: "There is a growing consensus that because of homology in behaviour and nervous structure all vertebrates, thus also fish, have subjective experiences and so are liable to suffer." Consider also what Victoria Braithwaite writes about fish: "I have argued that there is as much evidence that fish feel pain and suffer as there is for birds and mammals – and more than there is for human neonates and preterm babies" (Braithwaite, 2010: p. 153). She also notes (Braithwaite, 2011), "We now know that fish actually are cognitively more competent than we thought before – some species of fish have very sophisticated forms of cognition... In our experiments we showed that if we hurt fish, they react, and then if we give them pain relief, they change their behavior, strongly indicating that they feel pain." Fish have also been observed punishing other individuals who steal their food (Raihani *et al.*, 2010).[4]

So, how "low" can we go when line drawing and excluding various species from moral consideration because they don't feel pain or because they're not "all that smart" or "all that emotional"? Not "very low" at all if we pay attention to what solid scientific research has shown us over and over again, namely that line drawing is bad biology and ignores what we know about a vast and diverse array of other animals.

Tigers in Bangladesh: sentience, euthanasia, and speciesism in the real world

On 30 March 2011, I received an email from Christina Greenwood, Project Manager of the Sundarbans Tiger Project (2011), about a conservation project in Bangladesh dealing with tiger–human conflict, overseen by the Wildlife Trust of Bangladesh and the Zoological Society of London. It arrived at the best and worst of times. On the one hand, it couldn't have come at a better time, as I was reading some of the essays for the roundtable from which this book emerged and struggling with how to reconcile my own inability to come to terms with what I would like the world to be like and what it really is. On the other hand, it came as I was rushing about to leave town and I didn't need more on my plate. But it served as an important wake-up call that moved me out of my comfort zone and the net effect was that it came to be a most welcomed email.

The reality of the situation in Bangladesh made it a perfect case study because it brought into focus many of the issues with which my colleagues and I wrestle each and every day that all center on the daunting question, "Who lives and who dies?" Greenwood wrote that she and her colleagues were working on a tiger conservation

---

[3] http://en.wikipedia.org/wiki/Bird_intelligence
[4] www.time.com/time/printout/0,8816,1952458,00.html

project in Bangladesh and one of the issues they face is tiger–human conflict. Around 50 people per year are killed by tigers when they enter the forest for fishing or wood collection, and when tigers stray into villages adjacent to the forest villagers beat them to death or put out poison bait (in contrast, Bakels, this volume p. 161, notes how people in other cultures also kill tigers who trespass into villages but that "the dead body is not scolded or ridiculed, but treated with respect").

Greenwood wondered if tigers who come into villages should be euthanized, especially the injured ones? Would this really be the most humane choice since there were no obvious options? Of course, Greenwood and her team wanted to do the most ethical thing for both the tigers and the people. However, it's a very tricky situation in that the villagers are very poor, veterinary care for injured tigers is extremely limited, there aren't any rehabilitation centers, and zoos are in bad condition and can't take in any additional animals. Collaring animals and learning more about their movement patterns might help in the future (Siddique, 2011) but decisions had to be made immediately.

I struggled with what I would likely do in the real world (or what Lori Gruen, personal communication, calls the "non-ideal world"). Here is what I hurriedly wrote: "I'm leaving for Amsterdam tomorrow so let me give you a brief summary of some of my ideas about this because there are no 'easy' or 'fast' answers – in the best of all possible worlds, which this is not, such conflicts would be avoided or there would be enough money to take care of the problem tigers – good zoos or rehab centers with proper veterinary care – clearly this isn't the case – so the way I see it, cutting through the chase, is that the tigers are ultimately likely to suffer different sorts of deaths because there's no money to care for them in ways that could give them a 'good' life – life in a zoo would be horrible, there are no rehab centers, they can't get the proper veterinary care, and if they're left to their own they'll be poisoned – so it's a matter of our having to make life and death decisions for them, something we do for billions of other animals... while I know some might/would disagree with me, and I wish I had more time right now to write more, I see that in some/many instances in the situations with which you're faced, euthanasia – mercy killing to avoid prolonged suffering and likely/sure death – would be the more ethical option – of course each situation should be considered on its own – case by case – but it *seems* to me that in the dire situations and options you describe, euthanasia would be more humane and ethical then putting a tiger(s) in a zoo or leave him or her on their own to die in the wild or be poisoned – I deeply wish it weren't so..."

What if the tigers were ants, fish, birds, or rats? Compassionate conservation to the rescue

After I sent my email I sat at my desk for more than an hour wondering if I had done the best I could and every day since I have wondered what other options might be available in this situation, one that is mirrored in many other places around the world. What if this were about a less charismatic species or individuals we thought to be insentient? Would we still struggle with the decision to euthanize, kill, these individuals? Would Greenwood and her team be working on this project? We squash ants, overfish, killing billions of individuals without blinking or thinking about how they suffer when we do so, and poison millions of birds and brown rats as if their lives didn't matter to them (see for example, Stolzenburg, 2011). Nonetheless, the fact is that we're not even very good at protecting charismatic species (Bennett, 2011). And they're not the only species who really matter. Consider what renowned conservation biologist Edward O. Wilson (n.d.) has written: "If all mankind were to disappear, the world would regenerate back to the rich state of equilibrium that existed ten thousand years ago. If insects were to vanish, the environment would collapse into chaos."

Speciesism in the real world directly informs the practical strategies we choose to use. For example, we need to be able to identify those characteristics of an individual or species that warrant keeping them alive or allowing them to suffer or die, and when we factor in ecological variables this becomes a difficult practice. Nonetheless, we treat "higher/more valuable/special" species differently from and better than "lower/less valuable/not all that special" species – although these designations, based on our attempts to draw lines separating species, are fraught with error and ignore nature including evolutionary continuity (Bekoff and Gruen, 1993; Bekoff, 2010, 2013a).

Questions and issues such as these keep me awake at night, and I'm sure I'm not alone, tossing and turning and wrestling with reality. So just what is the most compassionate thing we can do? Is killing in the name of conservation a compassionate option? Should individuals be traded-off for the good of their own or other species or for ecosystem integrity? Can we really put a price on an animal to reflect their value? What does compassionate conservation entail? Can we reconcile the differences between those people interested in individuals and those more interested in species, populations, and ecosystems (for detailed discussion see Aitken, 2004)? How can we "rewild" our hearts and bring animals back into the picture where they truly belong (Bekoff, 2013b)? How can we build and maintain corridors of compassion and coexistence? How can we cross disciplines and work together? How can we walk the talk so that something actually gets done? Can we do better?

These questions were dealt with at the first international meeting on compassionate conservation held in Oxford, England (Compassionate Conservation, 2010) and followed up with a session at the 2011 Asia for Animals meeting held in Chengdu, China. The demands of the real world require us to expand our personal and professional comfort zones and think and act "out of the box" because we simply cannot do everything that needs to be done. Compassionate conservation is the perfect catalyst for such a move. If we continue to redecorate nature, and we surely will, we must do so with compassion and caring for the animals who we are displacing and killing.

We're cognitive relatives yet moral strangers: matters of mind inform matters of welfare

One of my major goals is to make the case that the time has come to expand our compassion footprint (Bekoff, 2010) and get rid of dualisms –"them" versus "us" – that simply do not work. Even the Great Ape Project (Cavalieri and Singer, 1993) was speciesist (Burghardt, 1997; Bekoff, 1998a; Corbey, this volume; Dunayer, this volume) but it was important for practical reasons to start somewhere, and it was thought that beginning with great apes would likely meet with the least resistance. I argued that we should rename the Great Ape Project as the Great Ape/Animal Project. I agree with Burghardt's (1997) concern about the original Great Ape Project: "As one who believes the true test of our respect for other animals lies in our treatment of venomous snakes and large carnivores, I (too) am wary of a creeping speciesism inherent in the proposal set forth here." Research shows we're "just one of the group," as ecopsychologist David Abram (personal communication) says. And, as Judith Benz-Schwarzburg and Andrew Knight (2011) note in their review of animal cognition and its moral relevance, many animals are our "cognitive relatives yet moral strangers." Surely we can do better in how we treat them?

It is easy to expand sentience and broaden our circle of concern for all animals, including those called pejoratively and incorrectly "lower" animals. These animal beings exist and deserve respect but it's a major concern that what we now know about insects, invertebrates, fish, mice and other rodents, chickens and other birds, for example, has *not* led to any revision of the federal Animal Welfare Act in the United States (Bekoff, 2010). These species are not protected from invasive research. However, the Treaty of Lisbon (n.d.) passed by member states of the European Union that came into force on December 1, 2009, recognizes that "formulating and implementing the Union's agriculture, fisheries, transport, internal market, research and technological development and space policies, the Union and the Member States shall, since animals are sentient beings, pay full regard to the welfare requirements of animals, while respecting the legislative or administrative provisions and customs of the Member States relating in particular to religious rites, cultural traditions and regional heritage."

The scientific community has created a set of professional standards that is supposed to guide scientists in how to develop and conduct their research to, as much as possible, preserve animal welfare, and in theory the federal Animal Welfare Act protects animals used in research in the United States. But these have so far been inadequate safeguards. Only about 1 percent of animals used in research in the United States are protected by this legislation, and the legislation is sometimes amended in nonsensical ways to accommodate the "needs" of researchers. For instance, here is a quote from the Federal Register (volume 69, number 108, Friday June 4, 2004): "We are amending the Animal Welfare Act (AWA) regulations to reflect an amendment to the Act's definition of the term *animal*. The Farm Security and Rural Investment Act of 2002 amended the definition of *animal* to specifically exclude birds, rats of the genus *Rattus*, and mice of the genus *Mus*, bred for use in research."

It may surprise some people to hear that birds, rats, and mice are no longer considered animals, but that's the sort of speciesist logic that epitomizes federal legislators: because researchers are not "allowed" to abuse animals, the definition of "animal" is simply revised until it only refers to creatures researchers don't need.

We need to be patient as science collects evidence and we cannot continue to pick the traits we use for comparison for self-serving convenience. We must shift the burden to the naysayers – the dualists – and force them to "prove" that speciesism represents reality. We should care about these misrepresentations because there are numerous practical and moral consequences. Of course, they can't.

Overcoming specious speciesism with compassion and optimism: the need for a new social movement

The late theologian Thomas Berry (1999) stressed that our relationship with nature should be one of *awe*, not one of *use*. Individuals have inherent or intrinsic value because they exist and this alone mandates that we coexist with them. They have no less right than we do to live their lives without our intrusions, they deserve dignity and respect, and we need to accept them for who they are. Deep ethology (Bekoff, 1998a), studies of animals that take us not only into their minds but also into their hearts, is a beginning as we expand our compassion footprint. Deep ethology will help us overcome speciesism by broadening the array of animals who receive protection from invasive research and other forms of abuse including our seemingly innate tendencies to fill every inch on Earth and in the air and in water.

We're all over the place and it's arrogant to think we can pick and choose where we have impacts, employing speciesistic criteria for convenience, for we have impacts everywhere. Many different animals display compassion for others, so it's natural to call upon compassion to alleviate the suffering of others. *It is natural and "animal" to be fair, good, compassionate, empathic, and moral* (Bekoff, 2010, 2011, 2013a, b; see also Bekoff and Pierce, 2009; de Waal, 2009; Keltner, 2009; Corning, 2011) and optimistic (Sharot, 2011).

Academics, advocacy, and activism go hand in hand. It is impossible to be neutral on issues of animal or environmental protection. Many scientists like to think science is objective and that they don't have an agenda and that they have no obligation to interact with non-researchers, but this is rarely the case (Bekoff, 2010). Indeed, it shouldn't be. In his wonderful and bold book, *A World of Wounds*, renowned biologist Paul Ehrlich (1997) wrote: "Many of the students who have crossed my path in the last decade or so have wanted to do much, much more. They were drawn to ecology because they were brought up in a 'world of wounds', and want to help heal it. But the current structure of ecology tends to dissuade them... Now we need to incorporate the idea that it is every scientist's obligation to communicate pertinent portions of her or his results to decision-makers and the general public." I could not agree more.

It's also the case that "more science" will not necessarily make us more compassionate or optimistic (Bekoff and Bexell, 2010). It's who we are. So, let's tap into our moral inclinations to make the world a better place for all beings. Science alone doesn't hold the answers to the current crisis nor does it get people to act. As historian Lynn White (1967) wrote in his classic essay "The historical roots of our ecological crisis" – "More science and more technology are not going to get us out of the present ecological crisis until we find a new religion, or rethink our old one." More than four decades later this claim still holds: we don't need more science to know we need a new mindset and social movement that centers on compassion and being proactive. The "putting out the fires" mentality and speciesistic thinking simply haven't worked and cannot work.

We must remain open to new discoveries although science is not all that convincing for many people, including policy-makers. Along these lines, Michael Shermer (2011: 4) writes, "the majority of our most deeply held beliefs are immune to attack by direct educational tools, especially for those who are not ready to hear contradictory evidence. Belief change comes from a combination of personal psychological readiness and deeper social and cultural shift in the underlying zeitgeist, which is affected in part by education but is more the product of larger and harder-to-define political, economic, religious, and social changes" (see also Clayton and Myers, 2009; Cooney, 2011; Bekoff, 2013a,b). Joe Zammit-Lucia (2011) says it poignantly, "Conservation is all about people" (see also Schulz, 2011).

Compassion begets compassion and there's a synergistic relationship, not a trade-off, when we show compassion for animals and their homes (Bekoff, 2010). There are indeed many reasons for hope (Goodall, 1999; Goodall *et al.*, 2009). In the future we must harness our basic goodness and optimism, and work together as a united community. We can and must look to the animals for inspiration and evolutionary momentum (Bekoff and Pierce, 2009).

First do no harm: the importance of humane education

Rather than continuing to wantonly ignore, destroy, and redecorate nature, it is important that we expand our compassion footprint and recognize that individuals count and that emotional and moral intelligence have evolved in many other species. A good place to begin to overcome the rampant effects of speciesism is with children by supporting programs in humane education and conservation education that stress and encourage coexistence and peaceful relationships among all beings (e.g. Verbeek, 2008). And a good rule of thumb to stress in our encounters with our own and other species is "First do no harm."

So, what would a global moral imperative look like as we broaden our taxonomic concerns? Guiding principles would be (1) do no intentional harm, (2) respect all life, (3) treat all *individuals* with respect and dignity, and (4) tread lightly when stepping into the lives of other animals. Let's give it a try. There are numerous animals and habitats worth protecting and saving so let's get on with it now, because time is not on our side.

It is essential to recognize other animals for whom they are. Colleen Patrick-Goudreau (2013) says it well:

My hope is that we can navigate through this world with the grace and integrity of those who need our protection. May we have the sense of humor and liveliness of the goats; may we have the maternal instincts and protective nature of the hens and the sassiness of the roosters. May we have the gentleness and strength of the cattle, and the wisdom, humility, and serenity of the donkeys. May we appreciate the need for community as do the sheep and choose our companions as carefully as do the rabbits. May we have the faithfulness and commitment to family of the geese, the adaptability and affability of the ducks. May we have the intelligence, loyalty, and affection of the pigs and the inquisitiveness, sensitivity, and playfulness of the turkeys. My hope is that we learn from the animals what we need to become better people.

While some may scoff that this is "too anthropomorphic" and not scientific, they are wrong. Much of what she writes is now supported by solid research and there is no doubt that future research will broaden not only the number of other animals we recognize to be sentient, intelligent, emotional, and moral beings but also will reveal additional "surprises" about what they know and feel. Following up on Darwin's ideas about evolutionary continuity, it is clear that it is *not* being anthropomorphic to say we know other animals have deep and rich emotional lives. Rather, we're identifying commonalities and then using human language to communicate what we observe. It's simply bad biology to rob other animals of their emotional lives. Anthropomorphism is not anti-science.

When we ignore *who* other animals really are we ignore nature. We must use what we know about animal minds to develop a new and inclusive ethic that blends respect, caring, compassion, humility, generosity, kindness, grace, and love to motivate wide-ranging action on the behalf of other animals. Animals depend on our goodwill for their well-being. A new paradigm based on the principle "first do no harm" will allow us easily to expand our compassion footprint by paying careful attention to what animals want and need from us. It's long overdue.

We all must reflect on how all too conveniently we ignore and redecorate nature. The animals are constantly telling us what they want and need and we well know that their desires and preferences are much like our own – to avoid pain, and to feel safe and secure in their homes and surroundings. Their manifesto, simply put, is treat us better or leave us alone (Bekoff, 2010). As we save other animals, we'll be saving ourselves.

The time has come to debunk the myth of human exceptionalism based on bad science and speciesism. It's a hollow, shallow, and self-serving perspective on who we are and who "they" are. Of course we are exceptional in various arenas as are other animals. Perhaps we should replace the notion of human exceptionalism with species exceptionalism, a move that will force us to appreciate other animals for who they are, not who or what we want them to be. Indeed, it might be more appropriate to concentrate on *individual exceptionalism* because it is *individuals* who matter when discussing their unrelenting abuse in education and research and for food, clothing, and entertainment.

Specious speciesism undermines our collective efforts to make the world a better place for all beings and undermines a social movement concerned with social justice for all beings, human and non-human alike. Part of the strategy for rewilding our hearts (Bekoff, 2013b) will be to get out of this misleading, false, and destructive mindset and get rid of speciesism once and for all. It is that misleading and harmful.

2 The rights of sentient beings
Moving beyond old and new speciesism

Joan Dunayer

The "state of the art" egg factory comprised four windowless warehouses imprisoning a total of half a million hens. Allowed to enter the building with the youngest hens (five months old), I saw the birds crammed nine to a cage, unable to walk or even lift a wing. The hens were nearly as tall as their wire cage (see Figure 2.1). Row after row, four tiers of cages extended into the distance, disappearing into the dimly lit haze. From manure pits directly below, huge mounds of excrement saturated the air with eye-stinging ammonia. Cagemates shared a single water nipple and were forced to squeeze past one another to reach the food trough in front of their cage. In bursts the birds gave frantic cries. With a dazed look, they stared outward, as if into empty darkness.[1] Over time, I knew, many of the hens would develop deep, infected wounds (Johnson, 1991: p. 123). Many would get caught in the cage wire and die unable to reach food or water (Singer and Mason, 2006: p. 38). Because they would remain nearly immobilized and continuous egg production would rob their bones of calcium, many would lose the ability to stand or even sit upright (Cutler, 2002: p. 480). Many would suffer broken bones (Marcus, 2001: p. 108). Until their death, all would experience relentless suffering.

Such suffering is entirely unnecessary. Humans don't need to eat eggs or any other food from non-human animals. Non-humans need basic rights, including rights to life and liberty, to protect them from human abuse. Humans who would withhold such rights show a form of bigotry as unjustifiable as racism or sexism: speciesism.

Old speciesism

Psychologist Richard Ryder (1992) coined the word *speciesism* in 1970. Although he didn't explicitly define the term, he stated that speciesists don't "extend our concern about elementary rights to the non-human animals" (Ryder, 1992: p. 171). Similarly, philosophers Peter Singer (1990: p. 6) and Tom Regan (2001: p. 170) define speciesism as bias toward humans and against all other animals. That definition is too narrow.

[1] The egg factory was Country Fair Farms in Westminster, Maryland. I visited the facility on January 25, 1993.

The Politics of Species: Reshaping our Relationships with Other Animals, eds R. Corbey and A. Lanjouw. Published by Cambridge University Press. © Cambridge University Press 2013.

Racism includes bias toward or against any number of races – for example, bias toward whites and Asians or bias against blacks and Native Americans. Analogously, species-ism includes bias toward or against any number of animal species, such as bias toward humans and other apes, bias toward primates and cetaceans, bias against all non-mammals, or bias against all invertebrates. What Ryder, Singer, and Regan call "speciesism" actually is only one type of speciesism – the oldest and most severe form, which I refer to as "old speciesism." Old-speciesists don't believe that any non-humans should receive as much moral consideration as humans or have basic legal rights.

Old speciesism relies on a double standard. According to one old-speciesist argu-ment, non-humans shouldn't have legal rights because they can't enter into contracts. Young children and numerous human adults with mental disabilities also can't enter into contracts; nonetheless, they possess legal rights.

Commonly, old-speciesists point to the Bible as authorizing humans to enslave and kill other animals. Even if we overlook the irrationality of believing that the Bible represents divine communication, Biblical endorsement fails as a defense of old speciesism because the Bible also endorses human slavery and, in many situations, murder.

Some old-speciesists argue that only humans deserve rights because humans are more intelligent than other animals. Even if intelligence is defined as the type of intelligence typical of humans, many non-humans are more intelligent than many humans. According to tests designed for humans, the gorilla Koko has an IQ of about 90 (Patterson and Gordon, 1993). Many humans have much lower IQs. Koko understands spoken English and communicates in American Sign Language, using a vocabulary of more than a thousand words (Patterson and Gordon, 1993). Some human adults have no language ability. Even by conventional human standards, the average pigeon or rat possesses greater learning capacity and reasoning ability than many humans with mental disabilities. In many ways a mature catfish is more cognizant than a newborn or senile human. Any intelligence criterion that excludes all non-humans also excludes many humans.

Another old-speciesist claim is that only humans need rights because they have a greater capacity for suffering than non-humans. With respect to both physical and psychological suffering, that claim is untenable. In some ways, non-humans are appar-ently more sensitive to pain than humans. For example, the nerve endings in a rainbow trout's skin react to noxious pressure at a lower level than those in human skin. Trout skin has a threshold comparable to that of a human cornea (Lynne U. Sneddon, 26 August 2003, email to author). When handled, a trout may feel as you would if someone were pressing on your cornea. Humans with congenital analgesia can't feel any pain at all.

Many non-humans clearly experience severe psychological suffering, such as intense fear and grief (Dunayer, 2001). For example, many birds and non-human mammals deeply, permanently mourn the death of a close companion (King, this volume). Humans with a psychological condition termed "blunted affect" apparently feel little emotion. Having some sense of control can make adversity more bearable (McMillan, 2005b: p. 191). Non-human victims of inescapable human abuse can't make sense of their plight, change their circumstances, or foresee an end to their situation. These factors may exacerbate their suffering.

Missing the irony, some old-speciesists contend that humans deserve special moral consideration because they're morally superior to other animals. Non-humans rarely kill or seriously injure other animals except to survive, as when a predator kills prey. When non-humans do needlessly harm others, they might not recognize the harm as needless or harmful. Like young children and mentally incompetent adults, they can't reasonably be held accountable. In contrast, humans routinely torture and kill other animals without need, even for recreation or entertainment (as in sportfishing and bull-"fighting"). Also, many humans abuse other humans. For example, child abuse and wife beating are widespread. Buyers of animal-derived food or clothing willingly participate in the needless infliction of suffering and death for trivial reasons such as convenience, taste preferences, and vanity. When it comes to cruelty, humans have no rivals. Sociopaths completely lack empathy. Sadists take pleasure in causing suffering. And if morality is limited to self-conscious morality, neither human infants nor many human adults with severe mental disabilities are moral. They still have rights.

According to some old-speciesists, humans warrant more protection than other animals because humans have more value to others. Many non-humans are greatly valued by their kin or other companions. Some humans are valued by no one. Compare a dog loved by a large human family, or bear cubs loved by their mother, to an unloved human recluse or orphan.

Nor do humans necessarily have more value to others in the sense of having a more positive effect. Overall, humans tend to have negative impact on other humans, with whom they compete for opportunities and resources. Also, most humans have an extremely negative impact on non-humans, whom they harm both directly (e.g., in vivisection and animal "agriculture") and indirectly (e.g., by supporting non-human exploitation and destroying non-human habitat). Many humans provide care to other individuals and otherwise contribute to their communities, but so do many non-humans. Unlike most humans, free-living non-humans also contribute to their ecosystems. For example, insects of numerous species pollinate a host of plants that provide food to countless other animals (including humans). In terms of their lives' objective value to most other beings, humans probably rank lowest of all animals.

If old-speciesists value all humans more than all non-humans, that valuation simply reflects their personal preference. We're all entitled to our preferences; however, we're not entitled to translate our preferences into privileges for some at others' expense. Most humans favor members of their own family and ethnicity. Favoritism isn't a just basis for rights. Indeed, human rights act as safeguards against bias. So would non-human rights.

In sum, arguments for favoring all humans over all non-humans are easily refuted. In the case of any capacity (e.g., sensitivity, intelligence, or morality) used as a criterion for rights, at least some non-humans possess the capacity as much as, or more than, some humans. Also, many non-humans surpass many humans in their value to others. Old speciesism is indefensible.

New speciesism

As previously explained, the Singer–Regan definition of speciesism is too narrow in that it limits speciesism to old speciesism, bias against all non-humans. The definition is too narrow in another way: it limits speciesism to bias based solely on species membership (Regan, 2001: p. 181; Singer, 2003), excluding bias based on species-typical characteristics. According to Singer, speciesism "refers to discrimination on the basis of species, not to discrimination on the basis of cognitive capacities" (quoted in Raha, 2006: p. 19). However, Singer's cognitive criteria for a right to life are themselves based on species. He advocates a right to life only for animals as rational and self-aware as a normal human beyond earliest infancy (Singer, 1993: p. 101). Why a normal human? Why not a normal octopus or crow? Clearly, Singer's criteria are human-centered and human-biased – speciesist.

Singer (1990: p. 21) contends that giving preference to human life over non-human life isn't speciesist if the preference is "based on the characteristics that normal humans have" and not on species per se. To fully recognize the falsehood of that claim, consider comparable statements about race and gender: giving preference to the lives of whites over the lives of blacks isn't racist if the preference is based on characteristics typical of whites rather than on race per se; giving preference to the lives of men over the lives of women isn't sexist if the preference is based on characteristics typical of men rather than on gender per se.

Singer's philosophy exemplifies what I've termed "new speciesism." In contrast to old-speciesists, new-speciesists believe that moral and legal rights should extend beyond humans. However, they advocate rights for only some non-humans, those who seem most human-like. Believing that most humans are superior to all non-humans, new-speciesists see animalkind as a hierarchy with humans at the top. Typically they regard chimpanzees, dolphins, and other highly cerebral non-human mammals as more important than other non-humans. They also rank mammals above birds; birds above reptiles, amphibians, and fishes; and vertebrates above invertebrates.

Singer advocates rights to life and liberty only for humans, other great apes, and perhaps other mammals (Cavalieri and Singer, 1993: p. 4; Singer, 1993: p. 132). In his view, all non-mammals and possibly many mammals "do not qualify for a right to life" because they have no sense of their past and future (Singer, 1993: pp. 119–20). Singer's premise that non-mammals have no memory of their past or expectations for their future is factually incorrect. Some crows drop shell-encased fruit onto roads and wait for motor vehicles to run over the shells, breaking them open (Shumaker et al., 2011: p. 207). Why do the crows wait if they lack a concept of their future? Toads who are stung by a honeybee subsequently avoid bees (Kavaliers, 1988). Their reaction is learned. Once minnows have been attacked by a pike, or merely seen other minnows being attacked, they flee at a pike's scent. They don't flee at this scent until they've had some experience of pike predation (Kelley and Magurran, 2003). When seeking a shelter, mantis shrimps avoid an empty cavity that contains the odor of a mantis shrimp who previously defeated them in a competitive encounter but enter a cavity that contains the odor of a mantis shrimp whom they previously defeated (Griffin, 1992: p. 199). How could they do that without remembering their past, anticipating their future, and having a sense of self? Fruit flies "run toward odors previously associated with sugar" and "run away from specific odors that they previously experienced with electric shock" (Waddell and Quinn, 2001: p. 1284). (I'm opposed to all experimentation that harms sentient beings; I reluctantly cite vivisection results to convince readers who place special stock in such evidence.) Like humans, fruit flies have both short-term and long-term memory (Waddell and Quinn, 2001). Across phyla, animals remember and anticipate.

According to Singer (1993: pp. 89–98, 2003), a typically human sense of past and future entails long-term goals and such goals are relevant to a right to life. In reality, there's nothing inherently better about long-term goals, which may be selfish (e.g., the goal of amassing wealth), have overall negative consequences for others (e.g., the goal of wielding unjust power), or simply remain unachieved (resulting in frustration and disappointment). Highly future-oriented individuals may be less benign and/or less happy than individuals whose goals are more immediate. Why, then, propose long-term goals as a criterion for a right to life? I see no reason other than speciesism: a typically human sense of past and future is best because it's typical of humans.

Singer also links strong social ties to the value of a life and to the rights to life and liberty that he endorses for humans, other great apes, and possibly other mammals (Cavalieri and Singer, 1993; Singer, 2001–2002). Because of humans' "close family ties," he claims, "it is not speciesist" to believe that "the killing of several thousand people" is "more tragic" than "the killing of several million chickens" (Singer, 2001–2002). Actually, it's deeply speciesist. Chickens (Burton, 1978: p. 72), other birds (Eiseley, 1959; Barber, 1993, pp. 81–6), and at least some fishes (Brown, 1984) are known to form strong personal relationships, but Singer ignores non-mammalian social ties. Also, in his comparison the total number of humans who either die or grieve the loss of a family member doesn't come close to the number of chickens who die, so Singer gives far more weight to humans' loss of a loved one than to chickens' loss of their very lives (including their capacity for strong social ties).

Singer's entire social-ties criterion is speciesist. First, like long-term goals, social ties don't equate to a richer life. Social animals need social ties to be happy; asocial animals don't. Also, social ties can involve negative feelings such as stress, anger, envy, and shame. The end of loving relationships causes grief. Second, individuals who are loved, respected, or otherwise valued by some may be hated or despised by others. Third, although group behavior can be highly constructive, it also can be extremely destructive. Only social beings engage in ostracism, maintain power hierarchies, gang up on individuals or minorities, or wage war. Finally, Singer indicates that he considers relationships between humans automatically most important. He states, "Notoriously, some human beings have a closer relationship with their cat than with their neighbours" (Singer, 1993: p. 76). Why should a human and cat who live together *not* have a closer relationship than humans who are merely neighbors? Singer's social-ties criterion is simply another criterion based on humanness.

Singer (1993: p. 61) also claims that a capacity for "abstract thought" is a non-speciesist criterion for a right to life. By itself, that criterion involves intellectual elitism but not necessarily speciesism. However, within the context of Singer's argument, the criterion reflects speciesist bias toward the "mental powers of normal adult humans"

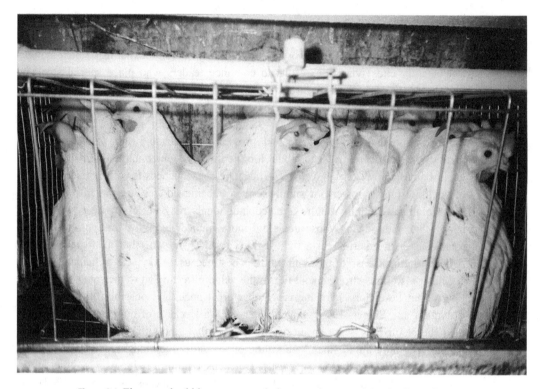

Figure 2.1 Five-month-old hens, crammed nine to a cage, at a Maryland egg factory. Photo by Joan Dunayer.

(Singer, 1993: p. 60). Given that Singer doesn't advocate a right to life for any non-mammal, his failure to acknowledge abstract thought in non-mammals indicates the speciesist (and incorrect) assumption that a non-human's capacity for abstract thought largely depends on the individual's biological relatedness to humans. As demonstrated in experiments, non-mammals such as African gray parrots (Pepperberg, 1988), pigeons (Wright *et al.*, 1988), and (evolutionarily distant from humans) honeybees (Giurfa *et al.*, 2001) are capable of abstract thought. In any case, a capacity for abstract thought doesn't signify a happier or more valuable life. A brilliant physicist may be intensely unhappy and highly destructive. Sensual pleasures can be at least as gratifying as intellectual ones. Is the pleasure that a dog feels running through a meadow, a lizard feels basking in the sun, a condor feels soaring at a great height, or a human feels smelling a rose more or less than the pleasure a human experiences when formulating a theorem? No one can say. Also, the answer partly depends on the circumstances and the individual. Like many other humans, Singer values abstract thought more than other ways of thinking. Offering no valid reason for his view that abstract thought makes a life more valuable, he merely expresses his own bias toward yet another characteristic typical of humans.

Similarly, Singer (1993: p. 61) contends it is not speciesist to consider complex communication, also characteristic of humans, relevant to a right to life. As with his other criteria for rights, Singer doesn't address the fact that complex communication isn't limited to mammals. In English, the African gray parrot Alex would request various foods and toys; solicit information; identify more than a hundred objects by name, shape, material, number, color, and size; and express emotions such as frustration, regret, and love (Pepperberg, 1994). Another African gray parrot, N'kisi, reportedly has an English vocabulary of approximately a thousand words, which he uses correctly and creatively in both familiar and novel contexts (Kirby, 2004). N'kisi also modifies verbs to form past, present, and future tense (Kirby, 2004).

Further, like non-human mammals, non-mammals have their own forms of communication, which humans are just beginning to decipher. For example, alligators and crocodiles communicate through infrasound, at too low a frequency for our unaided ears (Garrick and Lang, 1977). Electric fishes "read" one another's discharges, which vary according to the signaler's age, species, individual identity, and intentions, such as courtship or rivalry (Bleecker, 1975). Honeybees possess symbolic language. Having found food, honeybee workers return to their hives and communicate its location to their sisters. They dance a message of direction and distance that varies with the time of day and the food source's unique location (Griffin, 1992: pp. 182–5).

Most importantly, there's no reason other than bias to select complex communication as a criterion for a right to life. We would be equally justified in selecting perceptual powers that humans lack and asserting that such powers make a life more valuable. Many non-mammals see ultraviolet light, have a nearly 360-degree visual field, or perceive the direction in which light waves are vibrating. Many sense infrared energy or variations in the intensity of the Earth's magnetic field. Through infrasound, homing pigeons can detect thunderstorms, waves breaking against the shore, and winds passing over mountains at distances up to thousands of miles (Barber, 1993: p. 67). Compared to humans, some sharks are millions of times more sensitive to electrical fields – so sensitive that they can locate small fishes hidden in sand by sensing the electric potentials of their heartbeats (Griffin, 1992: pp. 26, 38). It's presumption – and speciesism – to rate the richness of other animals' lives according to what constitutes richness within most human lives.

Without the support of evidence or logic, new-speciesists believe that human lives generally are richest and most important. Singer (1993: p. 86) acknowledges that his criteria for rights are characteristics typical of humans. With regard to each criterion, he treats the capacities of "normal adult humans" as the gold standard. According to James Rachels (1990: p. 209), another proponent of new speciesism, the seriousness of killing non-humans depends on the extent to which "they have lives similar to our own." Whereas old speciesism is based strictly on membership in the human species, new speciesism is based on characteristics typical of humans. Both forms of speciesism are rooted in bias toward humanness.

Non-speciesism

Paola Cavalieri (2001: p. 70) has noted that speciesism includes "any form of discrimination based on species." As I've shown, such discrimination can be based either on species membership or on the characteristics typical of one or more particular species.

Therefore, for full clarity I define speciesism as follows: a failure, on the basis of species membership or species-typical characteristics, to accord any sentient being equal consideration and respect. Whereas old-speciesists advocate rights only for humans, and new-speciesists advocate rights only for those non-humans who seem most like humans, non-speciesists reject human-biased criteria for rights and advocate rights for all sentient beings.

Sentience, a capacity to experience, is necessary and sufficient for rights. Why? By definition, sentient beings can experience suffering and well-being. Also, sentient beings lose all opportunity for pleasant experiences when they die. (Insentient things such as plants don't lose anything when they die because they're completely unaware whether alive or dead. They have moral relevance only insofar as they affect sentient beings.) Therefore, all sentient beings need legal protection against humans who would needlessly deprive them of their well-being or lives. The whole point of laws is to protect interests. All (and only) sentient beings have interests.

Much suffering has nothing to do with amount of human-like intelligence. Beatings hurt and lack of food causes distress whatever an individual's IQ. Most animals will suffer from imposed immobility. Social animals will suffer from isolation. Curious ones will suffer from monotony. Freedom from deprivation and pain is no more relevant to humans than to any other sentient beings. The same is true of a right to life.

The question then becomes "Who is sentient?" The scientific consensus is that all vertebrates are sentient – for example, can experience pain. Substance P is a chemical associated with pain in humans. The human body also produces opiates, neurochemicals that reduce pain. Humans feel less pain when they receive opiate drugs, such as morphine. In keeping with evolutionary theory, extensive evidence indicates that all vertebrates have substance P activity, produce opiates, and feel less pain when they receive opiate drugs (Kavaliers, 1988; Machin, 2001; Sneddon, 2003).

Like all vertebrates, most invertebrates are bilaterally symmetrical (have a lengthwise body axis that divides them into left and right sides) and possess a brain, defined as a primary nerve center in the head. Among others, these invertebrates include crustaceans, mollusks, insects, arachnids, earthworms, and flatworms. Abundant evidence indicates that all invertebrates with a brain are sentient.

Bilaterally symmetrical invertebrates of numerous species are known to produce substance P and natural opiates. These animals include crustaceans (e.g., crabs, lobsters, and shrimps), mollusks (e.g., octopuses, squids, and snails), and insects (e.g., flies, locusts, and cockroaches) (Fiorito, 1986; Harrison *et al.*, 1994). Also, crustaceans, mollusks, and insects show less reaction to a noxious stimulus when they receive morphine. For example, morphine reduces the reaction of mantis shrimps to electric shock, land snails to a hot surface, and praying mantises to electric shock (Harrison *et al.*, 1994). Further, crustaceans, mollusks, and insects act as if they experience pain. Dropped into boiling water, lobsters show struggling movements, not reflex reactions (Singer, 1990: p. 174). Electrically shocked, marine snails pull in the part of their body that was shocked; release mucus, ink, and opaline; withdraw their gill and siphon; breathe more rapidly; and move away (Kavaliers, 1988). Attacked, insects produce sounds, alarm pheromones, and repellent secretions (Eisemann *et al.*, 1984). When a heated needle is brought close to their antennae, assassin bugs "react violently" and try to escape (Wigglesworth, 1980: p. 9). One of the world's foremost entomologists, Cambridge University professor V. B. Wigglesworth (1980: p. 9), stated, "I am sure that insects can feel pain."

Earthworms, flatworms, and leeches produce natural opiates (Fiorito, 1986; Harrison *et al.*, 1994). When earthworms receive morphine, their efforts to escape aversive pressure are less intense (Kavaliers, 1988). Earthworms writhe when impaled on a fishing hook. Flatworms learn to avoid foods and routes associated with electric shock (Rollin, 1992: p. 64). Initially, leeches crawl in response to a shock but not a touch; after a touch and shock have repeatedly been paired, they crawl in response to a touch alone, indicating they've learned to associate a touch with a shock (Sahley and Ready, 1988). Stuck with a pin or pinched with forceps, leeches coil or writhe (Kavaliers, 1988). Because of the evidence that worms can suffer, Bernard Rollin (1992: p. 78), who is a physiologist as well as an ethicist, includes worms among beings who have "interests."

Radial invertebrates, who radiate out from a center, don't have any apparent brain – at least, as traditionally defined – but do have a nervous system. These animals include comb jellies; cnidarians such as hydras, jellyfishes, sea anemones, and corals; and echinoderms such as sea urchins and sea stars (formerly called starfishes). Are radial invertebrates sentient? Hydras produce substance P (Fiorito, 1986). Hydras, jellyfishes, and sea anemones show escape behaviors, such as withdrawing from harmful chemicals (Kavaliers, 1988). Sea anemones are capable of associative learning apparently related to pain. For example, California shore anemones react to electric shock, but not bright light, by folding their tentacles and oral disk. After bright light has been paired with shock, the sea anemones react this way to the light alone (Haralson *et al.*, 1975). Studies indicate that a sea star's nerve ring acts as a control center – that is, like a brain (Landenberger, 1966).

Of all organisms whom biologists currently classify as animals, only the following lack a nervous system: sponges, rhombozoans (tiny worm-shaped parasites), orthonectids (microscopic worm-shaped parasites), and placozoans (which resemble amoebas but are multicellular). It's highly unlikely that these organisms can experience pain or anything else. When I speak of "animals" or "non-humans," I'm excluding these taxonomic anomalies and referring only to creatures who possess a nervous system. (Biologists don't classify viruses, bacteria, or protists such as amoebas and paramecia as animals. Like plants, those organisms have no nervous system.)

In sum, there's strong evidence that all creatures who possess a brain are sentient and growing evidence that all creatures with a nervous system are sentient. All nervous systems have a shared ancestry and numerous physiological similarities. The capacity to feel confers survival advantages. Evolution would be inexplicably discontinuous if only humans, only mammals, or only vertebrates could suffer. Because it's so important not to exclude any sentient being from moral and legal rights, every creature with a nervous system should receive the benefit of any doubt and be regarded as sentient.

For protection against humans, all sentient beings need legal rights, including a right to life. As previously remarked, humans kill other animals for all sorts of reasons other than need. They kill non-humans for profit, information, sport... They kill out

Figure 2.2 This calf is about to be confined in a hutch typical of the "dairy" industry. Only twenty minutes old, she has already been separated from her mother. She will be exploited for her milk or slaughtered for "veal." Photo taken in Spain in 2010 by Jo-Anne McArthur, We Animals.

of revulsion, groundless fear, and mere annoyance. They readily exterminate insects, small rodents, and other small animals, even when benign measures, such as improved sanitation, would resolve any serious conflicts.

Under non-speciesist law it would be illegal for a human to intentionally kill a non-human except in particular situations. For example, it would be legal to euthanize a non-human experiencing apparently incurable suffering. Overall, it also would be legal to kill a non-human who directly, immediately threatened someone's health or safety. Such defense would include killing internal parasites and killing an animal, such as a tiger or poisonous snake, who posed an immediate threat of serious injury or death, except that it would not be lawful to interfere with predator–prey relationships among free-living non-humans. Humans couldn't legally kill a non-human in testing or experimentation, even for some potential health benefit to others. Nor could they legally kill a non-human for that individual's pelt, flesh, or other body parts, except under extraordinary circum-stances, such as the otherwise-imminent starvation of someone stranded without access to adequate plant food. Humans needn't wear cow skin, sheep hair, "fur," or any other animal-derived material. They can be healthy without consuming animal-derived food. Indeed, abundant evidence indicates a vegan diet is safest and healthiest for humans

(Robbins, 1987; Brody, 1990).[2] Also, raising plant foods directly for human consumption is far more resource-efficient and far less damaging to the environment than animal "agriculture" (Robbins, 1987; Singer and Mason, 2006: pp. 231–40). Except in certain rare and desperate situations, all exploitive killing by humans – such as hunting, fishing, and slaughter – would be illegal.

Non-speciesist law also would accord all sentient beings a legal right to liberty – physical freedom and bodily integrity. With the temporary exception of most non-humans captive at the time of emancipation, non-humans would live unconfined and free from humans. The law would allow humans to remove small animals such as mice, rats, insects, and spiders from inside human-built structures, provided that the removal didn't involve intentional injury. The law also would allow humans to briefly restrain or subdue (e.g., tranquilize) a non-human, such as a lion or bear, to counter an immediate threat of serious injury or death. Otherwise, humans couldn't legally take non-humans captive, even briefly.

It would be illegal for a human to torture any non-human. It also would be illegal for a human to intentionally maim, batter, or otherwise injure a non-human except in someone's immediate, direct defense (excluding interference with natural predation). Vivisection, the pelt industry, horse and dog racing, zoos and "aquariums," circuses with non-humans, food-industry enslavement, and all other institutions and practices that violate non-humans' right to liberty would be unlawful. Legally, taking milk from a cow or venom from a snake would constitute assault, as would the forced insemination of turkeys, pigs, dogs, or any other non-humans, including members of endangered species. Humans couldn't legally breed non-humans for any purpose.

The number of "domesticated" animals would rapidly decline. Upon non-human emancipation, dogs, cats, and other domesticated animals living with loving, responsible human companions would stay with those humans. Liberated from exploitation and other abuse, other domesticated animals – such as chickens freed from egg factories, rats freed from vivisection laboratories, and cats freed from "shelters" – would receive any needed veterinary care, be euthanized if experiencing apparently incurable suffering, and otherwise be cared for at sanctuaries and private homes. Undomesticated non-human captives would be set free if they could thrive without human assistance (after any necessary rehabilitation) and if appropriate habitat existed. If not, they would be permanently cared for at sanctuaries. As much as possible, these sanctuaries would provide natural, fulfilling environments. While captivity was being phased out, humans would use benign measures (such as "spaying/neutering") to prevent captive non-humans from breeding. Eventually, virtually all non-humans would be undomesticated animals living in their natural habitats.

A non-human right to property would help ensure the preservation of such habitat. Non-speciesist law would prohibit humans from appropriating or damaging non-human property. Non-humans would own what they produce (eggs, milk, honey, pearls...),

---

[2] Vegans consume very little saturated fat and zero cholesterol. Whereas veganism promotes human health and longevity, consumption of animal-derived food increases the risk of diabetes, osteoporosis, arteriosclerosis, heart disease, and various cancers (Robbins, 1987; Brody, 1990).

what they build (nests, bowers, dams, hives...), and the natural habitats in which they live (marshlands, forests, lakes, oceans...). For example, honeybees would have a legal right to the honey they produce to nourish colony members. A dam built by beavers would legally belong to its creators and their descendents. Humans couldn't legally drain lakes, bulldoze woodlands, or slash and burn rainforest. All non-humans living in a particular area of land or water would have a legal right to that environment, their communal property. It would be illegal for humans to destroy or dramatically alter any "undeveloped" habitat. Land currently inhabited by non-humans and humans could remain cohabited, but humans wouldn't be permitted to encroach farther into non-human territory (e.g., by building more houses on land occupied only by non-humans). If humans failed to practice zero or negative population growth, they could replace houses (especially, needlessly large homes) with high-rises, thereby housing more humans on the same amount of land.

Non-human rights would involve new restrictions on human behavior. However, every expansion of human rights has likewise involved new prohibitions against unjust behavior, such as discrimination on the basis of race or sex. Non-speciesist law would prohibit unjust behavior on the basis of species. All animals would have legal rights to life, liberty, and property. Overall, the law would prohibit humans from exploiting non-humans or otherwise intentionally causing them harm. Apart from a transition period, non-human "participation" in human society would cease. Non-humans would be protected from human interference.

Conclusion

The standard definition of speciesism is too narrow because it limits speciesism to bias against all non-humans and to bias based solely on species membership. That definition describes only the most obvious and severe form of speciesism, old speciesism. Old-speciesists limit rights to humans. Old speciesism is illogical, based on a double standard. Apart from membership in the human species, every old-speciesist criterion for rights that would exclude all non-humans also would exclude many humans. Privileging members of one's own species simply because they belong to one's own biological group is sheer favoritism, which is not a just basis for rights.

New speciesism is bias toward some animal species and against all others. New-speciesists would limit rights to relatively few non-humans, those who seem most human-like. The new-speciesist criteria for rights are characteristics typical of humans, even though those characteristics don't equate to a richer, happier, more valued, or more constructive life.

Sentience, the capacity to experience, is the only defensible criterion for inclusion among rights-holders. Because sentient beings lose everything when they die, and because they can experience suffering and well-being, they need legal protection against humans, who have a great propensity to cause needless harm. Non-speciesists advocate rights for all sentient beings.

In keeping with evolutionary theory, a tremendous amount of evidence indicates that all creatures who possess a brain – all vertebrates and all bilaterally symmetrical invertebrates – are sentient. Growing evidence also indicates the sentience of radial invertebrates, who possess a nervous system but no brain. Therefore, it's logical, fair, and empirically justified to give all creatures with a nervous system the benefit of any doubt regarding sentience and accord them basic rights.

Under non-speciesist law, all sentient beings would have rights to life, liberty, and property. With few exceptions, such as immediate defense, it would be illegal for humans to intentionally kill any sentient being or deprive that individual of freedom or bodily integrity. It also would be illegal for humans to appropriate or intentionally damage non-human property, which would include what non-humans produce, what they build, and the natural habitats in which they live. Apart from a transition period during which humans would care for non-humans then in captivity, all non-humans would live free from humans.

Now let's return to the hens in the egg factory. I hope you see that their exploitation epitomizes speciesism. Needless. Cruel. Unjust. In a non-speciesist society, humans couldn't legally breed, confine, or exploit chickens. The egg industry wouldn't exist. Nor would any other legal form of human abuse of non-humans.

3 Indexically yours

Why being human is more like being here than like being water

David Livingstone Smith

My interest in the ethical issues surrounding the use of non-human animals grew out of my investigations into the phenomenon of *dehumanization* – the tendency to conceive of groups of people as creatures that are less than human. Dehumanization is a common feature of war, genocide, slavery, and other atrocities (Smith, 2011). Its purpose is to disinhibit violence against the dehumanized group by excluding them from the universe of moral obligation.

Dehumanization raises deep metaphysical and ethical questions about the human/ non-human binary. Answering these questions is also crucial for understanding the normative dimension of our attitudes toward other species. Given this focus, my approach in this chapter will be rather different than that of contributors to this book such as Bekoff, Dunayer, and Marino, who seek to motivate or defend a particular normative stance toward non-human animals. I will neither take for granted nor defend a position on the question of whether or to what degree non-human animals ought to be given moral consideration. Instead, I aim to interrogate the metaphysical framework that sets the parameters for discussions about their moral status. I will argue that both advocates and detractors of the view that non-human animals have moral standing misconstrue the nature of the human/non-human distinction by conflating it with the distinction between *Homo sapiens* and other animal species.

Does "human" pick out a biological category?

I begin with an observation about semantics. Two terms that loom large in the literature on animal welfare are *human exceptionalism* and *speciesism*. "Speciesism" is a derogatory term (analogous to "racism" and "sexism") for the view that species distinctions are morally relevant. Most often, "speciesism" is used to refer to the moral privileging of our *own* species (sometimes called "human speciesism"). As such, it entails the doctrine of human exceptionalism, which is the view that human beings are categorically distinct, in a specifically moral sense, from other animals. Notwithstanding their apparent interdefinability, these two terms are part of two different vocabularies. "Speciesism" is parasitic on the biological category of "species." It refers to the moral

The Politics of Species: Reshaping our Relationships with Other Animals, eds R. Corbey and A. Lanjouw. Published by Cambridge University Press. © Cambridge University Press 2013.

privileging of certain *biological kinds*. In contrast, "human exceptionalism" uses the vernacular term "human" and owes nothing to the biological lexicon. The definitional tie between these two expressions suggests that there is some principled relationship between them. Of course, it is widely accepted that there is such a relationship; it is generally supposed that "human" straightforwardly refers to the species *Homo sapiens*, and therefore that "human" is a name for a biological kind. If this assumption is correct, then it provides a clear basis for demarcating humans from non-humans. Only *Homo sapiens* are human, and all other species are non-human. But what if it isn't correct? If "human" and "*Homo sapiens*" are not equivalent terms, then this upsets certain suppositions about the human/non-human dichotomy as well as the moral implications that supposedly flow from them.

What is it to be human? Minimally, to be human is to be "one of us." But this response begs the question of the category of creatures to which "us" refers. The *Oxford English Dictionary* informs us that to be human is to be "a member of the species *Homo sapiens* or other (extinct) species of the genus *Homo*." It is not clear whether the second clause of the disjunction is intended to refer to the whole of genus *Homo* or some proper subset of it. If the *OED* lexographers had it in mind that the human is either *Homo sapiens* or genus *Homo*, then "human" denotes a genuine biological kind, although exactly what kind of biological kind remains uncertain, but if the lexographers had in mind that the human is only a subset of the species that make up genus *Homo* (for example *Homo sapiens* and *Homo neanderthalensis*), then "human" denotes an odd, gerrymandered category rather than a genuine biological kind.

Whereas the dictionary gives us three options (*Homo sapiens*, some proper subset of genus *Homo* that includes *Homo sapiens* among its members, or all of genus *Homo*), the scientific literature presents us with even more. Although for the most part paleoanthropologists identify humans with *Homo sapiens*, or with the genus *Homo*, some restrict it to the subspecies *Homo sapiens sapiens* or enlarge it to include all of the hominin lineage (for a range of views, see e.g., Leakey and Lewin, 1993; Falguères *et al.*, 1999; Potts, 2003; Schmitt, 2003; Lewin and Foley, 2004; Mikkelsen, 2004; Pollard, 2009). These differences of opinion are not due to the scarcity or ambiguity of empirical evidence. They are due to the complete absence of such evidence – or, to put the point with greater precision, the absence of any conception of what sort of evidence would settle the question of which primate taxa or taxon should be counted as "human". Biological science can specify, with a reasonable degree of accuracy, the taxon to which an organism belongs, but it cannot tell us whether an organism is a *human* organism. The epistemic authority of science does not extend to judgments about what creatures are human because "human" is a folk category, not a scientific one.

Some folk categories correspond more or less precisely to scientific categories. To use a well-worn example, the folk category "water" is coextensive with the scientific category "H_2O."[1] In the philosophical jargon, water is said to be reducible to H_2O, which means that H_2O is nothing over and above water, and therefore anything that is

---

[1] For a dissenting view, see Hendry (2006).

true of water is also true of H_2O. Not every folk category is reducible to a scientific one. Consider the category "weed." Weeds do not have any biological properties that distinguish them from non-weeds. In fact, one could know everything there is to know biologically about a plant, but still not know that it is a weed. In this respect, the category "human" has more in common with the category "weed" than it does with the category "water," because it is false to say that an entity is human if and only if it possesses certain determinate biological attributes. If this sounds strange to you, it is probably because you are already committed to one or another conception of the human (for example, that all and only *Homo sapiens* are human). Claims like "an animal is human only if it is a member of the species *Homo sapiens*" are stipulated rather than discovered. Neither you nor anyone else has sifted through the available data (what data?) to emerge with the finding that humans are *Homo sapiens*. Rather, in deciding that all and only *Homo sapiens* are humans, you are expressing a preference about where the boundary separating humans from non-humans should be drawn (Clark and Willermet, 1997; Corbey, 2005; Bourke, 2011).

Essentially human

Classifying organisms as human is not a morally innocent exercise in descriptive taxonomy. Attributing humanity carries immense moral weight, and denying it to a creature diminishes its moral status. As Kant famously articulated the principle in 1798, human beings are "altogether different in rank and dignity from things, such as irrational animals, with which one may deal and dispose at one's discretion" (Kant, 1974: p. 9). He assured his readers seven years later in the *Groundwork of the Metaphysics of Morals* that non-human animals "have only a relative worth, as means, and are therefore called things" in contrast to creatures classified as human beings, who are ends in themselves (Kant, 1993: p. 30).

Classifying an organism as a specimen of *Homo sapiens* carries no such moral weight, because scientific taxonomies do not in themselves have normative implications. They only acquire such implications when folk taxonomies are superimposed upon them. Putting the matter somewhat differently, *Homo sapiens* uniquely merit the highest degree of moral consideration if and only if all *Homo sapiens* are human. Now, scientists *qua* scientists at least agree that the category of the human includes *Homo sapiens*, even if it also may include other hominin taxa. That is, they hold that being a member of our species is sufficient for being human. In contrast, non-scientific conceptions of the human are far more flexible. It has often been the case that only a subset of the members of our species is counted as human. Being a member of *Homo sapiens* is perhaps regarded as necessary, but is by no means sufficient, for being human.

The imperfect fit between the folk category of the human and the biological category *Homo sapiens* has significant moral implications. Consider the problem of abortion. As Judith Jarvis Thomson points out, "Most opposition to abortion relies on the premise that the fetus is a human being, a person, from the moment of conception" (Thomson, 1971: p. 47). The idea is if the fetus is a human being then it falls into a moral category that makes taking its life impermissible.[2] Questions about the moral permissibility of slavery also often turned on whether slaves were human beings. We see this very clearly in the writings of seventeenth-century Anglican clergyman Morgan Godwyn, who reported that American colonists held a "disingenuous position" that "the Negros, though in their Figure they carry some resemblances of Manhood, yet are indeed no men," and that they advocated "Hellish Principles. . . that Negros are Creatures Destitute of Souls, to be ranked among Brute Beasts and treated accordingly" (Vaughn, 1995: p. 66). Similar considerations apply in cases of genocide, as is exemplified by the following remarks by Reichsführer-SS Heinrich Himmler, which appeared in a 1942 publication entitled *The Subhuman*. "As long as there have been men on the earth," he expounded, "the struggle between man and the subhuman will be the historic rule; the Jewish-led struggle against the mankind, as far back as we can look, is part of the natural course of life on our planet. . . ."

Just as the night rises against the day, the light and dark are in eternal conflict. So too, is the subhuman the greatest enemy of the dominant species on earth, mankind. The subhuman is a biological creature, crafted by nature, which has hands, legs, eyes and mouth, even the semblance of a brain. Nevertheless, this terrible creature is only a partial human being. . . Not all of those, who appear human, are in fact so.   (Himmler, 1942: p. 1)

Psychological essentialism

For one group of people to conceive of another group as less than human they must make a distinction between how a person appears and what they really are "beneath the surface." This distinction allows a further distinction to be made between *appearing* human and *being* human that is at work whenever groups of people are dehumanized (Smith, 2011). The Nazis held that although Jews look human, they are not truly human. Their humanity is merely skin deep, concealing a dangerously non-human core. A similar idea was expressed in the religiously infused vocabulary of American colonists, who believed that Africans are actually "Creatures Destitute of Souls" with only "resemblances" of humanity. These two examples are instances of an immensely more widespread pattern. It has been quite common for people to count all and only members of their own ethnic group as human beings (Smith, 2011). Yale University sociologist William Graham Sumner was the first person to write extensively about this phenomenon, and he coined the term "ethnocentrism" to denote it. "As a rule," Sumner remarked, "it is found that native peoples call themselves 'men.' Others are something else – perhaps not defined – but not real men. In myths, the origin of their own tribe is that of the real human race" (Sumner, 2002: p. 13). Both ethnographic and historical researches confirm Sumner's insight that *Homo sapiens* have often excluded other *Homo sapiens* from the human community (Figure 3.1).

---

[2] The quotation from Thomson nicely exemplifies the confusion between the concepts of persons and humans that is endemic to this literature (and indeed, endemic to moral philosophy generally) – a confusion that stems primarily from a lack of clarity about the concept of the human. Generally speaking, "human" is (wrongly) taken to be a biological category, while "person" is taken to be a moral category. For an interesting discussion of the use of the term "human" in the abortion controversy, see Tooley (1972).

Figure 3.1 "Bolshevism means drowning the world in blood." This German propaganda poster from 1919 represents the Bolshevics as wolves. It thus excludes them from the moral community and legitimizes violence against them by denying that they possess a human essence. Similarly, we exclude other animals from the moral community by conceiving of them as essentially different from ourselves. Library of Congress.

The distinction between how others appear and what they really are is one part of a broader pattern of thinking that we use to make sense of the variety of living things. Psychologists have found that we tend to conceive of organisms as divided into kinds and that we think of each of these kinds as constituted by a causal essence that is possessed by all and only members of the kind. This way of carving up the natural world is known as *psychological essentialism* (e.g., Hirschfeld, 1998; Gelman, 2003; Xu and Rhemtulla, 2005). As cognitive anthropologist Scott Atran (1998) observes:

> There is a commonsense assumption that each... species has an underlying causal nature, or essence, that is uniquely responsible for the typical appearance, behavior, and ecological preferences of the kind. People in diverse cultures consider this essence responsible for the organism's identity as a complex, self-preserving entity governed by dynamic internal processes that are lawful even when hidden. This hidden essence maintains the organism's integrity even as it causes the organism to grow, change form, and reproduce. For example, a tadpole and frog are in a crucial sense the same animal although they look and behave very differently, and live in different places. (Atran, 1998: p. 548).

From an essentialist perspective, individual organisms may or may not be *true to their kind*. If an organism is true to its kind, that is, if it is a stereotypical specimen of its kind, its essence is fully realized in its appearance. Organisms that are not true to their kind have a deformed or atypical appearance, which is explained by something's preventing their essence becoming fully realized. For example, on the essentialist view, an animal is a porcupine just in case it possesses a porcupine essence. Stereotypical porcupines are brownish quadrupeds, covered with sharp quills, etc. It is possible for a porcupine to fail to manifest any of these stereotypical attributes while remaining a porcupine: it might be an albino porcupine, or a three-legged porcupine, or a porcupine without quills. It might even be possible for a porcupine to resemble some other animal (say, a beaver) more than it does a stereotypical porcupine, and yet remain a porcupine in virtue of having a porcupine essence.

Psychological essentialism is an *intuitive theory*. Intuitive theories are tacit, informal theories that are so thoroughly imbedded in our everyday thinking that we take them for granted. The essentialist view of natural kinds, understood as an intuitive theory, can be usefully contrasted with a viewpoint that I call "psychological empiricism." Psychological empiricism is the view that "we have a natural tendency to classify things by their superficial, observable properties, such as color and shape" (Kornblith, 1993: p. 63) and that all of our conceptual categories – however refined – ultimately rest on our perception of such properties. If people were psychological empiricists about living things, they would conceive of something as a member of a biological kind if and only if it satisfies the description of a stereotypical member of that kind. Clearly, the notion that something might have stereotypically human appearance and yet not be human, or fail to have a stereotypically human appearance yet still be human, is inconsistent with psychological empiricism.[3] Contrast this with psychological essentialism. Insofar as we are psychological essentialists, we believe that one is human just in case one possesses

---

[3] Of course, we are psychological empiricists about some kinds, for example, the kind denoted by the expression "things that are more than six feet tall."

an essence that is unique to human beings. From an essentialist perspective, an entity might be human even if it does not have a stereotypical human appearance, and it might conform to the stereotype and yet nevertheless not be human.

Psychological essentialism is deeply bound up with the categories of things that philosophers call "natural kinds." It is difficult to frame an entirely satisfying account of natural kinds, and there is a large philosophical literature on this topic. However, the following definition by Koslicki conveys the basic idea.

> Kinds are categories or taxonomic classifications into which particular objects may be grouped on the basis of shared characteristics of some sort. Natural kinds are best construed, not so much as kinds found within nature, but rather as classifications that are in some sense not arbitrary, heterogeneous, or gerry-mandered.   (Koslicki, 2008: p. 789)

Natural kinds are contrasted with non-natural kinds. Non-natural kinds are artifacts of our classificatory practices that do not correspond to the structural fault lines of a mind-independent world. Weeds are a non-natural kind, because they exist only in virtue of certain social conventions, but *Pteridophyta* (ferns) are (arguably) a natural kind because, unlike weeds, their existence as a botanical group is insensitive to our taxonomic conventions. Scientific theories purport to be theories about natural kinds, and scientific laws and empirical generalizations purport to be about the regularities displayed by natural kinds. Scientific explanations privilege natural kinds because of their explanatory and predictive power. One can infer a great deal about every member of the kind from observations of a single member of the kind, and use this to predict the behavior of other members of the kind. Consider a gold ring. When we describe it as a ring, we describe it as a member of a non-natural kind, but when we describe it as a piece of gold we describe it as a member of a natural kind. Knowing that the item is a ring does not allow us to make very many inferences about it (we can make inferences about its decorative function, that it is designed to be worn on a human digit, but not much more than that). But knowing that it is made of gold tells us quite a lot about it (for example, that its atomic mass is 196.96655 amu, its melting point is 1064.43°C, that it conducts heat and electricity, and that it dissolves in aqua regia).[4]

Chemical elements like gold are defined by their essences, which are represented by their atomic numbers. But having an essence is not a *sine qua non* for being a natural kind as described by science. Biological species are often regarded as paradigmatic natural kinds, notwithstanding the fact that species do not have essences.[5] Just as it is important to distinguish between psychological and philosophical versions of essentialism, it is also important to differentiate scientific from intuitive concepts of natural kinds.

Sometimes, scientific taxonomies mesh together with folk taxonomies with remark-able precision. For instance, the biologist Ernst Mayr found that the 136 bird species discriminated by natives of the Arfak mountains of New Guinea corresponded almost

---

[4] I ignore the fact that gold rings are normally made of alloys.

[5] The claim that species are kinds is controversial. Many philosophers consider them to be individuals rather than kinds (Ghiselin, 1974; Hull, 1978; Kitcher, 1984; Stamos, 2003).

exactly to the 137 species identified by scientific taxonomists (Mayr, 1988). But it is often the case that intuitive and scientific frameworks come apart. Consider the intuitive concept of race. It seems *obvious* that the human types commonly referred to as races exist. But the intuitive concept of race sits poorly with what science tells us about the distributions of genes in populations. It is not that science denies that the bare notion of race is coherent. There is nothing implausible about the claim that there are inbred populations that are biologically distinct from others in non-trivial ways. What is implausible (and indeed false) is that the groups conventionally described as "races" are examples of such populations. There is another significant difference at work here – one that accounts for the discontinuity between scientific and intuitive conceptions of race. Even if it were found that the populations conventionally regarded as racially distinct were discovered to be relatively genetically homogeneous, the folk-taxonomic notion that human races individuated by phenotypic characters that arise from a hidden racial essence (typically imagined to be located in their blood, and transmitted down their bloodline) is inconsistent with the way that biologists think about populations. There is a deep *metaphysical* difference between intuitive and scientific theories of natural kinds even when (as in the case described by Mayr) they pick out identical or nearly identical taxa.

The concept of the human is an intuitive, essentialized natural kind concept. It is a living kind concept akin to the concept of a biological species, but it is neither coextensive with any biological taxon nor is it consistent with the metaphysical underpinnings of biological taxonomy. Being human is more like being a weed than it is like being water, even though we are psychologically disposed to think of humans as more like water than like weeds.

An epistemological puzzle

The unobservability of causal essences gives rise to an epistemological problem for essentialism. If non-humans can appear to be human, as psychological essentialism entails, then how is it possible to tell humans and non-humans apart? The European colonists who believed Africans to be less than human also believed that having a stereotypically African appearance is a reliable indicator of subhumanity. The problem posed by this criterion is that it does not entail that anyone who is black must have stereotypically African features ("If P then Q" does not entail "P if and only if Q"). This failure of entailment explains why racists have always been haunted by the prospect of black people "passing" as white. The writer Lillian Smith provides an excellent illustration of this. Smith, who grew up in a small Georgia town early in the twentieth century, recalls that one day a child (who is described as "very white indeed") was spotted by white women living with an African American family. Presuming that the little girl, who was named Janie, had been kidnapped, the sheriff took her into custody, and entrusted her to the Smith family for fostering. One day, the Smiths received a surprising telephone call from an African American orphanage informing them that Janie was really black, and that the African American couple from whom she had been taken were in fact her adoptive parents. When Lillian's mother explained to her that Janie would have to return to "Colored Town," Lillian, who had grown very close to Janie, remonstrated "But she's white!" And then the following dialog ensued:

"We were mistaken. She is colored."
"But she looks –"
"She is colored. Please don't argue!"
"What does it mean?" I whispered.
"It means," Mother said slowly, "that she has to live in Colored Town with colored people."

(Smith, 1949: pp. 27–8)

Similarly, although Nazis had a stereotypical image of the Jew (whose leering, bearded, hook-nosed, visage frequently appeared in racial propaganda), they were aware that most Jews were physically indistinguishable from ordinary Germans. Consequently, they searched for a deeper, uniquely Jewish essence that was inaccessible to naïve observation. In common with many race essentialists, Nazi scientists assumed that the Jewish essence must be located in their blood. "Think what it might mean if we could identify non-Aryans in the test tube!" commented one German biologist, "Then neither deception, nor baptism, nor citizenship, and not even nasal surgery could help!... One cannot change one's blood" (Koonz, 2003: p. 197). Nazi race theorists' inability to discover reliable biological markers of Jewishness did not motivate them to abandon the notion of a Jewish race. Instead, they replaced the biological concept of race with a spiritual one, the essence of which is undetectable by the investigative methods of natural science. "The Jewish race is above all a community of the spirit," remarked Hitler in a letter dictated to his secretary Martin Bormann, "presenting sad proof of the superiority of the 'spirit' over the flesh" (quoted in Heinsohn, 2000: p. 412).

At a more rarified level, Kant's moral philosophy came up against the very same problem (Latta, 2011). As I have mentioned, Kant made an absolute distinction between human beings, who are ends in themselves, and non-human animals, which are not. This is supposed to explain why we have moral obligations to human beings, but not to other creatures. What is it that is supposed to account for the special moral status of our fellow humans? In Kant's view, humans are morally considerable because they are rational, and they are rational because they possess freedom of the will. If all and only humans have freedom of the will, then to determine whether or not any being is human – and therefore apt for moral consideration – one need only determine whether they possess freedom of the will. But certain of Kant's other philosophical commitments foreclose the possibility of ever making this determination. Kant believed that our minds have access only to the world *as it is presented* to the mind (the phenomenal world), and that we cannot know what the world is like *in itself*, independent of the constraints that are necessarily imposed by the structure of our consciousness (the noumenal world). Kant held that one of the constraints imposed by our minds is that true freedom is inconceivable. In the phenomenal world (the only reality that we are able to comprehend) uncaused events are inconceivable. So, in the phenomenal world, strict determinism reigns. Freedom of the will lies not in the phenomenal realm, but in the noumenal world that is forever beyond our ken. It follows from these considerations that although human beings are essentially free agents, our cognitive imprisonment in the phenomenal world renders this impossible for us to perceive. Latta (2011) points out in a penetrating discussion of this topic that Kant's contemporary Gottlieb Fichte homed in on the problem, inquiring "how do I know which particular object is a rational being?"

> How do I know whether the protection afforded by that universal legislation befits only the white European, or perhaps also the black Negro; only the adult human being, or perhaps also the child? And how do I know whether it might not even befit the loyal house-pet? As long as this question is not answered, that principle – in spite of all its splendor – has no applicability or reality.
>
> (Fichte, 2000: p. 75).

"Human" as an indexical term

Fichte's question "How do I know…?" presumes that there is something to be known, i.e., that there is some fact of the matter about what beings are human. I have shown that this is a questionable assumption. In the remainder of this chapter, I will motivate quite a different notion of ascriptions of humanity. We have seen that the word "human" has been given bafflingly diverse denotations. The traditional response to this state of linguistic affairs has been to press one or another concept of the human, and to assert that competing views are mistaken (for example, one might try to counter the racist's claim that black people are not human by pointing out to him that blacks are also *Homo sapiens*). To take this stance is to be a realist about the attribute of humanness, but there does not seem to be any non-question-begging criterion that can support the realist position: a person who insists that being a human being is identical to being a member of a certain biological genus and species cannot evidentially or logically justify this independent of the prior assumption (which is not justified either) that "human" is a biological species category.

I think that it is more fruitful to understand "human" in quite a different way; namely, that *referring to someone as human is referring to them as a member of one's own natural kind*. If I am right, "human" is most perspicuously understood as an indexical term – a term that gets its content from the contexts in which it is uttered.[6] Just as the word "now" names the moment at which it is uttered, the word "here" names the place where it is uttered (or the place where the speaker is pointing), and the word "I" names the person uttering "I," the word "human" names the speaker's natural kind.

When thinking about indexical terms, it is helpful to distinguish between their meaning and their content (Kaplan, 1989). For example, the word "here" means "the place where I am." Its content is fixed by wherever the speaker is located. However, "the place where I am" is ambiguous. It might mean *this very spot*, but it might also mean *this room, this city, this country*, etc. Because of this ambiguity, we draw on other contextual factors to interpret what, exactly, is meant by utterances of words like "here."

---

[6] By "uttered" I mean any tokening of the term, or its associated concept.

There is something similar at work in the case of "human," construed as an indexical term. Suppose that "human" means "my own natural kind." This might pertain to any of the many natural kinds to which the speaker regards herself as belonging. We have already seen that "human" is used to refer to various primate taxa to which the speaker belongs and that it has also been used to refer to the speaker's racial or ethnoracial group. Conceiving of "human" as meaning "my own natural kind" goes a long way toward explaining why a statement of the form "x is human," in the mouth of a fireman, does not mean "x is a fireman," while the same statement in the mouth of a biologist might mean "x is a member of the species *Homo sapiens*" or in the mouth of a Nazi might mean "x is an Aryan."

One shortcoming of this analysis is that it does not tell us why "human" picks out one taxonomic rank rather than another. Why does "human" in the mouth of a committed Nazi have the content "Aryan" rather than, say, "mammal" or "vertebrate?" Recall that in the paleoanthropological literature "human" is for the most part equated with *Homo sapiens* or genus *Homo* and in folk taxonomies (insofar as they differ from scientific taxonomies) being human is equated with membership of one's ethnoracial group. One possibility is that there is a single rank that is privileged across the board – a rank occupied by species and genus in biological taxonomy and ethnoraces in folk taxonomies. This conjecture finds some support in investigations by cognitive anthropologists who argue that all cultures use systems of nested biological categories to carve up the natural world. These taxonomies consist of from three to six ranks. The most important of these is the *generic species* rank, so named because it roughly corresponds to the ranks of genus and species in the Linnaean system. Generic species are especially salient groups of organisms, groups that "stand out as beacons on the landscape of biological reality, figuratively crying out to be named" (Medin, 1989: p. 53). By definition, the biological ranks of genus and species correspond to the generic species rank in folk taxonomies. Similarly, ethnoracial groups stand out as "beacons" in the social landscape, and are therefore treated as generic species. To borrow a felicitous expression from the anthropologist Francisco Gil-White, ethnoraces look like species to the human brain (Gil-White, 2001).

The notion of generic species can be interpreted as a special case of a more general folk-taxonomic concept. If organisms are identified as members of a generic species because of their salience, then it is their salience that is doing the work. In many contexts groups of organisms that are especially salient are also species or species-like. However, what is salient is, at least to some degree, a function of the observer's interests and context. What is salient to a hunter-gatherer in the highlands of New Guinea is unlikely to coincide with what is salient to a paleoanthropologist studying human evolution. In the latter case, the most salient group of organisms might consist of the entire hominin lineage rather than anything that is species- or genus-like. It seems most advantageous, then, to interpret "human" as referring to the maximally salient natural kind to which the speaker regards herself as belonging.

There is no contradiction between this view and a moderate social constructivist account of races. It is plausible that the form of racial thinking reflects a deep feature of our cognition, whereas the content of ethnoracial categories is historically contingent

(Machery and Faucher, 2005). For colonial slaveholders, "white" and "black" were especially salient categories, which were naturalized, essentialized, and slotted into an intuitive folk taxonomy. These Europeans identified themselves as belonging to the intuitive natural kind that they referred to as "white" and consequently reserved the term "human" for those who were members of this kind. The striking transmutation of European settlers' attitudes toward Native Americans is quite revealing in this connection. Native Americans were initially regarded as white. However, as territorial conflict between Europeans and Native Americans increased, settlers began to describe Indians as "red" or "copper colored." As the racial characterization of Native Americans shifted, colonists began to enact anti-miscegenation laws to prevent interbreeding between themselves and Native Americans. This process of otherization gained momentum during the eighteenth century, and by the beginning of the nineteenth century "the stereotypical color carried a host of unfavorable connotations that prevented the Indians' full assimilation into the Anglo-American community and simultaneously precluded their acceptance as a separate and equal people... Even relatively sympathetic spokesmen now believed the Indians to be permanently different" (Vaughn, 1995: p. 33). As might be expected, the racial distancing of Native Americans went hand-in-hand with their dehumanization. For example, Francis Parkman, a noted American historian, wrote in his 1847 book *The Oregon Trail: Sketches of Prairie and Rocky Mountain Life* that:

For the most part a civilized white man can discover very few points of sympathy between his own nature and that of an Indian... Nay, so alien to himself do they appear that... he begins to look upon them as a troublesome and dangerous species of wild beast, and if expedient he could shoot them with as little compunction as they themselves would experience after performing the same office upon him.   (Parkman, quoted in Berkhoffer, 1979: p. 96).

Concluding remarks

I have argued that ascribing humanity to others is equivalent to claiming that these others belong to the same natural kind as oneself. But consider the statement "I am human." It looks like on the indexical account this statement must be interpreted as expressing the thought "I am a member of the natural kind that I am a member of," which is a vacuous tautology. But utterances of "I am human" are often anything but empty. During the heyday of the civil rights movement in the United States, African American protesters carried placards inscribed with the words "I am human." Likewise, on a recent visit to the Middle East I saw graffiti scrawled by a Palestinian on the security barrier dividing Israel from the West Bank stating "We are human beings." And there is the memorable moment in David Lynch's film *The Elephant Man*, when John Merrick cries out to his persecutors "I am not an animal! I am a human being!" These are more than just tautologies. Bearing in mind the specific context of these statements, they are not only consistent with but also provide support for the indexical account. The three examples that I have provided, which I take to be representative of contexts in which "I am human" is sincerely uttered, all have profound moral implications. Construing "I am a human being" as "I am a member of taxonomic category x" comes nowhere near to capturing the moral gravity of utterances, and indeed completely misses their point. When African American protesters carried signs stating "I am human" they were not attempting to specify the taxonomic category to which they belonged. Instead, they were appealing to their white oppressors to recognize that that both whites and blacks are members of the same natural kind. In this context, "I am human" communicates the message "Although we appear to be different, I am a member of the natural kind that *you* call 'human', and therefore deserving of the respect that you accord to your own kind."

If it is true, the indexical account of what it is to ascribe humanity to others upsets conventional views of the nature of the human/non-human boundary. It indicates that we do not morally discriminate against non-human animals on the basis of the biological species to which they belong, but rather because we conceive of them as non-human. This does not entail that species membership is irrelevant to judgments about the moral considerability of organisms, but it does imply that considerations of species membership play only a contingent role in such judgments. The species to which an organism belongs is relevant to judgments about its moral standing only insofar as the person making the judgment regards herself as belonging to natural kind *Homo sapiens*.

It is worth noting that although considering other kinds of animal as non-human makes it easier for us to harm them, it is probably not necessary and certainly not sufficient for this. With respect to the sufficiency condition, if animals are excluded from the circle of moral consideration, it is not just because they are thought of as non-human, it is because they are believed to be *sub*-human (less than human). In this respect, there is a striking resemblance between the way that we habitually conceive of other animals and the way that we conceive of those members of our own species whom we dehumanize. This suggests the intriguing possibility that research into dehumanization might benefit from drawing on research into the moral marginalization of other animal species, and, conversely, that investigations into the ramifications of human exceptionalism might be enriched by the study of our propensity to dehumanize our fellow *Homo sapiens*.

4 Apeism and racism
Reasons and remedies

Edouard Machery

Racism, xenophobia, and other prejudices against human groups have unfortunately not disappeared from contemporary societies, and, as the 1994 Tutsi genocide in Rwanda reminds us, such prejudices can still lead to disasters. On the other hand, in only 50 years, racism has strikingly decreased in at least some regions of the world. Focusing on the USA, explicit racism and anti-Semitism (racist and anti-Semitic prejudices that people are aware of harboring and that they can verbalize) have massively decreased since the 1950s (e.g., Devine and Elliot, 1995), which is not to say that explicit racism has entirely disappeared and that there are no other, implicit forms of racism and other prejudices (Kelly *et al.*, 2010b; Nosek *et al.*, 2011). For instance, in 1947 the department of physiology at Washington University in St Louis did not find it inappropriate to publish an advertisement for a faculty position stating that "preference will be given to [a faculty candidate] who is a gentile." In a 2011 Gallup poll (June 20, 2011), 94% of the people polled said that they would vote for an African American for president[1], compared to only 77% in the early 1980s, 50% in 1960, and 38% in 1958, when the question was asked for the first time.[2]

While the network of causes that led to this decrease of explicit racism is complex and not well understood, this decrease resulted in part from social policies, such as affirmative action, laws forbidding various forms of racial zoning, etc. Psychologists and social scientists have examined the efficiency of some of these social policies in great detail, as well as the efficiency of various strategies by which individuals can attempt to control their own biases (what I will call "individual strategies"), and much has been learned about them (for a recent review and discussion, see Kelly *et al.*, 2010a).

In this chapter, I will attempt to identify policies that could change people's callous attitudes toward apes on the basis of what is known about the efficiency of policies against racism.[3] I call these wrong attitudes "apeism" by analogy with speciesism,

---

[1] www.gallup.com/poll/148100/Hesitant-Support-Mormon-2012.aspx. Accessed 05/24/2012.

[2] www.gallup.com/poll/3400/longterm-gallup-poll-trends-portrait-american-public-opinion.aspx. Accessed 05/24/2012.

[3] The word "apes" is used in several different ways. Some use "apes" to refer to all hominoids (humans, chimpanzees, bonobos, gorillas, orangutans, and gibbons). More commonly, however, "apes" is used to

The Politics of Species: Reshaping our Relationships with Other Animals, eds R. Corbey and A. Lanjouw. Published by Cambridge University Press. © Cambridge University Press 2013.

racism, and sexism. As a French citizen living in the USA, I was astonished to find out that experimentation on apes was still legal in the USA, and I was also surprised to discover that my condemnation of this practice was not universal. My interest in the strategies that could be used to undermine apeism grows out of this recognition that too many people care very little about the well-being of apes.

So, the goal of this chapter is practical: I will recommend some policies, and I will also argue that other policies, some of which are commonsensical, are unlikely to be successful. (Similar points could be made with respect to people's attitudes toward several other species, including cetaceans.)

Here is how I will proceed. I will first describe what I take apeism to be, and I will briefly argue that it is wrong. I will then consider three policies that have been proposed to undermine racism.[4] For each of them, I will discuss whether it can be extended to undermine apeism, and I will assess the prospects of this extension.[5]

Several caveats should be noted. Which strategy is most appropriate to change people's attitudes toward apes may vary across groups (children, academics, etc.) and across cultures. In particular, much of this article is written from a Western perspective (for other perspectives, see Bakels, this volume; Riley, this volume). The attitudes of those people living in greater contact with apes or monkeys may be quite different from the apeism discussed here. Furthermore, while the strategies described below could perhaps be usefully combined, I will not examine how this could be done.

Apeism

What is apeism?

I will be using "apeism" to refer to an indifference toward the welfare of apes: someone is an apeist if and only if she has no or little concern for apes' welfare. Undermining apeism consists in increasing people's concern for or, equivalently, decreasing their indifference toward, the welfare of apes. (I will not attempt to specify what apes' welfare consists of.)

Indifference toward the welfare of apes can manifest itself in a variety of ways. One important manifestation is an incapacity or unwillingness to empathize with apes' suffering and a lack of sympathy for their poor well-being. In addition, apeism manifests itself when people fail to take into account the welfare of apes while making decisions that affect them. These include where to go on vacation, which social-reform

---

refer to non-human hominoids: chimpanzees, bonobos, gorillas, orangutans, and gibbons (the last being sometimes excluded). This is how "apes" will be used in this chapter.

[4] Other strategies could be investigated in future work. In particular, recategorization (thinking about oneself and the members of other races as belonging to a single group, e.g., human beings) could naturally be extended to apes: people could learn to think about themselves and apes as being sentient beings.

[5] For the sake of space, I will not review the body of research about the determinants of people's attitudes toward the natural world (sometimes called "conservation psychology"), but the reader could usefully consult the following website: www.conservationpsychology.org/, including the bibliography at www. conservationpsychology.org/resources/articles/. Accessed 05/24/2012.

movement to support (such as the recent push for prohibiting experimentation on apes[6] or the movement for granting legal personhood to apes and other non-human animals[7]), whether and to which charities to make donations, what books to give children, etc.

One can be a speciesist – that is, one can believe that species membership is a relevant moral difference (Singer, 1975; see Bekoff, this volume; Dunayer, this volume) – without being an apeist. One can have some concern for the welfare of apes (even a lot of it) while also thinking that species membership is relevant for moral judgments and decisions. Furthermore, there is no contradiction between being a speciesist and thinking that one should have some concern for the welfare of apes. As a result, defending the wrongness of apeism does not commit oneself to defend the controversial claim that speciesism is wrong. On the other hand, if someone rejects speciesism, then she commits herself to having some concern for the welfare of apes, and she is thus committed to reject apeism. Whether or not she lives up to this commitment is, of course, another matter.

Apeism is wrong

Apeism is not merely an indifference toward the welfare of apes, it's a callous indifference; it is wrong. Its wrongness follows from many ethical theories, which, even when they grant different types of moral status to different species (perhaps because membership in different species correlates with the possession of different cognitive capacities or different forms or degrees of agency), typically entail that non-human animals are justified objects of moral concern or that they have some moral rights (for discussion, see Chapters 1 and 2 of Gruen, 2011). Failing to take their interest into account or violating their moral rights would thus be wrong for most ethical theories. (Noteworthily, then, the wrongness of apeism is easier to defend than the wrongness of speciesism.)

Of course, some ethical theories disagree with the claim that non-human animals are proper objects of moral concern or have moral rights (e.g., Carruthers, 1992). For Immanuel Kant, only rational beings – where rationality is a matter of being able to act and to reflect on the reasons for one's actions and beliefs – have rights, and only humans count as rational. Rational beings are persons, while everything else – objects, plants, and non-human animals – are things, which persons can use in any way they want to reach their goals (provided this use does not violate other persons' rights). Thus, Kant writes in "Conjectures on the beginnings of human history" (cited in Korsgaard, ms: p. 2):

When [the human being] first said to the sheep, "the pelt which you wear was given to you by nature not for your own use, but for mine," and took it from the sheep to wear himself, he became aware of a prerogative which, by his nature, he enjoyed over all animals; and he now no longer regarded them as fellow creatures, but as means and instruments to be used at will for the attainment of whatever ends he pleased.

---

[6] www.nytimes.com/2011/12/16/science/chimps-in-medical-research.html?hpw&gwh=09F269688E0B60D6 A774335315AF0F9A. Accessed 05/24/2012.

[7] www.nonhumanrightsproject.org/. Accessed 05/24/2012.

On the other hand, Kant also holds that we have a duty to treat non-human animals humanely, but not because these have rights or are objects of moral concern. While, for Kant, people don't owe non-human animals anything, they owe themselves to treat animals well. As a consequence, even for Kant, a complete indifference toward the welfare of apes would be wrong (see also Hume, 1777/1975: p. 185).

Furthermore, Kantians need not follow Kant in denying moral rights to non-human animals. Korsgaard (2005, ms), in particular, argues that rationally choosing involves viewing oneself, *qua* animal for which things can be good and bad (and not *qua* rational being), as an end in itself and thus as having moral rights. Thus, one acknowledges that all sentient creatures have moral rights and that one has duties toward them (see Korsgaard, ms, for a discussion of non-human animals' political rights).

What amount of concern for the welfare of apes is appropriate is a difficult ethical question, and ethical theories are unlikely to agree on this. In what follows, I will bracket this question, assuming throughout that it is wrong to be indifferent to or to have little concern for the welfare of apes.

Contact

Contact and racism

At least since Allport's (1954) groundbreaking work, psychologists and social scientists have examined in detail whether, when, and why contact between different racial groups undermines racism (see Pettigrew, 1998; Pettigrew and Tropp, 2006; Kelly *et al.*, 2010a; Hodson, 2011). They have typically distinguished two forms of intergroup contact. The simplest hypothesis about intergroup contact – "the warm-body hypothesis" (Sampson, 1986) – proposes that simply increasing the frequency of interracial interactions reduces racism.

Psychologists have long been skeptical of the warm-body hypothesis because field studies of the outcomes of desegregation (e.g., in schools) revealed that increased contact did not always decrease prejudice or undermine stereotypes. To explain these findings, they have proposed that some social and psychological conditions must be satisfied for contact to reduce prejudice and undermine stereotypes. In developing this idea, Allport's (1954) formulation of the contact hypothesis in his celebrated book, *The Nature of Prejudice*, has been particularly influential: the participants in interracial interactions should have equal standing, they should have a common goal, they should play distinct, complementary roles for accomplishing this goal, and participation should be sanctioned by authority. These four conditions are well illustrated by the jigsaw classroom method in educational settings (Aronson and Patnoe, 1997). According to this method, classroom projects must be so structured that students from different racial backgrounds cooperate to reach the assigned common goal. Members of each race fill different, equally important, roles and the value of the interracial cooperation is identified and praised by the teacher. Decades of research have now shown that, when the conditions specified by Allport and others are met, contact reduces explicit racism, while more recent research suggests, somewhat tentatively, that it may reduce implicit racism too (for discussion, see Kelly *et al.*, 2010a).

Contact and apeism

If contact undermines people's biases only when they engaged in collaborative, authority-sanctioned projects, the idea that contact could also be used to undermine apeism would be a non-starter. Humans and apes do not, possibly cannot, and perhaps should not engage in the kind of authority-sanctioned, cooperative, complex inter-actions envisaged by Allport (1954) and others.

From a practical standpoint, because the number of apes is small and, sadly, dwin-dling, only a few people – mostly scientists studying apes and people in Africa and Asia living near ape populations – have the opportunity to have any direct, sustained interaction with apes.

Furthermore, apes may not have the cognitive capacities, motivation, emotional dispositions, or lifestyles that are required to engage in the kind of interaction described by Allport and others. Chimpanzees do not seem to be motivated to engage frequently in collaborative activities, possibly because the product of these activities (e.g., food) is rarely shared equally among chimpanzees (e.g., Hamman *et al.*, 2011; Melis *et al.*, 2011; see, however, Hare *et al.*, 2007 for a difference between chimpanzees and bonobos). For instance, in contrast to what is observed in human children, collaboration does not lead chimpanzees to share food (Hamann *et al.*, 2011). The relatively solitary lifestyle of orangutans does not seem conducive to the kind of interactions described by Allport and others, neither does dominant gorillas' lack of tolerance for other males.

There are also some relevant cognitive differences between humans and apes. Allport and others have insisted that, to undermine biases, interactions should be normatively sanctioned. Viewing interactions in normative terms (as being done properly or improp-erly) comes very easily to humans, including young children (e.g., Rakoczy *et al.*, 2008), but it is doubtful that the same is true of apes. Furthermore, some comparative psychologists doubt that apes are able to engage into what is sometimes called "we thinking" or "shared intentionality" – that is, conceiving of themselves as being part of a larger group and of their goals as being coordinated with others' goals. Thus, Tomasello and Herrmann (2010: p. 5) write that "the key difference is that humans have evolved… social-cognitive skills and motivations geared toward complex forms of cooperation – what we call skills and motivations for shared intentionality." Tomasello and colleagues may be exaggerating chimpanzees' unwillingness and incapacity to engage in collaborative efforts – in many of their own experiments, chimpanzees do engage in collaborative activities – but it is true that from a very early age on human beings are distinctively motivated to engage in such efforts.

Finally, apes may be better off if they are left alone in their natural habitat. Moral considerations relating to apes' welfare would thus prevent us from endorsing the proposal to undermine apeism by means of interspecies contact.

Fortunately, Pettigrew and Tropp's (2006) important meta-analysis of the research on contact recently revealed that even "warm-body" contact reduces racist prejudices (although to a lesser extent than the structured interactions described by Allport and others). While it may be impossible to engage in collaborative, authority-sanctioned interactions with apes for the reasons just given, other forms of contacts are definitely possible.

First, contact with apes can be limited to observing them ("observational contact"). In this case, contact does not involve any interaction, even less the kind of structured interactions frequently discussed in the literature on contact and racism. Zoos, research centers, and sanctuaries already make this form of contact possible in non-natural settings, and ecotourism makes it possible in apes' natural habitat. In the USA, zoos and aquariums already attract more people than football, basketball, hockey, and baseball *combined* (Vining, 2003). Documentaries shown on TV or on the web are an indirect form of observational contact (i.e., a type of contact that does not require physical proximity and actual face-to-face interactions).

Another form of indirect contact could be called "asymmetric contact." In this form of contact, people engage in actions that influence the lives of others, are given some feedback about the consequences of their actions, and modify their behavior as a response to this feedback. The recipients of these actions need not be aware that others are so acting. For instance, middle or high school students could follow the fate of a group of, for example, chimpanzees or the fate of particular chimpanzees within a group (see below for the importance of viewing apes as individuals), and see how this group or these individuals benefit from their actions (e.g., donations).

A third form of contact is worth mentioning: "imaginary contact." In race-related imaginary contact, people imagine interacting with the members of another racial group. Crisp and Turner's (2009) literature review suggests that, at least in a laboratory setting, just asking people to imagine engaging in interracial interactions is sufficient to reduce racism and other negative attitudes toward outgroups: intergroup attitudes seem to be improved, more positive characteristics are attributed to outgroup members, and contact-induced anxiety is reduced. Applied to apes, people could imagine interacting with apes. Since people often know little about apes this form of interaction should probably be supervised. For instance, middle or high school students could imagine engaging in some form of interaction with apes under the supervision of their teachers. Well-crafted, accurate novels, movies, and documentaries could make it easier to imagine such interactions.

Pessimism about contact

Various considerations cast doubts on whether extending the contact strategy from racism to apeism would be useful. First, it is dubious whether some of the forms of contact mentioned above can undermine apeism and, if they can, whether we should promote them. While the relevant literature is not extensive and incomplete (Swanagan, 2000; Vining, 2003), it is unclear whether zoos and aquariums have a positive influence on visitors' attitudes toward non-human animals and conservation (see below for a possible partial explanation). In particular, Marino and colleagues (2010) have questioned the validity of recent studies suggesting that zoos have such an influence (see also Rhoads and Goldsworthy, 1979 for older concerns). Some zoos and entertainment parks also provide misleading information about the animals they host. For instance, Michael Mountain (2010) notes that SeaWorld does not highlight the intelligence of cetaceans. In addition, apes in zoos and research facilities harbor many symptoms of anxiety and mood disorders, including hair plucking and feces eating (Birkett and Newton-Fisher, 2011; Ferdowsian et al., 2011; McGrew, personal communication). On moral grounds, one may object to an anti-apeism strategy that requires sequestering apes in environments that harm them. While it is less questionable than zoos, ecotourism is also not without criticism, and should comply with some best practices guidelines (e.g., Homsy, 1999; Macfie and Williamson, 2010; Annette Lanjouw, personal communication; see Lanjouw, this volume).

Second, no research bears on the intriguing idea that asymmetric contact can influence people's attitudes. On the other hand, the capacity of warm-body contact to reduce biases gives ground for optimism. And, as noted above, research suggests, perhaps surprisingly, that even imaginary contact undermines racism.

Third, while warm-body contact does reduce racism, contact has a much larger effect when the conditions identified by Allport and others are satisfied, and we should not expect a substantial reduction of apeism from the forms of contact envisaged above. Still, one may retort, any reduction would be good.

Fourth, and most importantly, it is unclear whether contact is the proper strategy for reducing apeism. Contact does not seem to reduce all types of prejudice and bias: males and females have been in contact for eons, but sexism is still alive and well. One may speculate that the persistence of sexism results from the fact that the interactions between males and females do not meet the conditions spelled out by Allport and others. But, if this were right, this would not bode well for the proposal of reducing apeism by means of the forms of contact discussed above. Alternatively, and perhaps more plausibly, the explanation for the persistence of sexism despite constant contact between males and females is due to the differences between sexism and racism. While the effect of contact on attitudes is probably mediated by several mechanisms (e.g., producing positive norms for social interactions), contact reduces racism in large part because it reduces the anxiety and fear that one racial group may have toward another (e.g., Voci and Hewstone, 2003). Because anxiety and fear play at best a minor role in sexism, contact turns out to be unable to undermine this form of prejudice. This plausible explanation again bodes poorly for the proposal to extend the contact strategy to apeism since it is dubious that apeism is due to anxiety toward and fear of apes. Rather, it seems to be due to a lack of empathy and sympathy and to a poor moral imagination.

It may be worth trying to implement some of the forms of contact described above, in particular because interviews of conservationists have highlighted the role of personal experiences, which require some form of contact, in the development of their attitudes toward nature (Vining, 2003). On the other hand, there are reasons to be skeptical of the capacity of contact to lead people to care for apes' well-being, and more promising strategies should be investigated.

The Enlightenment Strategy: informing and convincing

Racism and the Enlightenment Strategy

Racism is often defined as involving the possession of erroneous beliefs about racial groups (e.g., Appiah, 1990), and even those who have questioned the definition of racism in terms of false beliefs (e.g., Garcia, 1996) agree that racism is often paired with such beliefs, which causally sustain the persistence of racism. Some of these erroneous beliefs are factual, while others are moral. They include essentialist beliefs about membership in racial groups – e.g., that there are race-specific genes or that some differences between races are genetic and immutable – as well as beliefs about the worth (including moral worth) of these groups – e.g., that whites are intellectually superior to blacks.

It is thus natural to propose undermining racism by providing information that would lead people to correct their own erroneous beliefs and by convincing them by argument and reason. I call this policy "the Enlightenment Strategy." Philosopher of science Naomi Zack, for instance, has repeatedly advocated such a policy vis-à-vis racism:

The remedy for this type of racism has been to correct unsubstantiated generalizations and educate unintentional racists toward a realization that their actions harm non-whites. (Zack, 1998: p. 43)

Apeism and the Enlightenment Strategy

Lay people have little knowledge of apes' cognitive capacities and ways of life (e.g., Ross *et al.*, 2011). They do not know much about the threats that bear on apes' survival or about humans' responsibility for these threats (e.g., for the destruction of apes' natural habitat). They may also have problematic moral beliefs that would not withstand criticism. Such erroneous or problematic factual and moral beliefs may either be essential to apeism or, at least, may causally sustain people's indifference toward apes' welfare.

It would thus be natural to propose extending the Enlightenment Strategy to apeism: erroneous factual beliefs about apes should be corrected by providing information about their cognitive capacities and ways of life and about the threats bearing on them, and problematic moral beliefs should be challenged by argument. It is probably very tempting for academics and educators to endorse this strategy. After all, improving people's knowledge and correcting their misconceptions is part and parcel of academics' and teachers' task.

Such a policy can be put into practice relatively easily since it merely expands upon existing educational efforts. Documentaries can be (and are already) shown on TV and in schools or universities. Books aimed at a broad audience, including children, more technical, academic books aimed at teachers and opinion makers, and textbooks aimed at students can be (and are already) written. Public conferences or more restricted workshops can be (and are already) organized. Past educational efforts directed at other forms of prejudice (sexism, racism, anti-gay prejudices) can be used as models for putting the Enlightenment Strategy into practice.

Pessimism about the Enlightenment Strategy

There are several reasons to be skeptical of the capacity of the Enlightenment Strategy to undermine apeism. The counterparts of some of the erroneous factual beliefs that are central to racism are true in the case of apeism. Racists often believe that people's race is determined by a biological nature or essence, which also fixes their psychological or behavioral capacities. Such beliefs are undoubtedly false. By contrast, humans and, for example, chimpanzees do belong to different species, and they do have different psychological and behavioral capacities for biological reasons to do with inheritance.

Furthermore, people's false factual beliefs about apes may not be as central to apeism as the relevant erroneous beliefs are to racism. The possession of erroneous beliefs does not seem to be definitional of apeism, and it may not even cause people's indifference toward apes' welfare. Apeism seems to be in part a failure of people's emotional sensitivity – a limit to people's capacity to care – and this failure may have little to do with what people believe.

Even if false beliefs were as important for apeism as they are for racism, research on motivated reasoning in social and cognitive psychology (Kunda, 1990) suggests that people are unlikely to modify the beliefs they have emotionally invested in even when they are presented with inconsistent information. People are motivated to ignore or to be skeptical of arguments and information that undermine these beliefs, and to search for any kind of evidence that would support them. Standards of evidence shift depending on whether the arguments and evidence support or undermine people's emotionally laden beliefs. In a series of experiments reviewed by Kunda (1990), participants were presented with some body of evidence. Those who were motivated to disbelieve the conclusion supported by the evidence tended to be more skeptical of this body of evidence than those motivated to believe the conclusion. For instance, participants told that they had a low IQ were more likely to view studies establishing the validity of IQ tests with suspicion (Wyer and Frey, 1983). After reviewing more evidence fitting this pattern, Kunda (1990: p. 490) concluded that "taken together, these studies suggest that the evaluation of scientific evidence may be biased by whether people want to believe its conclusions." Skepticism about human-caused global warming provides a sad, real-life illustration of this phenomenon.

One may object that the discussion of motivated cognition paints a too pessimistic picture of what can be done to undermine apeism. Because racist beliefs are connected to people's emotions (Faucher and Machery, 2009), motivated cognition should be a concern for those who propose to fight racism by disseminating knowledge (Kelly *et al.*, 2010a). However, the case of the beliefs tied to apeism is different: because these may not be similarly emotionally laden, people may be more willing to modify their beliefs when provided with evidence and arguments. Furthermore, the target of education may be less people's false or problematic beliefs about apes than their ignorance about apes. To the extent that the Enlightenment Strategy is meant to remedy people's ignorance rather than change their beliefs, motivated cognition may not be an issue.

While these two objections make valid points, their significance should not be overstated. For many people, the belief that human beings are very different from all other animals may be emotionally laden, for instance because of its connection to religious convictions, and this belief may be causally involved in people's indifference toward apes' welfare.

One should not conclude from this discussion that disseminating knowledge about apes and correcting factual and moral misconceptions are bound to be useless to undermine apeism. Rather, the moral is that on their own evidence and arguments are unlikely to have a large effect on apeism. Evidence and arguments may not be sufficient to convince people to abandon their erroneous factual beliefs and their problematic moral beliefs about apes. As Gordon Allport (1954: p. 485) once said: "Information seldom sticks unless mixed with attitudinal glue." Even if people could be so convinced, it is unclear whether apeism would then be undermined since beliefs may not be essential to, or causally involved in, apeism. Thus, more promising strategies should probably be preferred.

Beyond species categories: individualizing apes

Beyond racial concepts: individualizing humans

Research on interracial interactions has shown that people's prejudices decrease when they come to view others as individuals – Maria, Todd, Kim, etc. – instead of members of particular racial or ethnic groups – Hispanics, whites, Asians, etc. (Wilder, 1981). Making a concerted effort to see members of another race as individuals, rather than instances of a category, lessens the scores of white individuals on tests of implicit racial bias (Wheeler and Fiske, 2005; Lebrecht *et al.*, 2009), and it leads to less activation of the amygdala when people are presented with pictures of members of other races (Wheeler and Fiske, 2005). It also increases people's ability to recognize and distinguish faces of members of another race – that is, it reduces what is known as the "other-race effect" (Lebrecht *et al.*, 2009). These effects of individualization result from the fact that explicit and implicit biases are more likely to be triggered when others are categorized as members of racial groups instead of being treated as individuals.

Individualizing apes

Individualizing apes – conceiving of apes as individuals instead of as members of particular species such as chimpanzees or gorillas – could take two forms as a policy. First, it could be an individual policy – that is, a form of cognitive habit that each of us tries to acquire – on a par with intellectual policies or cognitive habits one can develop to avoid racist or sexist thoughts and emotions. Second, it could be a social policy: scientists, scientific journalists and writers, and other opinion makers could depict apes as individuals for the lay public instead of as members of particular species.

Practically, individualizing apes involves distinguishing individuals from one another, e.g., by identifying them by names, as we do for human beings and pets

Figure 4.1 Michelle rescued Max as a piglet and bottle fed him until he was healthy. They're great friends and Max loves to romp with the dogs in the yard. New York, 2006. Naming and becoming familiar with individual animals is an effective strategy against discrimination. Photo Jo-Anne McArthur, We Animals.

(Figure 4.1). This also involves describing the particulars of each individual's life, as we would do for the life of a human being. Differences between the lives of particular apes, variability in their lifestyles, and personality differences should be highlighted. Negatively, individualizing apes involves refraining from characterizing apes' lifestyles, behavior, and cognition in generic terms, or, when this is necessary (e.g., when commonalities among chimpanzees or bonobos are under scientific examination), readers should be reminded of the existence of differences.

There are already numerous fine examples of this kind of approach, such as the description of Yeroen's, Nikkie's, and Luit's lives in de Waal's (1982) *Chimpanzee Politics* and Barbara King's (1998) obituary of Washoe in the January, 2008 issue of *Anthropology News*. Following Yerkes, Lori Gruen has described the lives of chimpanzees in Yale laboratories of primate biology (http://first100chimps.wesleyan.edu/). Each chimpanzee is named, its genealogy, when known, is described, and, when possible, its personality briefly sketched.[8] Many chapters in Robbins and Boesch's recent book,

---

[8] See also the portraits of the residents at Chimp Haven: www.chimphaven.org/meet-the-chimps/photo-gallery/. Accessed 05/24/2012.

Among African Apes, describe the lives of particular apes, and some of the pictures in this book name apes. Boesch's description of the death of Lefka is particularly touching:

Night life in the forest is a very special experience, with all the noises, such as the large hammerhead fruit bats making their courtship calls, the little bushbabies with their swift sewing-machine calls... However, with Lefka's laborious breathing near me, the experience of this night was different. Lefka had stopped breathing by 5:00 am, and in my heart, a piece of the magic of the Taï forest vanished. As the daylight entered the forest an hour later, I realized that a little chimpanzee had descended from a nearby tree. It was Léonardo, Lefka's tiny two-year-old brother. Seemingly, totally stunned by the second death within two days, he glanced at his dead brother and then, not knowing where the other chimpanzees were, headed north alone. I did not expect to see him again.   (Robbins and Boesch, 2011: p. 40)

It is also worth noting that people are already used to individualizing non-human animals: people tend to individualize pets, which are found in a majority of households in North America and in Europe (Vining, 2003) as well as in most cultures (though not in all). The proposed policy is merely to extend this habit to apes. Furthermore, accurate novels, movies, and documentaries could make it easier to view apes as individuals when their protagonists are individualized.

On the other hand, even excellent articles, books, and documentaries by scientists and science writers are typically at odds with this individualizing policy since they aim at depicting the behavior of chimpanzees in general (e.g., Boesch's, 2009, *The Real Chimpanzee*), or orangutans in general (e.g., Van Schaik's, 2004, *Among Orangutans*), etc. These writings and documentaries are bound to reinforce people's habit to view apes in generic terms instead of inviting to individualize them. The need to change the way apes are typically presented (as species exemplars instead of as individuals) is likely to be a stumbling block in the implementation of this individualizing policy, but hopefully an increasing number of writings and documentaries about apes will embrace it.

Grounds for optimism

Either as an individualistic or as a social policy, one may be optimistic that individualizing apes will undermine apeism. Individualization played a crucial role in the passage of the Animal Welfare Act of 1966. In 1965, journalist Coles Phinizy described the fate of Pepper, a Dalmatian who had been stolen, sold, and killed for experimental purposes, in the November 29, 1965 issue of *Sports Illustrated*. In 1966, *Life* published an article entitled "Concentration camp for dogs" that describes the horrific conditions of dogs at a Maryland dog dealer's farm with individualizing pictures of specific dogs.[9] These articles caused a national stir (*Life* received more letters about this article than about the Vietnam war), which contributed to the passage of the Animal Welfare Act.[10]

---

[9] http://tinyurl.com/7tnur68. Accessed 05/24/2012.
[10] www.nal.usda.gov/awic/pubs/AWA2007/intro.shtml. Accessed 05/24/2012.

Moral emotions as well as the capacities for empathy and sympathy are more likely to be engaged by individuals rather than by groups (e.g., Small and Loewenstein, 2003; Small *et al.*, 2007). Research on the identifiable victim effect has repeatedly shown that people's motivation to help, including their willingness to donate money, is greater when it is directed at specific individuals rather than at social groups (e.g., poor people, people in some specific countries, etc.). This is why relief agencies and non-governmental organizations often describe the plight of particular individuals.[11] Chimp Haven is already using a similar strategy, linking a request for donation to videos of its residents.[12]

The indifference toward apes' welfare is plausibly in part caused by our viewing them in generic terms. Because we tend to think about, for example, chimpanzees in general, or even about apes in general, instead of about particular individuals with distinctive fates, and because empathy, sympathy, and, more generally, caring are engaged by individuals more than by groups, we are disposed to be indifferent to apes. Remarkably, zoos, aquariums, and entertainment parks are likely to reinforce this disposition to think about apes and other non-human animals in generic terms since zoo animals are typically presented as exemplars of species ("a lion," "a giraffe," "a chimpanzee") rather than as individuals with particular lives. By contrast, learning to think of apes as individuals would plausibly result in greater interest in their welfare, exactly as we care for the well-being of pets.

More speculatively, learning to think of apes as individuals would perhaps make it easier to moralize people's attitudes toward their well-being. Moralizing consists in treating as morally wrong or right a behavior or an attitude that was previously neutral. Smoking, eating meat, and eating healthily have recently been moralized in parts of the USA (Rozin *et al.*, 1997; Rozin, 1999). Our intuitive moral sense seems to be geared toward making judgments about individuals rather than groups (which does not mean that it cannot be extended to making judgments about groups such as nations, cultures, or species): we spontaneously think of individuals as having rights and duties. As a result, it may be easier to judge that apes viewed as individuals instead of species exemplars have rights or deserve consideration. Failure to respect these rights or to take their interests into account, and more generally the indifference toward apes' welfare, may then be judged wrong. That is, treating apes as individuals may open the door for a successful moralization of people's attitudes toward them.

This is not to say that the individualization policy is guaranteed to succeed. In particular, people may not generalize their concern for the welfare of specific apes to the welfare of apes in general. This concern notwithstanding, individualization clearly has the potential to abate people's indifference toward the welfare of apes.

---

[11] See, e.g., Oxfam's website: www.oxfam.org/en/emergencies/west-africa-food-crisis/sahel-photos-food-crisis-foretold. Accessed 05/24/2012.

[12] https://donationpay.org/chimphaven/bestfriends.php. Accessed 05/24/2012.

Conclusion

People are too often indifferent to the well-being of great apes, an indifference that is morally condemnable. In this chapter, I have examined three possible strategies for undermining this indifference: contact, the Enlightenment Strategy, and individualization. While it may be worth trying to put the first two strategies into practice, they do not seem to be the most promising. By contrast, the third strategy – learning to view apes as individuals instead of as exemplars of their species – is not only a natural extension of our relation to pets, it is also the most promising strategy.

5 "Race" and species in the post-World War II United Nations discourse on human rights

Raymond Corbey

Discursive and analytical resources used to study or combat stereotypes can themselves be articulations of such stereotypes. In the following I will show this for the humanist post-World War II United Nations discourse on human rights and racism, which proclaimed a new, more inclusive demarcation of morally respectable beings by a continuing exclusion of others. This constitutes a formidable obstacle for a definition of rights and moral responsibility that includes non-human animals. The same problem faced the subsequent "Great Ape Project," which proclaimed an extension of the rights of human apes to non-human apes only.

From a background in both philosophy and anthropology I have always had a keen interest in historical roots and cultural backdrops to views of human and non-human animals, in particular the Western idea of human exceptionalism and unique dignity. In the following, I compile a tentative inventory of the main backgrounds, roots, and contexts of the implicitly speciesist post-World War II United Nations Declarations on Human Rights and Racism, ending with some reflections on the nature of such proclamations.

All humans equal

In the aftermath of the racist atrocities of World War II a new, anti-racist conception of humankind and human rights was proclaimed. *The Universal Declaration of Human Rights* (1948; published in 1952: United Nations, 1952; see Figure 5.1) states that "Recognition of the inherent dignity and of the equal and inalienable rights of all members of the *human* family is the foundation of freedom, justice and peace in the world... [All] *human* beings are born free and equal in dignity and rights. They are endowed with reason and conscience and should act towards one another in a spirit of brotherhood" (my italics). "Everyone has the right to life, liberty, and security of person, and to not to be held in slavery or servitude" (United Nations, 1952).

The *Statement on Race* (UNESCO, 1950) was the first of several subsequent declarations explicitly focusing on racism. The *Proposals on the Biological Aspects of Race* (1964), for example, claim that "[all] men living today belong to a single species, *Homo*

The Politics of Species: Reshaping our Relationships with Other Animals, eds R. Corbey and A. Lanjouw. Published by Cambridge University Press. © Cambridge University Press 2013.

Figure 5.1 Eleanor Roosevelt holding a copy of *The Universal Declaration of Human Rights*, a document that she helped create. US National Archives.

sapiens, and are derived from a common stock... Neither in the field of hereditary potentialities concerning the overall intelligence and the capacity for cultural development, nor in that of the physical traits, is there any justification for the concept of 'inferior' and 'superior' races" (Dunn *et al.*, 1975: p. 358).

Until then a Eurocentric double standard for "races" – one for "whites," another for "non-whites" – had predominated. Quote marks are in order, because developments in twentieth century genetics made it abundantly clear that biological variability within present-day humankind are a matter of very small and gradual differences caused by variable gene frequencies. There are no human "races" in terms of types with fixed essences that can be hierarchically ordered in terms of moral qualities, mental capacities, or motivational characteristics. All variability was now assumed to be cultural (see Stoczkowski, 2009).

The inclusion of non-European "races" in the "human family" was made possible by a persisting speciesist, *Homo*-centric standard for all other species as the foundation of society's moral and legal order. Humans were not to be treated like "beasts." Quote marks again. There are millions of animal species, among them thousands of mammal species and two hundred extant primate species, one of which is the extant human species. Yet this single species is set apart in opposition to millions of other species that are lumped together not just as "animals" – or more accurately as *other* animals – but as "beasts" – a term with negative connotations. It was in this pejorative sense that "lower races" were associated time and again, metaphorically, metonymically, literally, with (other) animals.

The biological homogeneity of humankind was presented as an argument for moral and political equality within that species. The inclusion of all members of the human species in a community of moral equals was thus made possible by the exclusion of the members of all other species. The human kind was uniform, which forbade dominion over some humans by others; it was unique in living nature, and so justified human dominion over and exploitation of other, disposable, commodified, animal species. Caucasian exceptionalism was combated while human exceptionalism persisted.

Human exceptionalism in European metaphysics

The post-World War II United Nations discourse on human rights and racism was, and still is, a broadly humanist one. It issues from a European tradition of exceptionalist metaphysical and moral views of (unique) human nature and dignity, and the (special) place of humans in nature. This tradition reaches back over 2000 years to both Greek philosophy and Judeo–Christian religious doctrines. This is the first root of the humanist United Nations discourse I would like to discuss. Time and again, from Plato and Aristotle through Augustine and Thomas Aquinas, through Descartes and Kant, up to substantial parts of present-day philosophy, human individuals were taken to be fundamentally different from individuals belonging to other animal species. As minds with subjectivity and agency they were also taken to transcend and stand apart from their own natural, animal-like bodies.

The decisive, "essential" difference – reflecting a specific, immutable metaphysical *essentia* in the sense of Aristotle, Scholasticism, and Roman Catholic orthodoxy – was humans' capacity to be reasonable, both in a cognitive and in a moral sense. In broad circles worldwide, including the overwhelming majority of legal and political systems, it is still seen this way. In a present-day phrasing, which is much indebted to Immanuel Kant and Enlightenment thought: humans are uniquely self-conscious, free-willing, and, therefore, morally responsible beings. In a leap, which I have never understood very well, this is usually taken to imply automatically the unique moral respectability – "dignity" – of humans themselves. Non-human animal subjectivity has always been, and continues to be, a blind spot, a conundrum for mainstream European thought.

This humanist discourse has loosened itself from, but still converges with, religious ideas in the Judeo–Christian tradition. Subscribed to by over two billion people, the latter sees humans as the only living beings with reasonable, rational souls, created in the image of God and therefore standing high above the rest of nature. As Pope John Paul II phrased it in an address to the Pontifical Academy of Sciences on October 22, 1996:

It is by virtue of his spiritual soul that the whole person possesses such a dignity even in his body... if the human body takes its origin from pre-existent living matter the spiritual soul is immediately created by God... theories of evolution which, in accordance with the philosophies inspiring them, consider the mind as emerging from the forces of living matter, or as a mere epiphenomenon of this matter, are incompatible with the truth about man. Nor are they able to ground the dignity of the person. With man, then, we find ourselves in the presence of an ontological difference, an ontological leap.   (Pope John Paul II, 1996)

The "new synthesis" of evolutionary biology and genetics

Ideological pressures toward assuming the unity of humankind, prompted by the racist excesses of World War II and drawing upon mainstream European thought, converged with an important development in twentieth-century biology. The New Synthesis between evolutionary theory and genetics replaced earlier scientific approaches to races in terms of a hierarchy of fixed types, so forming a second important influence on the United Nations Declarations on Human Rights and Races. Prominent biologists such as Julian Huxley and Theodosius Dobzhansky were actively involved with both the New Synthesis and with those declarations. They defended the biological and genetic homogeneity of the species *Homo*, which served as an argument for moral and political equality and the right to full citizenship of all humans, whatever their cultural background or physical characteristics.

Humankind was supposed to be variable only culturally. New synthesis biology saw the human species as a "grade," an adaptive evolutionary stage with a specific ecological niche that precluded the presence of similar species, and thus tended to stress the unity of the species. No two hominid species could live in the same ecological niche beause of competition. The new synthesis also tended to assume a linear, progressive development to more advanced stages (Corbey, 2012), still wrestling to be rid of the influence of early conceptions of evolution as directed by an inner drive toward a fixed goal along a pre-ordained path – rather than by Darwinian natural selection.

Nowadays various forms of metaphysical naturalism are growing ever stronger. They hold that natural sciences such as evolutionary biology or cognitive neuroscience understand nature and human nature. They replace essence with variation, higher purpose in cosmos and history with coincidence, top–down metaphysics with bottom–up ones. Remarkably this was not yet the way in which the New Synthesis, part of the naturalistic turn in twentieth-century thought, influenced United Nations' discourse: anthropocentrism lingered on in the latter's scientific humanism.

While in earlier views presumed biological differences had been taken to imply moral and political inequality, now similarities between all members of the species *Homo sapiens* were stressed. These similarities were taken to imply humans' moral and political equality and their right to full citizenship, whatever their cultural background or physical constitution (see Haraway, 1988). But the Caucasian yardstick for races was combated with a persistent human yardstick for species and the consequent exclusion of non-human species from moral respectability and from the "family of man." The pivotal role of that metaphor in the 1950s climate of opinion is illustrated by an exhibition of photographs showing people from 68 different societies, and entitled *The Family of Man*, that was mounted by the Museum of Modern Art, New York, in 1955 (Steichen, 1955). The exhibition travelled around the world for eight years, with stops in 37 countries.

The repudiation of biology in cultural anthropology

A third major influence on the human-exceptionalist, post-World War II discourse on human rights, was the cultural anthropology of that period, in particular in France and in the United States. In both traditions – the French one in the wake of Emile Durkheim, the American one in that of Franz Boas – cultural anthropology was seen as relatively autonomous with respect to the life sciences. It was supposed to deal with those aspects of humans that transcend their organic existence: linguistically expressed symbolic meaning and moral values, and how these structure society.

Many cultural anthropologists, under the influence of these two traditions, still think that the symbolic capacity of humans implies a rupture with nature and organic life. Humans are special, for they have entered into a different order of existence, that of symbolic language, reflexivity, and morality. Efforts to bring the symbolic and moral world of society and culture within reach of life-science perspectives such as behavioral ecology or gene-culture coevolution are repudiated (Carrithers, 1996; Corbey, 2005). "Culturalists" do not deny the role of biological and material constraints, but they see these as trivial; for them, symbolic meaning is the decisive factor. Cultural behaviors are not taken primarily to be objective matters of fact, but meaningful and appropriate in their specific contexts. As such, the argument goes, they partially, or even essentially, elude objectifying approaches, regardless of how germane such approaches may be to underlying biological or ecological processes. The distinctive quality of humans is not that they live in and adapt to a material world, like other organisms, but that they do so according to meaningful, culturally variable, symbolic schemes.

A major voice in the coming about of the declarations on race was that of the prominent French ethnologist Claude Lévi-Strauss, while American anthropologists in the Boasian tradition had their say too. In both cases we meet Immanuel Kant again, the most influential exponent of humanist Enlightenment philosophy. The Boasians and Franz Boas himself had been influenced decisively by the intellectual climate in late nineteenth-century Germany, where in a neo-Kantian setting philosopher Wilhelm Dilthey and others pleaded for the relative autonomy of hermeneutic, interpretive human sciences vis-à-vis the natural sciences. Boas, who studied in Germany, read Kant intensively; the same goes for two of his most influential pupils, Ernst Sapir and Alfred Kroeber. They saw culture as an extremely variable, relatively autonomous layer superimposed upon humankind's uniform biology. It therefore required a methodology different from and irreducible to that of the biological sciences. In a strongly relativist vein they defended the equality of all human cultures.

Lévi-Strauss concurred with both this dualism and this relativism in his own, idiosyncratic way. He too, through the work of Emile Durkheim and Marcel Mauss was influenced by the nineteenth century, in this case French, neo-Kantianism. Cultural anthropologists don't study nature, bodies, landscapes, and kinship in themselves, but examine instead the ways in which they are culturally perceived and symbolically categorized. Both disciplinary traditions, Boasian and neo-Durkheimian, focus on how humans as moral beings rise above their naturally selfish, Hobbesian animal individuality that is directly rooted in the biological organism.

People are not "animals"

A fourth root of negative attitudes toward other animals is a cluster of phenomena to do with cultural attitudes and the articulation of cultural identity. It is of a slightly different order than the first three. I will concentrate on modern European culture in these remarks, but I think the argument is relevant for many other cultural settings too.

Unfavorable stereotypes of "the" non-human animal in general and a number of specific animals in particular had a role to play in the cultural discourse and attitudes of Europeans. They provided models of "beastly," "uncivilized," "low" behaviors to be avoided by those who took themselves to be "civilized" – who behaved, ate, defecated, dressed, and made love in a manner they felt was correct. One's own body, bodily functions, and certain impulses were also perceived as "animal," and to be subdued. Expressions such as "you behaved like an animal," or "people should not be treated like beasts" in the post-Holocaust political discourse, as well as various forms of verbal abuse show how non-human animals served as forceful symbols of uncivilized conduct, connected with shame and disgust. They featured in articulations of cultural identity in terms of the exemplary alterity of the animal as such, while certain animals became paragons or "natural symbols" (Douglas, 1970b) of brutishness, associated with uneasiness and aversion regarding what was considered improper or unbecoming.

There is also an extensive literature on European citizens' tendency to extend these associations to peasants, the working classes, the colonized, various infamous professions, and the sexually "deviant," among others (Frykman and Löfgren, 1987; Blok, 2001). Various distorting and distancing mechanisms have added to the role of articulations of cultural identity in attitudes toward other animal species, for example misrepresenting them as insensitive or evil, or concealing cruel practices (see Twine, this volume). This facilitated their exploitation in a variety of ways and helped to maintain a moral order and dietary regime that favored humans. Beings that were seen as low and defiled were thus not only exploited materially, but also discursively, serving to express and deal with things human.

Finally, we should not forget that in corporally embedded cultural regimes of appreciation and feeling about living beings, evolved cognitive and motivational predispositions are at work. Recent empirical research suggests that all humans, species wide, have a domain-specific cognitive system for categorizing and reasoning about living beings. It tends to attribute fixed essences to living beings as their underlying causal nature and to construct hierarchies. Scientists disagree on the precise nature of this psychological equipment and its effects are hard to disentangle from culturally transmitted beliefs and attitudes, but this line of research is promising in terms of a better understanding of both racism and speciesism (Livingstone Smith, 2011, this volume).

The Great Ape Project

In 1993, inspired by the United Nations Declarations on Race and Human Rights, a group of academics from various disciplines argued for a further widening of the community of moral equals to include all great apes (Cavalieri and Singer, 1993: pp. 4–7). The Great Ape Project's "Declaration on Great Apes" claimed the right to life, individual liberty, and avoidance of suffering. It argued that there is no meaningful criterion of personhood that excludes non-human apes and no natural category that includes apes but excludes humans. This declaration has had a substantial bearing on changes in legislation regarding the legal status of non-human apes in a number of countries worldwide.

However, the fact that it left the status and personhood of other animals open to future reflection provoked criticism. Similarity to humans still seemed to be the standard against which non-human species should be judged. The inclusion of great apes was made possible by the exclusion of other species, analogous to the inclusion of non-Caucasian "races" at the expense of non-human species in the *Universal Declaration of Human Rights*. As one of the project's contributors conceded, "[we] need to change 'The Great Ape Project' to 'The Great Ape/Animal Project' and to take seriously the moral status and rights of all animals by presupposing that all individuals should be admitted into the Community of Equals" (Bekoff, 1998a; see Bekoff, this volume; Dunayer, this volume). The great apes could serve as a convenient starting point, a bridgehead to the animal world, and the "Declaration on Great Apes" could be extended to other sentient beings.

Ritual recognition

Present-day moral, legal, and political philosophy and discourse is rife with analogous cases of implicit, silent, taken-for-granted exclusion of other animals. A glaring example is the contemporary philosophical debate on recognition (*Anerkennung* in German), the act of acknowledging or respecting another being and its status, rights, and achievements. Among the most prominent recent contributors are Canadian philosopher Charles Taylor and German philosophers Jürgen Habermas and Axel Honneth. All three operate in the social contract wake of, among others, Jean-Jacques Rousseau, Immanuel Kant and G. W. F. Hegel. While there is much to be found on multiculturalism and cultural "others," one searches in vain for serious consideration – let alone recognition – of the status of non-human species among these thinkers (see, for example, Taylor, 1994).

Recognition in the sense of that debate is exactly what this book is about. Various views on ethics in connection with non-human animals are taken nowadays, some inspired by Aristotle, others by Immanuel Kant, yet others by Jeremy Bentham, to mention but three of the most influential. All three argue for a more inclusive definition of moral respectability. I will leave for now the difficult philosophical debate and end this chapter with some remarks on the ritual character of recognition, from an anthropological perspective. Here I think that the neo-Durkheimian tradition of research on categorization and ritual offers some interesting cues.

This tradition draws our attention to yet another aspect of the United Nations and Great Ape Project declarations, and the present book too: not so much of their content but rather of their performative aspect; less of what is said but of how it is said. I mean the difference between a passive description of reality like "the door is open" and an utterance changing something in reality, like " open that door!" or "I hereby open this meeting." The first type of statement is true or false, the latter one more or less effective or successful. Seen from this angle, what happens in the statements on rights we are dealing with is a ritual – solemn, public, formal – incorporation or proposed incorporation of beings in a community of, in this case, moral equals. It is a rite of passage, a ritual articulation and bestowal of a new identity on both the community and the beings involved.

In the 1920s Marcel Mauss (1990), in synchronization with Emile Durkheim, analyzed the exchange of gifts, services, or civilities as a profoundly moral activity, as the coming into being of society as a moral order. A gift asks for, or is itself, a countergift. There is a three-fold obligation implied in every gift: to give, to receive, and to give in return. When people started giving, Mauss argues, they laid aside the spear, and omnipresent conflict developed into contract. One takes from enemies, but gives to friends. The gift as a moral gesture is thus constitutive of human society – a profound philosophical thought. Bestowing a name or a title on someone is constitutive of the identity of the giver as well as the receiver. It further articulates the relationship between them, creating specific rights, duties, and attitudes on both sides.

This rather abstract line of thought has turned out to be heuristically fruitful in ethnographic research on the constitution of the identity of groups and individuals – particularly in terms of dignity – through ritual exchange. "Dignity" from the Latin word *dignitas*, carries such meanings as "the quality of being worthy or honourable; worthiness, worth, nobleness, excellence... honourable or high estate, position, or estimation" (*New Oxford Dictionary of English*, Pearsall and Hank, 1998). Various declarations on slaves, women, racism, all humans, non-human apes, all sentient beings, ritually recognize the dignity of these categories of beings, incorporating them within the community of moral equals. And here again, I tend to stress the performative, expressive, voluntaristic aspect of the declarations under consideration as not only statements on what there is – moral respectability – but also and in particular as an act of recognition, an emphatic attempt to bring about worth and worthiness.

Conclusion

I have discussed various roots of the humanist United Nations discourse on human rights and racism. This discourse was prompted by and explicitly referred to the Holocaust, the systematic state-sponsored murder of approximately six million Europeans during World War II. Remarkably, and controversially, detailed parallels have been drawn more recently by animal rights groups and prominent writers such as I. B. Singer, Marguerite Yourcenar, and J. M. Coetzee between the fate of animals in our global economy and that of the victims of the Holocaust. "[We] are surrounded by an enterprise of degradation, cruelty and killing", a fictive character in one of Coetzee's novels states in a lecture, "which rivals anything that the Third Reich was capable of, indeed dwarfs it, in that ours is an enterprise without end, self-regenerating, bringing rabbits, rats, poultry, livestock ceaselessly into the world for the purpose of killing them" (Coetzee, 1999: p. 21; see Figure 5.2).

The subjects of rituals of acknowledgment and incorporation change from appropriated, commodified, objects to acknowledged subjects. One of the most profound political thinkers and theoreticians on recognition, Jürgen Habermas, distinguishes two aspects and modes of action that are germane to his influential work on human

Figure 5.2 A print by, and courtesy of, Roger Panaman, aka Ben Isacat, a British animal rights activist, conservationist, biologist, and artist.

communication, democracy, and law, and for his cultural critique of industrial societies. On the one hand there is strategic action, which has to do with instrumental rationality, with optimizing the means to one's goals, and which can turn to violence. On the other hand there is reasonable, communicative action, which acknowledges – recognizes – the subjectivity of others, and enters into an open, reasonable communication on goals with them. Reason(ableness) is placed in opposition to oppression, giving to taking, inclusion to exclusion, respectful coexistence to appropriation, and otherness as enrichment to otherness as threat and lack. The challenge here, I believe, is to think *with* such theorists of recognition, using their sophisticated analytical resources, *against* them – against their too narrow, too Kantian view of subjectivity and respectability.

6 Addressing the animal–industrial complex

Richard Twine

> The modern animal industries tend to be regarded as technically inevitable and
> politically neutral, since it is only relations between humans which are considered
> political.
>
> (Barbara Noske, 1989: p. 22)

I am a sociologist with a major interest in the position of animals in the globalized
economy. I did extensive research on the economics and science behind the increasing
molecular commodification and production of animals for human consumption
(Twine, 2010a). Since this work I have turned to Dutch anthropologist Barbara
Noske's (1989) very useful concept of the "animal–industrial complex." This chapter
attempts to give it more definition. It increases its heuristic potential as a key
organizing frame of analysis and research collaboration in studying various domains
of human–animal relations. I will present various ways of refining its definition and
illustrate some intersections between different forms of animal use and with various
other "complex" concepts.

    While keeping the permeability of the boundaries of the animal–industrial complex
very much to the fore, the present chapter nevertheless has a focus on farmed
animals. Bringing the profoundly speciesist animal–industrial complex into a more
clearly delineated space of scholarly and public critical scrutiny is important for the
emerging field of critical animal studies. Although there is not always a distinct
boundary between critical animal studies and animal studies, the former places a
stronger emphasis on the "condition" as well as the "question of the animal" (see
Pedersen and Stănescu in Socha, 2012). It regards the early twenty-first century (most
obviously the emergence of climate change) as bringing much urgency to the
academic engagement with animals. Although some critical animal studies scholars
take the "critical" literally and draw upon the work of the Frankfurt School (e.g., see
Sanbonmatsu, 2011) it is usually understood in a more general sense of a politiciza-
tion of academic animal studies.

The Politics of Species: Reshaping our Relationships with Other Animals, eds R. Corbey and A. Lanjouw.
Published by Cambridge University Press. © Cambridge University Press 2013.

A techno-capitalist imaginary

Global farmed animal production is increasingly understood as both an emblematic and a problematic controversy in debates around the unsustainability of global capitalism. The ongoing expansion of this sector under the rubric of the so-called "livestock" revolution (notably in "developing" countries) sits in marked tension with the various sustainability question marks increasingly placed around these sets of globalized practices. This "revolution" is naturalized via an assumption of inevitable human population growth and the global dissemination of a hegemonic "human" that, by definition, consumes meat, and a lot of it. It truly is a revolution, at least in terms of economic, social, environmental, and interspecies relations. It represents both a considerable capitalization move by globally positioned "livestock" corporations and is effectively an attempt to normalize a wide range of identities, relations, and practices via the conduit of dietary change. Capitalist biopolitics typically operate via an assumption of human–animal hierarchy, but collectively resource humans and animals alike for capitalization often in the same places and at the same times (for example, farm workers and farmed animals; slaughterhouse workers and farmed animals).

An aside on terminology is in order before I continue. I put the word "livestock" within quotation marks, for it betrays a frame of thinking that thoroughly naturalizes the commodification of animals. I therefore prefer not to use it but do so occasionally, to critically address the particular discourse of the "livestock revolution" and self-titled "livestock"-related companies. My use of scare quotes around this word is intended to indicate a critical distance from its assumptions. I have also decided non-ideally to use the word "animal(s)" instead of "non-human animal(s)," which I have used previously. Both are problematic and I think that eventually the discourse and binary of "human" and "animal" must be challenged in more satisfactory ways.

Sustainability concerns include, at the very least, direct environmental pollution, greenhouse gas production, human health impacts of consumption and labor, effects on communities, land use priorities, water consumption, zoonotic disease, and ethical questions around the killing of animals. There is a further set of questions concerning how these issues are being elaborated by the molecular turn (biotechnology) and the ways in which novel breeding techniques are proffered by scientists and industry as the means to secure the livestock revolution (see Twine, 2010a) – to usher in a new era of productivity, which assumes the surmounting of pre-existing (especially biological and environmental) limits as a technological problem and inevitability. This is galvanized by an emerging Western policy discourse of food security that uncritically reiterates an urgent need to double food production in order to meet the demands of a growing human population (see Tomlinson, 2011).

In clear tension with other cultural calls for the *reduction* of the consumption of animal products, the assumption of a productivist livestock revolution also involves varied attempts to make animal agriculture more efficient in lieu of changes to consumption levels that could threaten pre-existing markets. Such attempts include manipulating genetics in animal breeding and also measures to capitalize upon the waste products of animal production, such as the Cargill corporation's scheme to generate energy from beef processing waste in Alberta, Canada. We can note simultaneously a vision of a "green" livestock revolution and one that adds significantly to the present approximate annual global kill figure of 56 billion land animals.

Genomics and biotechnology generally have been fetishized by animal or "meat" scientists and policy makers as the means by which to reinvent capitalism as a new more efficient and environmentally benign project often under the banner of the knowledge-based bio-economy (Cooper, 2008; Twine, 2010a). The latter in its very enunciation pretends to portend to ideas and not somehow also to material, bodily repercussions (see Kenway *et al.*, 2006). This fits very well within now standard accounts of "ecological modernization," which prefers not to elevate the question of growth itself (or indeed limits) to a central problematic of the economy.

For meat scientists, an under-theorized constituent of the animal–industrial complex, in alliance with corporations and policy makers, "developing" countries comprise a particular object of biopolitical scrutiny and potential resource. This is not only because their human populations are simply not (yet) eating "enough" meat and dairy products, and are therefore zones of under-capitalization, but because their populations of farmed animals are similarly conceived as biopolitically backward, possessing "inferior genetics."

With such examples, capitalist biopolitics suspends a moral human–animal binary in order to promote capitalization (see Holloway and Morris, 2007; Twine, 2010a: especially Chapter 3). It is arguably informed by the historical lineage of discursively placing various classed and racialized human others as "closer to animal." In this manner capitalism strategically employs contradictory accounts of human–animal difference that are enabling for various development and accumulation projects. Thus we can note attempts at knowledge transfer (often framed philanthropically) to improve the genomics infrastructure of developing countries, to train farmers therein, to assess herd genetic diversity, to catalog, to introduce biopolitically superior Western stock, but also, further up the commodity chain, to introduce more intensive farming technologies. Such policies incorporate an assumption not only that Western levels of meat/dairy consumption are superior and somehow more constitutive of normative "ways of being human," but also that Western production methods are superior.

This chapter explores this techno-capitalist imaginary by (re)turning to the concept of the "animal–industrial complex" (Noske, 1989): a significant component of the broader global food system. Through such a return this is a proposal to reinvent and rejuvenate Barbara Noske's work. I hope it will encourage further research and activism that can address urgent matters of animal, ecological, and human flourishing shaped by and within the practices of the complex.

By returning to her concept the aim is to tease out its original perceptive dimensions but also to add further rigor so that it can be employed less as a rhetorical term but actually come to be embodied by a delineated set of actors, relations, and usable definitions. While a hyperbolic sense of the concept has not been without use in the sense of a shared discourse between those politically interested in challenging its power, a more refined definition can provide much more focus, shape, and coherence. Moreover, this project can be valuable for understanding the context of the complex within broader relations of political economy, for a better appreciation of intersectionality, and in allowing the concept to do better political work for those engaged in its critique.

This notion of intersectionality has various histories. Originally it was a humanist concept born out of a concern to better understand the intersections of classed, raced, and gendered power relations. In affinity with ecofeminism, (critical) animal studies, and other post-humanists, I intend it here, obviously, also in a more-than-human sense (see Twine, 2010b) to consider how such categories are also mutually co-constructive with "species." A substantial body of work in ecofeminism that examines intersections of gender, nature, and animals of course pre-dates any proclamations of the fields of animal or critical animal studies. This ecofeminist work has also considered race and cultural difference although there have been notable recent publications in what could be termed post-colonial critical animal studies (e.g., Kim, 2007; Harper, 2010a,b; Deckha, 2012).

While this chapter advocates a revisiting of Noske's concept through an intersectional lens, the main hope is that it can initiate a collaborative process whereby the concept can be refined, mapped, and utilized by academics, activists, and activist–academics, as befits the goals of critical animal studies and cognate fields.

What is the animal–industrial complex?

The first discernible scholarly mention of the term is found in Noske's (1989) book *Humans and Other Animals: Beyond the Boundaries of Anthropology*, which devotes an entire chapter to discussing the concept. It has not been used very much since in academic work but seems to have at least in a limited sense entered critical discourse around human–animal relations. Two papers by Carol J. Adams (1993, 1997) are important exceptions to this lack of use. My concern is that its use and meaning may have become simply assumed and almost rhetorical; deployed monolithically to represent, but also to reduce, the myriad complexity of the multiple relations, actors, technologies, and identities that may be said to comprise the complex. I think one of the central tasks for the nascent field of critical animal studies has to be, via its championing of interdisciplinary analysis, to return to this concept, and to restore and tease out a sense of its nuance and complexity. This is not to say either that this is a small undertaking or, indeed, that research is not already being done to perform exactly this task, even if this might not explicitly be using the animal–industrial complex terminology.

Without any refinement the concept already piques our interest for two clear, overlapping reasons. The first revolves around Noske's intention to contextualize the use of animals as food not primarily within a rubric of inadequate ethical frameworks but as part of the wider mechanics of capitalism and its normalizing potential. Thus a large part of the answer to the question of "why do 'we' exploit so many animals" is found here in terms of the way in which "animal industries are embedded in a capitalistic fabric" (Noske, 1989: p. 22), combined with an assertion that the "main impetus behind modern animal production comes from monopolistically inclined financial interests rather than

Figure 6.1 Several cows hooked up to milking machines, shot from behind. This was at a family-run, "humane" organic dairy and veal farm in Spain, 2010. Photo Jo-Anne McArthur, We Animals.

from farmers, consumers, or workers here or in the Third World" (Noske, 1989: p. 22). This, as I read it, is not to reduce the human exploitation of other animals to a capitalist formation but to give due explanatory power to this economic realm in considering our present situation. Moreover this is not to deny a distributed and networked agency at play in the animal–industrial complex but to underline that corporate capital accumulation has been a significant factor in the emergence of globalized, industrialized animal production.

Corporate interests have had a direct shaping through marketing, advertising, and even flavor manipulation in constructing the consumption of animal products as a sensual material pleasure and one that can be an easily acquired identity resource for a wide range of consumers. Additionally, technological innovation in the production process is often sold to farmers as directly benefiting their business; yet the largest returns tend to be experienced much higher up the production chain.

Noske's attempt here to clarify whose interests are served by the emergence and intensification of the animal–industrial complex foregrounds political economy and intersects a critical analysis of human–animal relations with a critique of capitalism (Nibert, 2002; Best, 2009; Twine, 2010a; Sanbonmatsu, 2011). This is in line with the critical animal studies research agenda I subscribe to, including its commitment to interdisciplinarity (Best, 2009). Mainstream animal studies shares this interdisciplinarity but, regretfully, makes no special case for either political economy or the importance of capitalism in shaping human–animal relations.

Well taken, too, is Noske's emphasis on intersectionality. Her interest in political economy already signals an attentiveness to the relationship between the human exploitation of other animals and economic power relations. She specifically covers such issues as health risks to human workers in, for example, slaughterhouses; the impact of the complex on developing countries; environmental consequences; and antibiotic use. Elsewhere in her book she discusses intersections of gender, nature, and animality. It is significant that she also includes a section on animal research. While nowhere does she offer a clear working definition of the "animal–industrial complex" or even a schematic to show what it might comprise, she does draw our attention to important matters of scope and scale that are crucial to refining the concept. By gesturing to the issue of animal research, Noske essentially questions whether agribusiness can be said to delineate or exhaust the boundary of the complex. She answers this, correctly, with a resounding, no. Such attentiveness to the permeability of the boundaries of the complex is crucial for teasing out the material and ideological interconnections between different spatial and cultural contexts of human–animal relations. It also helps theorizing the interconnections between the animal–industrial complex and other sectors of the global economy.

Four other "complexes"

Although Noske does not reveal her sources it seems a reasonable assumption that her concept is in part influenced by the older idea of the military–industrial complex. While there is much popular and scholarly writing on the use of animals in war this connection is similarly not noted. The military–industrial complex concept emanates from the speechwriters of former US President Eisenhower's farewell address in 1961. He warned the country of the power of the emerging close relations between a government's military policy, the various armed forces, and the corporations that support the military. Some versions also included the role of academia and scientific knowledge.

The concept of a military–industrial complex is not dissimilar to the later idea of the "triple helix" (Etzkowitz and Leydesdorf, 1997), put forward to characterize the changing role of academia in the global knowledge economy in terms of its increasingly close relationship with government and the corporate sphere. This triple helix is itself enmeshed within the contemporary animal–industrial complex (see Twine, 2010a). Since the 1960s a partly analogous host of "industrial–complex" terms have been used in various ways. They potentially provide a useful means by which the material and semiotic overlaps between, on the face of it, apparently different parts of the global capitalist economy can be theorized. Next to the "military–industrial complex" there are at least three further examples to be noted here of other complexes.

A second is the notion of the "prison–industrial complex," defined as a set of bureaucratic, political, and economic interests that encourage increased spending on imprisonment, regardless of the actual need (Schlosser, 1998). This is a considerably more widely used and cited term than "animal–industrial complex." It could be drawn upon to highlight links between the incarceration of animals, of humans animalized by "race," class, or gender, and also of humans convicted of protesting the animal–industrial complex in the context of the increasing criminalization of some forms of activism (Potter, 2011). Furthermore, there are of course histories of prison inmates being used as subjects within scientific experimentation.

Animals are also used therapeutically in some prisons. Furthermore, as Fitzgerald (2012) points out, in the USA and Canada many inmates are employed in the slaughtering and "processing" of animal bodies. This is a source of cheap labor, which symbolically pairs animalized humans and animals or "nature" together in a similar manner to the phenomena of environmental racism (see Higgins, 1994). It also, perhaps, provides one economic rationale for the perpetuation of a specific prison population. Fitzgerald correctly alerts us to a particular problem of employing inmates in this way given other research findings that suggest a tendency toward psycho-social brutalization in such labor. Consequently such labor could seriously frustrate inmate rehabilitation.

Thirdly, the idea of an "entertainment–industrial complex," initially coined to refer to structural changes in the US film industry, has been in circulation since the 1980s (e.g., Christopherson and Storper, 1986). This concept could be broadened out to think through the intersections with the animal–industrial complex in the way that zoos, theme parks, and "pet" keeping have become considerable forms of profitable "entertainment." More generally it can be linked into other theoretical approaches, most obviously valuable, if problematic, research on the "culture industry" by members of the neo-Marxist Frankfurt School, especially Theodor Adorno and Max Horkheimer. Indeed Shukin's (2009: Chapter 2) skilful argument for an historically intersecting triangulation between animal capital, the film industry (the use of gelatin being perhaps the most obvious material enmeshment), and the automobile industry is a good example of both the sort of analysis required and the porosity within and between different "complexes."

Fourthly, sociological research has also engaged with the idea of a "pharmaceutical–industrial complex" (Abraham, 2010), which speaks to the "pharmaceuticalization" of Western societies: "the process by which social, behavioral or bodily conditions are treated or deemed to be in need of treatment, with medical drugs by doctors or patients" (Abraham, 2010: p. 604). There are at least four main points of overlap between the "pharmaceutical–industrial complex" and the animal–industrial complex:

1. A major point of intersection revolves around the use of animals as experimental subjects in the research and development of new drugs. The "pharmaceuticalization" thesis in a sense is a critique of Western societies for using drugs to try and treat what in effect are mental and physical illnesses with a considerable social, economic, and political etiology. This also calls into question those forms of animal exploitation that are especially caught up in this trend.

2. Many of the world's leading pharmaceutical companies including Abbott, Pfizer, and Ely Lilly/Elanco have considerable capital interests in also producing drugs for animals across various sectors. Such drugs are often represented under a rubric of

"animal health" but in the agricultural or sporting context health and welfare can often be proxies for the productivity of the animal body (Twine, 2007). Typically these are administered to farmed animals to alter meat or milk in some way. Sometimes this becomes more visible and prominent in the shape of human health scares.[1] A further important and recently well-publicized element of this overlap is the use of antibiotics in farmed animals as aids to productivity. There have now been calls (e.g., by the World Health Organization) to further regulate this use of antibiotics, not for the impact upon animals as such, but for their role in reducing the efficacy of antibiotics in humans.

3. The aforementioned pharmaceuticalization phenomenon also extends across species with, for example, the application of anti-depressant drugs to dogs. Although the 2011 US FDA approval of Ely Lilly/Elanco's canine anti-depressant known as Reconcile® (see www.reconcile.com) could be read as a welcome subjectification of companion dogs, it would be remiss not to interpret this in terms of familiar capital accumulation strategies and to underline the potential side effects to dogs (and others) of being administered such drugs.

4. A final point of overlap is that there is good evidence (e.g., for a review see Twine, 2010a: Chapter 7) to suggest that the consumption of animal products at levels now surpassed in many Western diets (and increasingly in non-Western ones) is productive of a wide range of health problems, which in itself is generative of pharmaceutical capital (and of course more animal experimentation). This is worth underlining since it also highlights intersections within the animal–industrial complex between the often separated issues of animal experimentation and meat/dairy consumption.

From these additional complexes we can note how capitalism creatively commodifies its own excess, its own ailments. Approaches to climate change mitigation via a carbon tax are the latest case in point. Most of these complexes outline examples where the mass production of various commodities arguably not conducive to human or animal well-being are anyhow pursued and capitalized upon. Thus war, human incarceration, meat and dairy production, and mental and physical illness have been transformed into profitable enterprises that, in turn, set up economic interests in their perpetuation. Although all of these have been naturalized by a familiar ideology that would root them in an assumed "human nature," they in fact reveal major tensions between corporate agency and democratic oversight. Contextualizing the animal–industrial complex within its broader milieu of other complexes is an important initial step that is reflexive to questions of scope and generative of intersectional and critical knowledge around our contemporary hegemonic economic and cultural forms.

It is worth pointing to a note of caution toward the notion of "complex" in the sense that it may suggest something akin to a conspiracy theory. On the one hand it points to relatively powerful alliances and networks in particular sectors of the global economy

---

[1] Although growth hormone in milk is a well-known example, more recently, in June 2011, Pfizer subsidiary Alpharma was told by the FDA to cease sales of Roxarsone (also known as 3-Nitro®) when a carcinogenic form of arsenic from the drug was found in the livers of broiler chickens. In the case of Roxarsone it has been used to improve both productivity (weight gain) of animals and the aesthetics (pigmentation) of meat.

that strive to maintain a hegemonic position. It underlines how capital accumulation and resource control may become the overriding rationale in spheres such as global food production or military conflict. It also shows how these relations are implicated in and used as conduits for geopolitical strategies. Yet it would be a mistake to overplay the degree of control or design those particular actors and networks are assumed to have. In other words, the discourse of the complex ought not to become fatalistic or deterministic.

The politics of food

Questions of scope also arise in the importance of cautioning the present approach away from a political vision (especially as it encompasses veganism) that (a) might reduce politics to consumption, and (b) might assume the consumption of all non-animal products to be morally benign. When in 2007 food writer Michael Pollan spoke of the "vegetable–industrial complex" in relation to *E. coli* scares, it served inadvertently as a reminder that any consideration of the animal–industrial complex must also consider its interdependence with the crop and vegetable industries; most notably, though not solely, in the form of animal feeds. There is thus no moral dichotomy here for a pro-vegan critical animal studies in terms of a simplistic valorization of non-animal products. This underlines the often-made point that veganism is more than "just a diet" and is better seen and practiced as a systemic and intersectional mode of critical analysis and a useful lived philosophy counter to anthropocentrism, androcentrism, hierarchy, and violence.

Similarly it is not surprising that the geographer of the global food economy Tony Weis has used the concept of the "temperate grain–livestock complex" (2007) – yet another complex. He uses this in an interdisciplinary manner to evoke the historical, spatial, and political economy dimensions of an important feature of global food relations. Weis outlines how during the twentieth century both grain and animal production came to be centred on a small number of domesticated species. He shows how increasingly, especially through the growth of soybean production, the "livestock revolution" and intensive high-input crop production became very much interconnected.

Thus any analysis of the globalizing dimensions of the animal–industrial complex cannot avoid factoring in economic, historical, spatial, and technological changes in plant crop production as crucial to what Weis refers to as the "meatification" of the global diet. By this process the global population has approximately doubled since the 1950s but meat production has increased almost five-fold in the same period (Weis, 2007: pp. 16–20). Including crops necessarily complicates the analysis of the animal–industrial complex. It encourages a focus on areas such as the local impacts of soy production, corporate practices, plant biotechnology, transnational food regulations and standards, as well as governmental agricultural policies. Furthermore it makes more likely the narrowing of the gap that is often noted academically and politically between the question/condition of the animal and that of ecology.

Figure 6.2 Factory farming: a sow separated from her piglets in a gestation stall. Photographed during an investigation with Igualdad Animal on a large pig farm in Spain, 2009. Photo Jo-Anne McArthur, We Animals.

Clearly it is not only the exploitation of animals for their meat, milk, eggs, and skin that has come under the rubric of capitalist processes but also such practices as the use of animals in sport, animal experimentation, the companion animal sector, and the zoo industry. If we are to take on, as I suggest we should, the animal–industrial complex as an organizing concept, then the continued mapping of these interconnections between overlapping sectors is a significant task therein. If these may be referred to as questions of scope, then Noske also points us toward the issue of scale. The rhetoric of agribusiness creates a promissory discourse that represents global "livestock" corporations as somehow benign and even philanthropic providers to the developing world (a discourse that has since been much recycled in the context of biotechnology). Noske (1989: pp. 29–34) begins to briefly outline some of the economic and consumption inequities that comprise the unmistakably global character of the complex.

However, I think we still lack a sophisticated understanding of how the complex endures across different regional, national, and global scales. Although the concept is clearly an attempt to construct a global representation of an important part of the global economy it should not be understood or researched monolithically. The animal–industrial complex will vary locally given historical and cultural variation in human–animal relations, in economic history, and in food traditions. This degree of complexity makes clear that a rhetorical use of the complex concept is wholly inadequate. It needs to be radically strengthened with expertise from, for example, sociologists, science and technology studies scholars, geographers, and critical economists. These disciplines have established traditions, theoretical and methodological approaches to examining, for example, the global food system, that are of obvious relevance to understanding the animal–industrial complex. Simultaneously there is a need for scholars in these disciplines who are not primarily focused on animals to become reflexive to the multi-levelled reasons for examining the animal–industrial complex.

The three meanings of "complex"

In thinking through how we might further add to Noske's animal–industrial complex concept I want to briefly consider three common uses of the word "complex," which I think can be heuristic overall. First, "complex" means "difficult to understand," complicated, hidden, and impenetratable. Second, it means a conglomerate, a network, structure, system, an association. Third, to have a complex is to have a psychological problem, to experience emotions such as anxiety, fear, disgust, and obsession, and to unreflexively exploit social psychological processes such as stereotyping, denial, and projection. It is not too difficult to see how these three meanings can interrelate in the context of the animal–industrial complex.

The complicated, partly-hidden character of the animal–industrial complex is bound up in the process of denial, which can be framed in the terms of sociologist Norbert Elias (1978). Elias argued that Western society underwent what he referred to as a "civilizing process" between the sixteenth and nineteenth centuries. This involved the emergence of a set of classed bodily and affective dispositions, visible in manners that can be read as the social construction of a new "human," relationally defined against constructed understandings of both animality and animalized human otherness. He illustrated how a multitude of embodied practices, such as eating habits, become emotionally charged with embarrassment and shame.

There is a wealth of literature during this time outlining the proper way to act in terms of etiquette and manners. This literature employs animal and classed others (such as peasants) as examples of how not to behave: "[The] use of the sense of smell, the tendency to sniff at food or other things," Elias writes, "comes to be restricted as something animal-like" (1978: p. 203). The meanings attached to various embodied practices in the sixteenth century changed and became in a sense "domesticated" by the nineteenth. The thresholds of embarrassment and shame had shifted.

In one sense this process was about concealing the "animal" in the "human," dealing with the shame of our connectedness to materiality. This theory speaks to what I have previously referred to as the "internally torn" model of the human, constructed around the idea of a human–animal binary and productively called upon to perform (intra) human difference (see Twine 2010a: p. 10). We can note here a disavowal of the "animal" within the human, but *also*, in part, an external banishment of human violence toward other animals. Thus, significantly, this "civilizing process" also involved the partial concealment of violence from everyday life and can be seen as bound up in the social sequestration of animal slaughter: a pivotal feature of the animal–industrial complex.

This may have been part of a general shift against the overt public display of violence as counter to emerging social norms rather than something more specific to human–animal relations. If so, then it also had the effect of quelling the potential disruptive potency of this violence to call into question the naturalization of human–animal hierarchy. This means that the sequestration of slaughter cannot be reduced to a conscious decision to conceal violence for political-economic ends, but was in fact embedded within a broader socio-historical trend. Nevertheless this new Western "human" consolidated its power over and against the "animal" and was able to an extent to keep its barbarism hidden from view and self-constitution.

These points underline a major contradiction of claims for Western civility and echo accordingly with the description by Zygmunt Bauman (1989) of modernity as "janus-faced." Yet it is precisely the affective processes of denial and projection that are apt to help explain both the possibility and precarity of such a social formation of the "human," inseparable from its histories of class, gender, "race," and species domination. This "human" then is precisely a "complex" in the third sense highlighted above. The affective processes are part of the day-to-day management of the possibility of the mirage of the "civilized human."

In what is now a very familiar argument in critical animal studies, and earlier from ecofeminist writing (e.g., Plumwood, 1993), the moral significance of harming those deemed inferior has been denied by projecting a whole interconnected cluster of discursive meanings onto "them," where, for example, the "animality" of "animals" is merely assumed, and the "animality" of various "humans" is (also) culturally produced (see Corbey, this volume; Livingstone Smith, this volume; Machery, this volume). These meanings are increasingly dated and difficult to defend in the face of, for example, ethological knowledge of other animals. Denial here is also applied to the material labor and active agency of the inferiorized (Plumwood, 1993). That denial is expressed spatially in the design of technologies and buildings that, in our case of animals, help to secure the relative invisibility of the violence acted out against their bodies.

Thus meanings one and three of "complex" can do the appropriate work to outline the inseparable material–semiotic dimensions of the animal–industrial complex. The second meaning of "complex," to mean network, conglomerate, or system is similarly useful for thinking how the animal–industrial complex achieves and sustains what we can refer to as "material–semiotic" hegemony. What networks are at play? How do they interconnect? How are particular speciesist norms naturalized, carried, and circulated? Why are such norms so successful at recruiting adherents? Relatedly, this second meaning of "complex" is also useful for addressing our interest in scope and scale. A pertinent question here is to ask what role the online and offline mass media play in consolidating the symbolic dominance of the animal–industrial complex (see Cole and Morgan, 2011), and to analyze how this is being resisted? A related research question is to probe the extent to which dominant assumptions and practices around the human–

animal binary have extended culturally and how might local resistances have expressed themselves?

This exploration of the three main meanings of "complex" offers an initial basic and succinct definition of the animal–industrial complex as a partly opaque and multiple set of networks and relationships between the corporate (agricultural) sector, governments, and public and private science. With economic, cultural, social, and affective dimensions it encompasses an extensive range of practices, technologies, images, identities, and markets.

I place agricultural in brackets only to highlight my personal interest in this chapter – a working definition of the animal–industrial complex must not, as I have argued here, be confined to this domain. However, empirically it is both useful and essential to narrow down the analysis and to focus on case studies of human–animal relations, while attempting to also draw useful broader connections. I see this as setting one of the most important research agendas for interdisciplinary critical animal studies.

Pointers for future research

When addressing the animal–industrial complex in terms of agriculture sociologists, science and technology studies scholars, geographers and critical political economists are all of relevance. As mentioned above, such disciplines have established traditions: theoretical and methodological approaches to theorizing the global food system. To end with, I will briefly discuss some of these, as pointers for future research.

The social-scientific study of food has very much become a growth area over the last 20 years, contextualized by a growing interest in environmental issues, the intersections of social inequality with food, and related questions of public health. One well-known area of research is based around analyses of the commodity chain. Instead of focusing on a particular sector of the food industry this approach follows the commodity as it moves through different stages of production and consumption. The aim is to understand what are referred to as "commodity stories," in order to better grasp moments of commodification and the complicated relationships between "people, places and commodities" (see Hughes and Reimer, 2004: p. 1). It is not surprising then that this work overlaps with an interest in the animal–industrial complex, even if that concept is not deployed.

The well-known geographer and political economist Michael Watts (2004), for example, has written in this tradition, specifically looking at the US hog (pig) and broiler chicken industry. Watts probes the extent to which the hog industry has followed in some respects and differed in others from the trajectory of the broiler industry in terms of, among other things, mechanization, contract farming, and the impact of the biology of the animals themselves. Although the tone is largely (politically) humanist, studies such as this are immensely useful to explicate the animal–industrial complex.

Analyses based around commodity chains represent one approach from Marxist political economy broadly construed. The emergence of actor–network theory during the 1990s offered an alternative conceptual and methodological approach to studying social phenomena that has been influential upon social scientists interested in food and agriculture. Some of these points of tension between political economy and actor–network theory revolve around how to conceptualize globalization, how to understand scale, and how to understand agency. There have been influential attempts to argue that actor–network theory offers more convincing approaches to the social scientific research agenda on food and globalization (e.g., Whatmore and Thorne, 1997) as well as perspectives that are more mutual and conciliatory between the two approaches (e.g., Busch and Juska, 1997; Castree, 2002).

It is not new to recognize that actor–network theory is of potential use to the study of human–animal relations (Whatmore and Thorne, 2000), in large part due to its insistence upon the agency of non-human actors within a given network. However, some of its other premises are also of interest to the analysis of the animal–industrial complex. As well as being critical of political economy approaches for assuming a passive nature, Busch and Juska (1997: p. 691) argue that they tend to reify corporate actors. Speaking of particular companies acting in this or that way somehow renders opaque that particular people are involved in making particular decisions and actions possible. They state that this is not a retreat to an asociological focus on individuals but a methodological call for analyses to focus upon how different actors achieve power within the context of specific networks, "a plea to begin to think of the globalization of agriculture in terms of the extension of actor networks" (Busch and Juska, 1997: p. 692).

Of course this does not mean that we no longer analyze corporations but that we think more specifically about what particular people within a corporation are doing, as well as how the corporation overall is linked into other networks. Similarly actor–network theory uses the idea of the actor–network to deconstruct the traditional social science distinction between micro and macro. Here globalization is understood as an extension of actor–networks made possible through the strategic enrolment of particular technologies that allow the power of "action at a distance" to occur (Busch and Juska, 1997: p. 694). Again the research focus is on how networks link together and endure, the extent to which they can be made open to scrutiny, on particular enabling technologies, and how weak points may be pinpointed. Although this merely provides a very brief flavor it is clear that the confluence of political economy and actor–network theory approaches to food and agriculture is very promising in attempting to better understand the animal–industrial complex. Other approaches to understanding social change such as practice theory will also likely prove useful (see Shove *et al,*. 2012). Such approaches can help guard against the repetition of potential methodological and conceptual mistakes though it remains important to bear in mind potential ontological and epistemological tension points between them.

Conclusion

In this chapter I have revisited the concept of the animal–industrial complex, picking up where Barbara Noske's initial outline came to a premature end, seeking to give it more definition, and making it truly heuristic. Focusing on the agricultural sector, I have analyzed it here as a partly opaque and multiple set of networks and relationships between the corporate sector, governments, and public and private science; with economic, cultural, social, and affective dimensions; and encompassing an extensive range of practices, technologies, images, identities, and markets.

This effort should also be situated within the wider context of attempts to outline the conceptual and methodological terrain for the future of critical animal studies as a promising research agenda. The approach here has been to explore how the animal–industrial complex relates to other "industrial complex" concepts, to consider three meanings of the word "complex" as a means to better define it, and to argue that when considering the complex we should situate our work within pre-existing social science research around, for example, global dimensions of food and agriculture.

Critically addressing the animal–industrial complex is not merely important to a small set of critically minded scholars but constitutes a reflexivity that is now vital to the entirety of humanity. This is perhaps most clear in the case of the contemporary "livestock revolution" and all that entails in terms of carbon emissions and violence against all (other) animals alike. Corporate and state promotion of the (over)consumption of animal products can be conceptualized, I think, as a form of material and cultural violence against humans as well. Given that business-as-usual carbon intensive social practices entail highly probable environmental disaster, and that the "livestock revolution" significantly exceeds business-as-usual levels of production and consumption, it represents an exceptional case for intervention, change, and mitigation.

Bringing the animal–industrial complex back into scholarly and social consciousness and conscience helps make this case. It is impossible to envision a transition to more convivial human–animal relations without a wholesale challenge to the animal–industrial complex. This effort stands then as a crucial contemporary form of knowledge production that argues this case by contesting persistent norms of "the human" and the ethical and spatial invisibility of "the animal."

Acknowledgments

A different earlier version of this chapter appeared in the *Journal for Critical Animal Studies* 10(1), 2012. The support of the UK's Economic and Social Research Council (ESRC) is gratefully acknowledged. The work was part of the program of the ESRC Centre for Economic and Social Aspects of Genomics (Cesagen).

Part II

Sentience and agency

7   Humans, dolphins, and moral inclusivity

Lori Marino

I am a behavioral neuroscientist by training and have studied intelligence, brain evolution, neuroanatomy, and self-awareness in dolphins and primates for over 20 years. My work led me to understand the very real ethical implications of my research findings. Ten years ago, after co-authoring a study in which we demonstrated that bottlenose dolphins (*Tursiops truncatus*) recognize themselves in mirrors, I came to understand how devastating captivity is for dolphins when the two subjects of that study, young male bottlenose dolphins, died after being transferred to other facilities. That experience and subsequent research into the world of captivity motivated me to recommit my scholarship to advocating for these and other animals who are exploited and abused. I've learned two important things during that period. First, there is no inherent conflict between science and advocacy. Second, being a scientist can make one a very powerful advocate for other animals. That is why I founded the Kimmela Center for Animal Advocacy, a new organization that brings together science, scholarship, and animal advocacy, helping to create a new professional path, the scholar-advocate, for scientists and others who want to apply their expertise to protecting other animals. It is my hope that by doing so we can transform our current relationship with other animals from one of exploitation of "resources" to one of respect on their own terms.

There is abundant scientific evidence for complex intelligence, self-awareness, and emotional complexity in dolphins. Dolphins possess large complex brains second in relative size only to those of modern humans. They have demonstrated prodigious cognitive abilities in such areas as language understanding, abstract thinking, and problem solving. They recognize themselves in mirrors, understand the relationship between their own body and that of others, and can reflect on their own thoughts, showing that their sense of self is not unlike our own. Moreover, their complex sociality and cultural traditions are well documented from ongoing field studies. These findings provide strong support for recognizing their status as individuals with basic rights comparable to those of humans. Yet, dolphins (and whales) continue to be treated as non-sentient objects, commodities, and resources. Egregious examples of this are the many dolphin and whale slaughters that occur around the world and the exploitation of dolphins in entertainment parks, research facilities, and the "dolphin therapy" industry.

The Politics of Species: Reshaping our Relationships with Other Animals, eds R. Corbey and A. Lanjouw.
Published by Cambridge University Press. © Cambridge University Press 2013.

In this chapter I will discuss the consequences of mistreatment for suffering in dolphins, the reasons for their continued abuse and exploitation, and new efforts to recognize dolphin personhood and its implications. I will argue that dolphins present an extreme challenge to our ability to consider similarity and difference simultaneously (a prerequisite for inclusivity) in our moral stance toward other animals.

Cetacean brains and psychology

The massive cetacean brain

Both absolute and relative brain size account for some of the variance in different aspects of intelligence among species (see Marino, 2006 for a review). Modern cetacean brains are among the largest of all mammal brains in both absolute and relative size. The largest brain on Earth, that of the adult sperm whale, at an average 8,000 g (Marino, 2009), is six times larger than the human brain.

Because brain and body size are positively correlated they are often expressed as an encephalization quotient or EQ (Jerison, 1973). An EQ is a value that takes body size into account by representing how large or small the average brain of a given species is compared with other species of the same body weight. Species with an expected brain size relative to body size have an EQ with a value of one. Species with brains larger than expected for their body size possess EQs greater than one, and so on. The EQ for modern humans is seven; modern human brains are seven times the size one would expect for a species with our body size. Almost all odontocetes (toothed whales, dolphins, and porpoises) possess above-average EQs compared with other mammals. Many odontocete species possess EQs in the range of four to five, that is, their brains are four to five times larger than one would expect for their body weights. Many of these values are second only to those of modern humans and significantly higher than any of the other primates (Marino, 1998). The EQs of mysticetes of baleen whales, it is important to note, are all below one (Marino, 2009) because of an uncoupling of brain size and body size in very large aquatic animals. However, mysticete brains are large in absolute size and exhibit similarly high degrees of complexity and progressive elaboration as the toothed whales.

The "new" cetacean cortex

Cetacean brains have been on an independent evolutionary trajectory from their closest relatives, the even-toed ungulates, for at least 52 million years (Gingerich and Uhen, 1998), and from the last ancestor with primates for about 92 million years (Kumar and Blair Hedges,1998). During that time they evolved a unique combination of features. However, the most important point to make about cetacean brains is that despite being different from primate (including human) brains in several ways, they are comparable in complexity.

The brain structure most relevant to intelligence and cognitive complexity is the cerebral cortex – the layered (often folded) sheet of neural tissue outermost to the mammalian cerebrum. The phylogenetically most recent part of the cerebral cortex, the neocortex, is differentiated into several horizontal layers of neurons forming vertical microcircuits and other architectural features that enable complex information processing. Functionally, the neocortex is the basis of thought, reasoning, awareness, and communication. The cetacean neocortex is very different from that of primates in that it does not contain all of the same layers as that of the primate brain and the layers are somewhat different in cellular morphology and architecture. These differences led to an earlier view of the cetacean brain as relatively simple and unspecialized, and generally lacking in the prerequisite organizational complexity needed for complex cognitive abilities (Kesarev, 1971; Gaskin, 1982). However, modern neuroanatomical techniques have demonstrated convincingly that the cetacean brain, and especially the neocortex, are at least as complex as that of other terrestrial mammals, including human and non-human primates (Hof *et al.*, 2005; Hof and Van der Gucht, 2007). Various regions of the cetacean neocortex are characterized by a wide variety of organizational features, i.e., columns, modules, and layers, that are hallmarks of complex brains.

There are specific cortical regions of the cetacean brain that are especially notable in their apparent degree of elaboration. The cingulate and insular cortices, both situated deep within the forebrain, in odontocetes (toothed whales) and mysticetes (toothless baleen whales) are extremely well developed (Jacobs *et al.*, 1979; Hof and Van der Gucht, 2007). The expansion of these areas in mammals is consistent with high-level cognitive functions such as attention, judgment, and social awareness (Allman *et al.*, 2005). Recent studies show that the anterior cingulate and insular cortices in larger cetaceans contain a highly specialized projection neuron, known as a spindle cell or von Economo neuron (Hof and Van der Gucht, 2007), which is thought to be involved in neural networks subserving aspects of social cognition (Allman *et al.*, 2005). Spindle cells, also found in human, great ape, and elephant brains (Hakeem *et al.*, 2009) are thought to play a role in adaptive intelligent behavior of the kind described in the next section.

Dolphin intelligence

Dolphin cognitive complexity and flexibility have been demonstrated abundantly in years of studies with bottlenose dolphins. These findings include the ability to learn a variety of governing rules for solving abstract problems, understanding televised representations of the real world, learning numerical concepts, and innovating motor behaviors. Dolphins are also one of the few species that can imitate arbitrary sounds and behaviors. In addition, they understand the semantic and syntactic features of a human-made symbolic language and understand and employ pointing as a referential gesture (see Marino *et al.*, 2008 for a review of these and other findings).

One of the hallmarks of cognitive sophistication on a par with that of our own species is self-awareness, which has been variously described as awareness of one's personal identity, a sense of "I," an autobiographical identity. Self-awareness is related to metacognition, the ability to think about, or reflect upon, one's own thoughts and feelings. There is substantial evidence for self-awareness and metacognition in bottle-nose dolphins and good reason to argue that these capacities are not limited to this cetacean species. Body awareness has been demonstrated by a bottlenose dolphin through an understanding of how symbolic gestures refer to her own body (Herman *et al.*, 2001). Also, awareness of one's own behaviors has been demonstrated through a dolphin's ability to repeat a behavior she just performed, in response to a "repeat" command, or to perform a different behavior if so instructed (Mercado *et al.*, 1998, 1999; Herman, 2002). In 2001, Reiss and Marino demonstrated conclusively that two captive bottlenose dolphins were able to recognize themselves in mirrors (Reiss and Marino, 2001) showing that they had a similar capacity for self-recognition as great apes and humans.

Additionally, bottlenose dolphins have demonstrated metacognition (the ability to think about their own thoughts): they were able to report on how uncertain they were about correctly completing a discrimination task (Smith *et al.*, 1995). This is a high-level capacity requiring them to consciously access their own memory and knowledge base and then make a critical decision based upon that awareness. Therefore, dolphins (and likely other cetaceans) possess rather uncommonly sophisticated aspects of self-awareness, shared, thus far, with humans, chimpanzees, capuchin monkeys, and rhesus macaques (Smith *et al.*, 1997; Call and Carpenter, 2001; Beran and Smith, 2011).

Social complexity and culture

Many dolphin and whale species live in large, highly complex societies with differentiated relationships (Baird, 2000; Connor *et al.*, 2000; Lusseau, 2007) that include long-term bonds, higher order alliances, and cooperative networks (Baird, 2000; Connor *et al.*, 2000) that rely extensively upon learning and memory. There is also evidence that individual role-taking exists in dolphin societies to facilitate cooperative relationships (Gazda *et al.*, 2005) and decision-making processes (Lusseau, 2006, 2007).

Field studies have documented impressive cultural learning of dialects, foraging sites, and feeding strategies in cetaceans. Culture, the transmission of learned behavior from one generation to the next, is one of the attributes of cetaceans that most sets them apart from the majority of other non-human species (Whitehead, 2011) and is likely underpinned by complex social learning abilities. Cultural attributes have been identified in many species of cetaceans but principally in those best studied: the bottlenose dolphin, the killer whale (*Orcinus orca*), the sperm whale (*Physeter macrocephalus*), and the humpback whale (*Megaptera novaeangliae*) (Whitehead, 2011). Even cultural tool use has been documented among bottlenose dolphins, who use sponges to probe into crevices for prey (Krützen *et al.*, 2005), among other examples. Therefore, there is abundant evidence that cetaceans not only are socially complex in terms of their relationships and societal dynamics but also culturally sophisticated in a way we are just beginning to understand.

The current scientific research on the intellectual and emotional abilities of dolphins and other cetaceans shows that they are self-aware, unique individuals with distinctive social roles and autonomy. As such, it has been suggested that dolphins and other cetaceans are "non-human persons" who qualify for moral standing as individuals with

Figure 7.1 Kiska, an orca whale, lives in solitary confinement in a tank at Marineland Theme Park, Ontario, 2011. Orcas in the wild have complex social lives and can travel up to a hundred miles in a day. Photo Jo-Anne McArthur, We Animals.

basic rights (White, 2007, 2011). Whether one agrees to recognize the formal classification of dolphins as persons or simply accepts the current evidence for person-like mental abilities in dolphins, the current way in which we regard and treat cetaceans, as non-sentient commodities, is entirely inconsistent with who they are and, as such, is morally indefensible.

Human abuses of cetaceans

Our professed affection for dolphins is somewhat belied by the fact that we still treat them as property, as non-sentient objects, as a means to an end. And indeed, they, like all non-human animals, are property legally. Tens of thousands of dolphins and whales are slaughtered each year around the world for meat and alleged cultural reasons. Thousands die each year as a result of human fisheries practices and anthropogenic degradation of the environment. And hundreds are held in captivity for our entertainment, scientific curiosity, use in military operations, and "therapy." Thus, the bottom line is, when it comes to our attitudes toward and treatment of dolphins and whales, they are still, in the end, thought of and treated as commodities.

Dolphins and whales as food and bycatch

The fact that most people who consider it cruel to harm a dolphin directly largely tolerate cruelty toward them at the hands of others, indicates that we see them as different enough to be excluded from moral concern on a par with other persons. Even today a constant battle wages between protection and conservation efforts, on the one hand, and commercial interests on the other. Various forms of dolphin and whale slaughter continue around the globe. These include the customary hunting of large whales by Greenland, Iceland, Japan, Norway, Canada, and other nations and the slaughter of dolphins and smaller whales in Japan, the Faroe Islands, the Solomon Islands, and other places. Dolphins and whales are slaughtered primarily for meat but are also considered competitors for fish and they are recurrent victims of incidental killing by the fisheries industry. The main method of intentional dolphin slaughter is dolphin drive hunting, also called dolphin drive fishing, a method of hunting dolphins and other small whales by driving them together with boats into a cove or onto a beach where they are killed with knives, harpoons, and other deadly instruments. Many die from the acute stress if not the actual butchering (www.savejapandolphins.org).

The pursuit, capture, and killing of whole groups of dolphins and whales, including families, is, by even the most relaxed standards, unimaginably cruel. But there are other more insidious, and related, forms of abuse and exploitation that generally go unrecognized and, indeed, are supported by the general public – that is, the abuse of dolphins in captivity for entertainment, research, military purposes, and even "therapy."

Dolphins and whales in research

Although a great deal of research on dolphins and whales is conducted in the field in a way that does not negatively impact the individuals under study, some research – particularly on cognitive abilities – has been and continues to be done with dolphins held in captivity. Research done in captivity affords a level of experimental control not easily achieved in the field setting. On the other hand, the results from studies of captive dolphins, particularly those that yield negative findings, may be limited in generalizability because of the psychological constraints and trauma associated with captivity (see below). Today there are very few dedicated dolphin research labs providing evidence of productivity. Almost all of the dolphins used in some of the key cognitive studies of dolphin intelligence and learning in the past two decades have died of conditions not common in wild dolphins; they have joined the many cetacean victims of a captive lifestyle. Therefore there is a current effort underway to create a new paradigm of research on dolphin cognition that excludes captivity and is shaped by the needs and desires of wild cetacean individuals who choose to interact with humans on their own terms (Marino and Frohoff, 2011). The future of cetacean cognitive and behavioral research is being paved by studies of wild and sociable cetaceans and habituated cetacean groups in the natural environment and, importantly, on their own terms.

Dolphins and whales in the military

Since 1960, the United States Navy has maintained dolphins and other marine mammals in captivity in order to use them in defense maneuvers, mine detection, and develop better submarine and sonar weapons. About 75 dolphins and small whales are held in the program, many of which were captured in the infamous Taiji, Japan drive hunts. In addition to using the dolphins in high-risk military situations, the Navy has, throughout the years, conducted invasive physiological research on dolphins in order to learn more about their brain function, sleep, and echolocation. This long history of often terminal military research on dolphins has been generally kept hidden from the public.

Dolphins as entertainment

By far the majority of captive dolphins and whales around the world are used for entertainment and recreation. Despite the claims of the zoo and aquarium industry, there is no evidence that dolphin and whale displays are educational or result in increased conservation attitudes or efforts (Marino et al., 2010). The truth is that dolphin shows and displays are commercial amusements equivalent to any theme park display. There is also a longstanding and intimate connection between the dolphin captivity industry and dolphin hunting around the world, and, most notably in Taiji, Japan (www.savejapan-dolphins.org). These kinds of associations reflect the fact that well-known dolphin circuses like SeaWorld and other entertainment parks are not as beneficent as they would like the public to think. Although couched in the guise of "education" and "conservation" there is no evidence for these benefits. Moreover, the dolphins and whales in these facilities still suffer shortened lifespans and the psychological and physical effects of confinement (Marino and Frohoff, 2011).

Dolphin displays and shows are often accompanied by swim programs where the public can pay to pet, swim with, or interact in some way with the animals. It is apparently no longer satisfying to the public to be able to watch dolphins and whales performing tricks. The new trend is to make physical contact with them. These kinds of activities grade into yet another form of dolphin exploitation called dolphin assisted therapy (DAT).

Dolphins as "therapists"

Dolphin assisted therapy is a popular form of animal-assisted therapy, marketing dolphin swims and interactions as a cure or treatment for a variety of psychological and physical illnesses. Children with such conditions as autism, developmental delay, attention deficit hyperactivity disorder, and musculo-skeletal diseases are often the target. Adults with everything from depression to multiple sclerosis to cancer are also exploited (Brakes and Williamson, 2007). Proponents of this modern-day "snake oil" claim that dolphins have healing abilities; curative mechanisms often cited are the supposed treatment effects of echolocation on disease processes and brain waves. None of these proposed explanations are supported by reputable data (Brakes and Williamson, 2007) and there is no existing evidence that DAT has any therapeutic value (Marino and Lilienfeld, 1998, 2007; Humphries, 2003).

Dolphin assisted therapy typically involves several sessions either swimming or interacting with captive dolphins along with more conventional therapeutic tasks. The standard cost of DAT, whose practitioners are not required by law to receive special training or certification, is exorbitant, often averaging $3,000 to $5,000 for a few brief swims. Dolphin assisted therapy has grown into a highly lucrative business with facilities all over the world, including the United States and, like the standard recreational "swim with the dolphin" programs, DAT is not regulated by any authority overseeing health and safety standards for either humans or dolphins. Because of this there are many risks to both humans and dolphins during DAT that include injury, disease transmission, opportunity loss for participants, and the capture and confinement of the dolphins (Marino, 2011).

The specific forms of exploitation and abuse described above all produce a unique set of ordeals for dolphins and whales. Yet they are all connected by a common thread – captivity and its devastating effects.

The effects of captivity on dolphins and whales

Ironically, the very characteristics of dolphins and whales that are so appealing, i.e., complex intellect, emotional sensitivity, and self-awareness, are the same that make them highly vulnerable to the psychological impact of captivity. Furthermore, because of the complexity and interrelatedness of dolphin and whale societies whole social groups can be destroyed even when a small number of individuals are captured (Reeves *et al.*, 2003; Lusseau, 2007). Captivity impacts social relationships, degrades autonomy through the imposition of an enforced schedule of activity and behavior, causes boredom produced by a relatively sterile and unchanging environment, induces frustration, and inhibits incentives and abilities to carry out natural behaviors such as hunting and traveling. The abundant evidence for stress, disease, and increased mortality in captive cetaceans attests to these effects.

Aberrant behavior

Many captive cetaceans display behavioral abnormalities indicative of distress and emotional trauma. These include stereotyped behavior, unresponsiveness, excessive submissiveness, hyper-sexual behavior (toward people and other dolphins), self-inflicted physical trauma and mutilation, compromised immunology, and excessive aggressiveness toward other dolphins and humans (see Frohoff, 2004; Stewart and Marino, 2009 for reviews). Birkett and McGrew (this volume) report on similar abnormal behaviors in chimpanzees who are confined in zoos or research labs. This is not surprising given the similar levels of intelligence, vulnerabilities, and types of deprivations endured by captive cetaceans and great apes.

Stress and disease

Stress derives from many aspects of captivity, including being transferred into and out of different pools and social groups without choice and being unable to resolve conflict in the way it is accomplished in the wild, that is, through dispersion and adequate social support. These factors eventually lead to reduced life expectancy (Waples and Gales, 2002). The U.S. Marine Mammal Inventory Report lists numerous stress-related disorders, such as ulcerative gastritis, perforating ulcer, cardiogenic shock, and psychogenic shock as "cause of death" (U.S. Marine Mammal Inventory Report, 2010).

Mortality

The effects of increased stress and disease in captive cetaceans are evident in shorter lifespans, lower survivorship, and higher mortality. For bottlenose dolphins, survivorship statistics in captivity are not statistically significantly higher than in the wild (DeMaster and Drevenak, 1988; Wells and Scott, 1990; Duffield and Wells, 1991; Small and DeMaster, 1995; Woodley *et al.*, 1997). However, there are numerous biases in these data; survivorship statistics from captive facilities often exclude periods of sharply increased mortality – those associated with capture and transfer. These biases can easily lead to artificially inflated survivorship data for captivity.

For orcas the discrepancy between captivity and the wild is glaring. The natural average lifespan for male and female orcas is 29 and 50 years respectively, with a maximum longevity of 60 and 90 years respectively (Olesiuk *et al.*, 1990; Ford *et al.*, 1994; Ford, 2009). In captivity most orcas do not survive much past the age of 20 years (Williams, 2001). DeMaster and Drevenak (1988) estimated the annual mortality rate for captive orcas at 7.0%, and two further studies, Small and DeMaster (1995) and Woodley *et al.* (1997) both estimated (captive) annual mortality rates at 6.2% (excluding calves), considerably higher than the 2.3% annual mortality rate figure for wild populations. Furthermore, the evidence for premature death in other dolphins and whales held captive, such as belugas (*Delphinapterus leucas*) is also mounting (Woodley *et al.*, 1997).

The evidence above shows the myriad ways that we directly or indirectly encourage the continued exploitation and abuse of dolphins and other cetaceans despite our professed affection for them and the abundant scientific evidence that they are similar to humans in terms of their level of awareness, autonomy, and uniqueness – their selfhood. In the next section I will explore a general psycho-social explanation for these apparent inconsistencies.

A challenge to moral inclusivity

The human mind is evolutionarily prepared to include members of our in-group in our moral circle. Those that appear to be part of an out-group are held outside our moral concern. History is replete with this psychological parsing of moral consideration within our own species; differences between groups of people form the foundation of discrimination, objectification, and abuse. Successful efforts to eliminate prejudice and exploitation rely upon emphasizing the similarities across groups over the differences. The psychological process of doing away with racism, sexism, and all other forms of intolerance is one that requires an ability to take into account similarities and differences simultaneously; we come to accept differences in the context of acknowledging that basic characteristics are essentially the same. The capacity to take into account both similarities and differences at the same time has proven to be difficult even when considering members of our own species. And it is a delicate process that disintegrates when one feels threatened. How much more difficult is it to perform the same mental feat for members of other species?

Those members of other species who enjoy *some* level of protection and moral consideration (pets, primarily) are typically those that we consign to be less autonomous individuals who need to be safeguarded and nurtured. This attitude has resulted in both positive and negative consequences for domestic pets. But there are few other animals that qualify to participate in this special relationship and most others are regarded as either pests or resources and sometimes a combination. Even animals that we label as charismatic megafauna, e.g., dolphins, polar bears, elephants, tigers, and chimpanzees are still, in the end, considered not on their own terms but as resources for our amusement, aesthetic appreciation, and for the ways they make the world more enjoyable and more beautiful for *human* generations to come. As such we make some effort to conserve these animals for our own sake but clearly exclude them from true moral concern as autonomous individuals in their own right.

Most people give lip service to the idea that dolphins and whales should be protected and conserved. But, as philosopher Thomas White points out, our relationship with them is fraught because we cannot comprehend their similarity to us in the face of their difference (White, 2011). Their intelligence, self-awareness, emotional sophistication, and social complexity mean that they are similar to us in that we both experience life as persons. But cetaceans (unlike great apes, for instance) look and move differently, lack changes in clearly recognizable facial expressions, communicate in strange modalities, live in a very different physical environment, and seem to possess a level of social cohesion foreign even to us. Therefore, cetaceans – probably more than any other animal – represent extremes of similarities and differences that challenge our ability to recognize them as moral equals. That is, probably more than any other animal, cetaceans are the most vulnerable and the most bewildering.

Our inability to take these two dimensions into account at the same time has resulted in extreme objectification, exploitation, and abuse at our hands. Although other animals (great apes, elephants, etc.) too are victims of these biases there are often enough similarities in the way they look and express themselves that provides a basis for recognizing their emotions, i.e., distress, fear, affection, excitement. For instance, it is difficult to understand how we can continue to tolerate the mass slaughter of dolphins and whales in the Taiji, Japan or Faroe Island drive hunts where thousands of individuals are herded together and hacked to death in water red with their blood. The dolphins call out to each other and thrash about in pain and panic. I would argue that the same atrocity perpetrated annually on great apes, or elephants, *in as visible and wholesale a way as is done in dolphin drive hunting,* would result in a more dramatic human response.

Unquestionably, great apes, elephants, and other animals are subject to horrific brutality and slaughter but we more readily recognize their emotional response, their cries for help, their body language – which are all more similar to our own than those of dolphins. While our recognition of the distress of other animals that are more similar to us moves us to help them, dolphins and whales are worse off. When dolphins are being slaughtered they continue to "smile." Those of us who know dolphins can recognize the sounds of panic and fear in their whistles and body language. But the ever-present "smile" of the dolphin and lack of facial expression even under the most horrendous of circumstances is a deception that appears to minimize our concerns for them. And we tolerate and even support more insidious types of abuse in the form of captivity because they do not exhibit all of the obvious signs of despair we recognize more readily in other primates, for instance. But the very personhood of dolphins makes them among the most vulnerable of victims to this and other forms of exploitation.

This is why dolphins represent an extreme challenge to moral inclusivity. We must find a way to internalize the notion that dolphins (and other animals) can be very different from us in many ways and yet on a par with us when it comes to the dimensions that are important for moral consideration. The captivity industry knows how difficult this challenge is and exploits it at every turn. For instance SeaWorld publishes online information booklets (Animal Info Books) on bottlenose dolphins, orcas, beluga whales, and other animals in their circuses (SeaWorld, 2011). These pamphlets and other information resources, such as their teacher's guides, are littered with inaccuracies – all aimed at biasing perceptions of dolphins and other cetaceans as interesting enough to pay admission to see but not so intelligent and like us as to give credence to any concerns about their captivity. Like the last bowl of porridge in the Goldilocks fairytale, dolphin intelligence is... just right.

Against a backdrop of the human need to create a separation between ourselves and the other animals, to uphold our so-called superiority, and to assume that other animals do not experience life with the same richness and intensity, dolphins and whales have the great burden of being so like and so unlike us at the same time that they are, simultaneously, among the most beloved and the most abused and vulnerable of all animals. Our response to this realization – as for all the other animals – will depend upon our ability to rise to the challenge of thinking about others in a more complex way than ever before. That is, to be as cognitive and emotionally sophisticated as our species claims it is. In our treatment of dolphins and other animals that claim is still in search of evidence.

8 The expression of grief in monkeys, apes, and other animals

Barbara J. King

The baboon Sylvia was close to her adult daughter Sierra, unusually so, even given the tight nature of the mother–daughter bonds found in these monkeys of Botswana's Okavango Delta. The two groomed each other to the near-exclusion of anyone else, and spent a significant portion of their time together.

Then Sierra fell victim to a lion attack. In the wake of this predation event, Sylvia changed her behavioral routine. For a week or two, she sat apart from the other baboons in her group, and initiated no social interactions. Primatologist and field researcher Anne Engh observed these changes and concluded that Sylvia was depressed: "Sylvia was high-ranking and intimidating," Engh told me, "so it didn't seem unusual that other females didn't approach her, but I was surprised that she had no apparent interest in interacting with anyone."

Was Sylvia mourning her daughter? Did her altered behavior amount to an expression of baboon grief? Based on other research done among the Okavango baboons, and on an emerging picture of the capacity for grief in a wide diversity of animals, I suggest that the answer to these questions is "yes."

Back in 1985 to 1986, when I myself was studying the yellow baboons in Amboseli National Park, Kenya, it wouldn't have occurred to me to look for grief behavior when one of the monkeys died. Even then I knew, of course, that non-human primates like monkeys and apes *thought* and *felt*. And Jane Goodall's (1990) description of the chimpanzee Flint's dying of sadness only a short time after his mother Flo died was already widely known.

In those days, though, my stance was (without my being aware of it) speciesist. I simply assumed that our closest living relatives, the chimpanzees and other great apes, were the cognitive and emotional powerhouses of the non-human animal world. Monkeys were infinitely intriguing creatures, but not, I thought, good candidates for behavior that depends on a combination of memory and emotional expression. And as for other mammals, birds, vertebrates, and invertebrates, they were barely on my scholarly radar, as they sometimes aren't for anthropologists who take up evolutionary questions.

My understanding is different now, almost three decades later. Bolstered by my own experiences with animals, by interviewing of others who live or work closely with

The Politics of Species: Reshaping our Relationships with Other Animals, eds R. Corbey and A. Lanjouw. Published by Cambridge University Press. © Cambridge University Press 2013.

animals, and by reading the literature about animals on the farm, in our homes, at the sanctuary or zoo, and in the wild, I no longer think that grief is limited to great apes and humans, or even to primates plus the "usual suspects" of elephants and cetaceans.

In this new and broader framework, Sylvia the baboon's behavior fits right in. Sylvia's depressed behavior was the catalyst for primatologist Engh to carry out a study of the physiological effects of bereavement among the Okavango Delta baboons. The monkeys are good subjects for this endeavor because their groups are organized around matrilines, groups of related females. The matrilines result because females stay and breed in their natal groups, whereas males transfer out at puberty to live and breed in new groups.

During a 16-month period in 2003 to 2004, twenty-six baboon deaths occurred at Okavango. Only three of the baboons had been visibly sick in some way. Of the remaining monkeys, ten were known by researchers to have been victims of a predator. The other thirteen were thought to have met the same fate, based on predator sightings by the scientists or alarm calls by the baboons.

We might expect that baboons experience significant stress from the presence of active predators in their habitats, and indeed Engh's research team found the physiological markers of it in the monkeys' bodies. Engh *et al.* (2006) collected fecal material from the baboons in order to measure circulating levels of the females' glucocorticoid (GC) stress hormones. They discovered that in the four weeks following a predation event in the group, females' GC levels increased measurably.

The GC levels of twenty-two females ("affected females") who had lost a close relative to predation were then compared to those of females ("control females") who had experienced no such loss. The affected females showed significantly higher GC levels. While predator attacks were witnessed by many adult females in the group, only the "bereaved" females showed significant GC increases.

Here we have qualitative information from the anecdote regarding Sylvia and Sierra, and data from the hormonal study that combine to showcase the emotional and physiological costs of grief in this group of monkeys. It's evidence that baboons, or at least these baboons, *feel* their relatives' deaths.

The bereaved females' elevated stress levels were temporary; soon the females began to increase the number of their grooming partners and the rate at which they participated in grooming. In monkeys, to groom and be groomed by a partner is a soothing social activity as much as it is a hygienic one. As Engh *et al.* (2006) put it, "bereaved females attempted to cope with their loss by extending their social network." Sylvia too widened her social circle again only a few weeks after the loss of her daughter Sierra.

Defining grief

The conclusions of Engh *et al.* (2006) add to a growing database that suggests individuals in a wide variety of species may mourn their losses in some way. Jane Goodall (1990) famously recounted the life-sapping grief of chimpanzee Flint at his mother Flo's death in 1972, at Gombe Stream in Tanzania. In the intervening forty years, evidence has accumulated for emotional responses to death in other wild chimpanzee groups (Teleki, 1973; Boesch and Boesch-Achermann, 2000) and indeed in other animals ranging from elephants and cetaceans to domesticated or companion animals (see Bekoff, 2007a; King, 2013).

The field of primate thanatology is still new. Definitions, methods, and hypotheses to be tested are only now being debated and refined. To date, monkeys' and apes' cognitive abilities have been researched in greater depth than have their emotional capacities. There's a sea change underway, however, and grief research may be the leading edge of research into monkey and ape emotion.

A growing number of animal-behavior scientists, led by some of the researchers and writers represented in this volume (e.g., Marc Bekoff and Steven Wise) assert freely that non-human animals feel grief and love, full stop. I agree and have adopted definitions of animal love and grief that, while conservative, remain open to refinement as our knowledge base encompasses more species and thus broadens.

Love: when an animal feels love for another, she will make an active choice to spend time with that individual, for reasons that may *include* but will also *go beyond* her survival-based needs. That is, she will go out of her way to be near to, and positively interact with, the loved one, going beyond only the realms of foraging, predator defense, mating, and reproduction.

The active choice to be together is a necessary condition to define "love," but it's not sufficient. The second component is this: should the animals become separated for some reason, including by the death of one of the pair, both animals, or the survivor if only one remains, suffer emotionally in some visible way. By "some visible way" I refer to refusing to eat or eat adequately, losing weight, becoming ill, acting listlessly, being uninterested in social interactions or other normal activities, and/or exhibiting body language that conveys sadness or depression. When one or more of these symptoms is seen in a survivor following an animal's death, the definition of *grief* has been met. In this scheme, then, the expression of animal grief becomes part of the criterion for identifying animal love.

Neither definition is without problems. Love, for instance, is likely to be underestimated using the definition I have offered. Strictly speaking, no infant mammal could be said to love his mother, because it's impossible for a scientist to tease apart the infant's survival needs (suckling for nutrition) from any non-survival-based reason he might choose to stay close to his mother. Given that emotional health isn't strictly separable from physical health and thus survival, it may be too simple to suggest a split between survival- and non-survival-related choices in the first place. Further, my definition probably can't distinguish reliably between animal friendship (see Seyfarth and Cheney, 2012) and animal love, which is a thorny task even as regards our own species.

Lastly, both love and grief will likely be underestimated on this scheme simply because it will be rare for observers to be present when animals become separated for any reason, including the death of one.

I envision a database of animal grief examples, in which some instances meet these conservative definitional criteria, and others don't but nonetheless point us toward expression of some key animal emotions. Such a resource would offer clues into the very process of how collectively we might rethink and revise the definitions proposed here.

Let's return first to monkeys, and then broaden out to consider grief in other primates and, finally, non-primate animals.

What *isn't* grief?

When a relatively new area of investigation is opening up that provides definitional challenges of the sort just outlined – basically, how do we know what an animal *feels* as opposed to reporting how that animal *behaves?* – it can be helpful to cite examples that *don't* meet the stated criteria.

Sylvia's behavior with her daughter qualifies as love, using the definition I have offered here, and her altered routine after her daughter's death qualifies as grief. Sylvia and Sierra clearly chose to be together even aside from basic survival needs, and upon Sierra's death, Sylvia's willingness to interact socially with her group mates altered markedly. By contrast, the statistical data from Okavango on hormonal profiles of "bereaved baboons" *don't* lead to a definitive conclusion of grief, because the social and emotional histories of the individual animals aren't reported. No judgment can be made along the lines that my suggested definitions require.

Monkeys' response to death is most often discussed in the literature through data about maternal corpse-carrying. When infants die, some mothers choose not to lay down the bodies but instead continue to carry them. Over a 3.5-year period, 14 female gelada baboons living in the Guassa highlands of Ethiopia carried dead infants (Fashing *et al.*, 2011). Most carrying episodes lasted between one and four days, with three females carrying their infants for significantly longer: 13, 16, and 48 days. In these extended cases, the infants' bodies gradually became mummified, and just as was the case with the same phenomenon in Japanese macaques at Takasakiyama (Sugiyama *et al.*, 2009), they emitted an unpleasant smell.

This maternal behavior by monkeys amounts to a *lack* of alteration in the mother's behavior, because the mother continues infant carrying despite the complete lack of response coming back to her. In the accounts I have read, the mothers do not appear depressed and there's nothing like a shrinkage of positive social interactions as there was with the Okavango baboons such as Sylvia.

If grief accompanies the act of maternal corpse-carrying in monkeys, we simply cannot know it from observation. Primatologists Dorothy L. Cheney and Robert M. Seyfarth (2007) take note of monkeys' lack of visible grief in their book *Baboon Metaphysics*. Even when monkeys carry the bodies of infants who are dying, they say, the mothers treat the infants pretty much the same way as they do their healthy infants.

Grief and great ape welfare

Great ape mothers too may carry the dead bodies of their infants for prolonged periods (e.g., Biro *et al.*, 2010), and in this case too there's no compelling evidence for the expression of grief. The emerging picture, however, suggests that mourning may be of

Figure 8.1 On April 4, 2012 the chimpanzee baby Ajabu died in Artis Zoo, Amsterdam, only twelve days old. Her mother put the little corpse aside on April 9. Only then were the other chimpanzees from the group allowed near it. The photo shows a member of the group inspecting the dead baby. Photo Roland van Weeren and Artis Zoo.

an emotionally richer nature in great apes than in monkeys, in line with the more elaborated nature of cognitive empathy and perspective-taking in these closest living relatives of ours (Preston and de Waal, 2002). At present, this is only a hypothesis. It may, in fact, be another speciesist perspective; only more comparative research will either falsify or support it.

Although still few, the most thorough examples we have of great ape response to death comes from chimpanzees. Nudged by the extraordinary observations in the wild, zoo scientists may pay keen attention nowadays to group behavior when a chimpanzee dies in captivity. When an old female chimpanzee fell ill at a Scottish safari park, for instance, zoo staff anticipated her death and filmed her group's actions. Because the apes were sheltered indoors in their heated winter quarters, visual access was quite good.

At the park, two mother–offspring pairs lived together as a group: Pansy, the dying female whose age was estimated to be in the 50s, and her daughter Rosie, age 20, plus Blossom, a female about Pansy's age, and her son Chippy who was 30. Pansy had been listless for some weeks when she began to show signs of labored breathing. Pansy's companions seemed aware that something was amiss; in the ten minutes before her death, they groomed or caressed her at what the observers judged a higher than usual rate.

Around the presumed moment of Pansy's death, Rosie, Blossom, and Chippy continued to be quite active, as a transcribed account by James Anderson and his colleagues (2010) makes clear. The numbers refer to time-stamp codes on the film; some observations in the sequence are omitted, but the words included are taken verbatim from the published report:

16:2421 Chippy crouches over Pansy's head then appears to try to open her mouth. Rosie moves toward Pansy's head.

16:2425 Blossom, Chippy, and Rosie simultaneously turn toward Pansy's head. Chippy and Rosie are crouched over Pansy's head. Chippy pulls Blossom's face down toward Pansy's.

16:2436 Rosie moves from Pansy's head toward her torso. Blossom moves away from Pansy. Chippy lifts and shakes Pansy's left shoulder and arm.

The chimpanzees continue to caress and groom Pansy. Then at

16:3656 Chippy jumps into the air, brings both hands down and pounds Pansy's torso, then runs across and off the platform.

Chippy's behavior may startle us – it certainly doesn't sound like the action of a chimpanzee steeped in grief! Yet his jumping and pounding actions aren't so different from what has been observed in some of the wild accounts. Was Chippy expressing anger or emotional upset? Or was he going about a sort of practical problem-solving approach, trying to elicit some kind of response from his immobile cage-mate? Both possibilities are likely, according to Anderson and his co-authors.

The behavior of Pansy's companions remained atypical through the night of her death and beyond, and clearly signaled some degree of emotional upset. (Data of this nature are unlikely to come from observations made in the wild because under natural circumstances, chimpanzees fairly soon move on from the corpse.) Pansy's survivors slept fitfully. Her daughter Rosie stayed near to her body. Unlike in the moments before her death, no one groomed Pansy's body, although Chippy attacked her corpse three more times during the night.

The chimpanzee survivors were "profoundly subdued" the next day, the scientists saw. In silence, the apes watched as zoo keepers removed Pansy's body. For the next five nights, none of the chimpanzees slept on the platform where Pansy had died, even though they had favored that spot in the past. Indeed, for weeks, they remained quiet, and ate less than usual. These chimpanzees were grieving.

Observations like these, in captivity as well as in the wild, matter not just for our understanding of great ape emotional capacities, but also for the lives of the apes themselves, including the way we choose to treat them even as we keep them captive in the biomedical or entertainment industries. Like many other chapter authors in this volume, I have argued against the use of chimpanzees biomedically (King, 2011), and remain concerned about their treatment in zoological parks.

One positive outcome of primate thanatology research is that it adds concretely to the established picture of great apes as creatures who may be deeply emotionally affected by what goes on around them. By analogy with what Birkett and McGrew (this volume) write about ape consciousness and suffering, I would point out that it's neither good science nor morally acceptable to insist that great apes don't grieve because we can't *prove* that they grieve. We can't *prove* that humans grieve either, even when people say that they are grieving. What we can do is look with open minds and hearts for behaviors that meet the criteria of love and grief, and see whether they are out there or not.

A second positive outcome of primate thanatology research is the gradual improvement in how zoo apes are now treated when one of their companions dies. As was evident in the example of Pansy at the Scottish park, apes may be offered a chance to spend time with a deceased companion and later to observe as the body is taken away. At Chicago's Brookfield Zoo, when the gorilla female Babs suffered from an incurable kidney disease, she was euthanized at age 30. The staff organized what they called "a wake" for Babs' companions.

Gorillas from different generations were present, some of whom were visibly emotional in the presence of the body. The event was described this way in a published account by the Associated Press (AP Worldstream, 2004): "Babs' 9-year-old daughter, Bana was the first to approach the body, followed by Babs' mother, Alpha, 43. Bana sat down, held Babs' hand and stroked her mother's stomach. Then she sat down and laid her head on Babs' arm... Bana rose up and looked at us and moved to Babs' other side, tucked her head under the other arm, and stroked Babs' stomach."

As mental images fill our heads of this gorilla goodbye, we recognize the grief of a child confronted with the utterly still body of a long-loved parent. Babs' other cage-mates approached her, too. Nine-year-old Koola brought close her infant daughter, a baby who had received Babs' affection during her young life. Ramar, the group's 36-year-old silverback male, however, stayed away from Babs.

Ramar's decision reflects his own nature, or at least his nature at that particular moment in that particular context; it is not some across-the-board tendency for emotional aloofness seen in gorilla silverbacks. A mature male acted quite differently when at the Franklin Park Zoo, the female Bebe was euthanized because of malignant masses in her body. Diane Fernandes, then curator of research at the zoo and now director of the Buffalo Zoo, told me (personal communication) about Bebe's mate's response: "We first let the male Bobby in with the body and he did try to revive her, touching her gently, vocalizing, and even placing her favorite food (celery) in her hand. When he realized she was dead, he began to call in this soft hoot but then started to wail and bang on the bars. It was clearly a demonstration of immense grief and it was very sad to watch."

Fernandes shows no hesitation in deploying cognitive and emotional terms for what Bobby experienced when Bebe died. She says that he *realized* and then *grieved* her death. As opaque as Bobby's thought processes must remain to us, the sequence of his behaviors does support this conclusion. The gift of Bebe's favorite food was apparently offered because Bobby thought, or hoped, that Bebe was alive, or because he wanted to somehow encourage her aliveness, to revive her with a sensory experience that she loved. When this strategy failed, and Bebe lay unresponsive, Bobby erupted into sadness.

The Franklin Park Zoo allowed Bobby some time alone with Bebe's body, and then allowed her three other gorilla companions to approach. These apes also, Fernandes recalls, touched Bebe's body "as if to rouse someone who was sleeping." Unlike Bobby, they did not vocalize. Perhaps they didn't make the cognitive leap that I suspect Bobby did, or perhaps they just expressed grief differently.

By focusing in this section on captive primates, I do not intend any across-the-board endorsement of the fact of their captivity. Whatever we may feel about the ethics of keeping animals in captivity at zoos, it can only be to the animals' benefit to apply emerging research on their emotional lives to how they are perceived and treated.

Beyond speciesism

I mentioned that I've expanded my circle of consideration recently into questions of grief in other animals. The first step for me involved elephants and cetaceans. The work of Cynthia Moss and Iain Douglas-Hamilton, who study wild elephants at separate sites in Kenya, documents expressions of elephant mourning in the wild (for a review, see King 2013: Chapter 5). Within their tight-knit matriarchies, elephants may surround a dying companion and, once death occurs, may manipulate and caress the bones of the loved one. Elephants do grieve.

Marino (this volume) makes a fascinating argument that compared to apes and elephants, dolphins are perceived as too different and thus aren't readily recognized as thinking, feeling creatures – as "moral equals." She writes that, comparatively, cetaceans "look and move differently, lack changes in clearly recognizable facial expressions, communicate in strange modalities, live in a very different physical environment and seem to possess a level of social cohesion foreign even to us." We may not, then, recognize cetacean emotion as readily as we recognize emotion in great apes and elephants.

I recognize the validity of Marino's point, yet also think that with dolphin response to death, we may find a clear window into cetacean emotion. In 2001 off the Canary Islands, rough-toothed dolphins were observed by scientists over the course of six days (Ritter, 2007). One animal, presumably the mother, was seen lifting and pushing a dead newborn, with escorts from two other adults. For five more days, even as the baby began to decay, the dolphins stayed with the infant corpse, defending it when a gull approached. The dolphins even swam with their backs underneath the baby's body. In total, 19 dolphins participated in some way, 15 who were thought to belong to the same social unit. The dolphins slowed their speed of travel and indeed barely moved for the whole six days.

Skeptics might suggest that the dolphins were attempting to awaken a lifeless baby. Based on everything we know about the cognitive and emotional lives of dolphins, however, and on the alteration in the dolphins' behaviors, I feel comfortable arguing that the scientists caught a glimpse of dolphin grief.

Now, if, based on unexamined assumptions, I stopped the "widened circle" from great apes at elephants, cetaceans, and monkeys, I'd not have made much progress from my speciesist stance of decades ago. In fact, I have recently collected and/or reviewed convincing evidence of animal grief from a much wider variety of species. In *How Animals Grieve* (King, 2013) I tell stories of mourning in domestic house cats, dogs, and rabbits, and in ducks and geese, when one of a long-bonded pair dies and leaves the other in a state of profound sadness. I raise the possibility of mourning in species as variant as Hawaiian sea turtles and wild Yellowstone bison.

The future of grief research

As I conclude this chapter, I focus once more on the behavior of monkeys and apes. Given what we now understand about intraspecific variation, it's clear that not all monkeys and apes will express grief when a relative, close friend, or group mate dies. We are dealing with animals who show variability not only in cognitive and emotional capacity, but also in personality and in nuances of behavior depending on the emerging circumstance of the moment. In some wild populations, animals may conserve energy and reduce the risk of predatory attack by hiding any emotional states they experience, and it will be quite difficult to tease out by observation alone this situation from a lack of emotion. Research that combines observation of visibly altered behavior with hormonal profiles of changing stress/emotional states may be the best bet in this regard.

In short, we don't yet know very much about grief in monkeys and apes. What we do know suggests that benefits will accrue both to animal behavior science and to animal welfare endeavors when grief is studied with a keen eye to variability in its expression across populations, age–sex classes, and individuals, and within the same individuals across time. Cultural tradition, individual life history, and developmental dynamics intersect to make animal social behavior highly plastic, and we should expect no less than this plasticity when we explore animal grief.

For me, the bottom line is how we can harness the exploration of animal grief to work *for* animals. Detailed qualitative as well as quantitative study of animal grief will bring into focus the degree to which a whole variety of primates, and as I have shown here non-primates as well, go about their lives as individuals who think and feel deeply in their relationships with others.

9 Great ape mindreading
What's at stake?

Kristin Andrews

There is nothing like travel for learning how others live. When I first entered an orangutan forest school in East Kalimantan, Indonesia, it wasn't clear who was more curious, me or the orphan orangutans who had lost their mothers due to palm plantations, the pet trade, or other horrors. That first day, I first met the 12 orangutan students through the bars of a cage where they spend the night before going out into the forest for the day's lessons in food processing, tree climbing, and social interaction. I took my first orangutan hand through the bars of the cage, and put my face close to the bars so he could sniff me. I was eye to eye with him, dark deep-brown eyes looking directly into my blue ones. Next, I met Cecep outside the cage, and when I bent down to say hello, he buried his head in my neck and gave me a loud kiss.

That first day I saw so many things. As a philosopher going into a Bornean forest, I came with a healthy skepticism. But that first day, I couldn't believe it – in only a few hours, I saw things that looked like social learning, joint attention, declarative showing, surprise, imitation, relationships, and lots of play. I had already known from reading that humans and other great apes are similar in many ways. We share an extended immaturity and intense infant–caregiver relationships, group living situations, cultural transmission of technology, and many emotions and cognitive capacities. Yet the communities of non-human apes are also very different from human communities. Humans build lasting tools, and store them to use later. We build permanent sleeping and living structures. We cook our food. We have courts of law and prisons and ethics books. There are vast technological differences between humans and non-human great apes.

After meeting Cecep and his cohort, I was even more interested in the question of what makes for such a difference between humans and other apes. On one account, one central difference between humans and other apes is that only humans develop the ability to mindread, the ability to see that others have beliefs that could be true or false, which permits joint attention and shared intentions. For example, Michael Tomasello takes mindreading (along with cooperation and having shared goals) to permit the development of cumulative culture, so that technological advances can spread through a society and future generations can continuously improve upon those advances (Tomasello, 2008). And Kim Sterelny suggests that our hominin ancestors thrived in an apprenticeship culture where naïve individuals were given the opportunity to learn

The Politics of Species: Reshaping our Relationships with Other Animals, eds R. Corbey and A. Lanjouw. Published by Cambridge University Press. © Cambridge University Press 2013.

from a master, and the master knew how to offer the apprentice the appropriate projects, tools, and materials as her skill sets improved – which of course was facilitated by a developing mindreading capacity (Sterelny, 2012).

An idea that emerges is that only humans can be teachers, builders, or judges because only humans understand their own reasoning patterns and the evidence they have for their beliefs, and only humans realize this about others in such a way that allows them to form the collaborative projects needed for complex culture. This requires mindreading capacities and realizing that some of our beliefs turn out to be false, or lack evidence, while others turn out to be true. Retaining the true beliefs and discarding the false ones is a generally advantageous evolutionary strategy. And it is traditionally taken to be a requirement for an *autonomous agent*, one who acts deliberately, and hence is responsible for her actions. Autonomous agents are the only sorts of creatures who can truly be moved by moral norms.

In this chapter I will offer an alternative account of the relationship between mindreading and the development of community social norms and individual autonomous agency. Mindreading is not needed for teaching, developing some cumulative culture, or agency. Rather, I will argue that the foundational role for mindreading is based in its usefulness for explaining behavior (Andrews, 2009a, 2012). It is the need to explain behavior that drove the evolution of mindreading, and this, in turn, required a prior understanding of social norms. Looking at the behavior of the non-human great apes in this light, it appears that the core elements required for agency are fairly well established in great ape communities. One consequence of this position is that mindreading should not be a requirement for either moral or legal standing.

The benefits of mindreading

Mindreading is one aspect of a theory of mind. *Theory of mind* refers most generally to the ability to attribute mental states of various sorts to others. The interest in theory of mind was for a long time focused on the ability to attribute false beliefs to others (Bennett, 1978; Dennett, 1978; Harman, 1978; Wimmer and Perner, 1983), but now it includes understanding others' perceptual states (such as what others see or hear; Hare *et al.*, 2000, 2003, 2006; Flombaum and Santos, 2005; Kaminski *et al.*, 2006; Melis *et al.*, 2006; Lurz, 2011), goals (Uller, 2004), intentions (Tomasello *et al.*, 2005), or knowledge states (Hare *et al.*, 2001; Kuroshima *et al.*, 2002, 2003; Kaminski *et al.*, 2008). What all these states share is their unobservable nature, so the story goes. Such mental states are purported to be theoretical entities whose existence must be inferred. While almost no one believes that apes can attribute propositional attitudes such as beliefs (Call and Tomasello, 2008), many claim that apes can attribute some of these other mental states, or as Hutto (this volume) puts it, there is evidence that apes can "mind minds."

Despite the growing number of states of mind being investigated, most parties agree that attributing these states of mind is beneficial for predicting behavior, and that mindreading is particularly advantageous, since it permits the attribution of states of mind such as belief states that the attributor knows may be false. Having this ability allows one to try to manipulate a competitor by giving him false beliefs, which would then lead the competitor to act contrary to his own interests (and, importantly, in the manipulator's own interests). The view has been recently described in this way: for mindreading to be advantageous, it must offer some benefit, and a benefit for a mind-reading animal is knowing what other animals' behavior will be (Lurz, 2011).

I aim to challenge the widespread assumption that the benefit of mindreading is in predicting behavior. Rather, since predicting behavior can be accomplished using a large variety of techniques that don't require thinking about what others think or believe, it is in explaining behavior that mindreading offers an advantage (Andrews, 2009a, 2012). When one has an explanation for someone's behavior, one knows how to act appropriately given that explanation, and nothing could be more advantageous than that. Of course, once one has explained someone's behavior, it is possible to generate plausible predictions about future behavior, or at least to limit the domain of expected behaviors. But it is the understanding of the other individual, rather than a simple ability to predict, that offers the advantage. And some explanations of behavior can only be given in terms of beliefs.

To illustrate, consider being approached by an individual who is talking to himself, acting in a jittery manner, and otherwise appearing strange. If you were to find this person aversive given his strange behavior, and move away, you haven't tried to understand him. Rather you have just consigned him to the out-group. But if you were curious, and tried to understand his behavior by examining him more closely, you might see that he was using a cool new gadget, and by learning something about his reasons for his behavior, you learn something about new technology. So that is a contemporary example. Consider now our hominin ancestors before the discovery of cooking. Meat was gained at great effort, and so was presumably highly valued. Fire is destructive and would have been constrained as much as possible. Putting the two together might seem like a terrible thing to do. If our hominin ancestors ostracized the meat-cooker, rather than trying to understand his motivation for cooking, they would have lost out on the increased nutrition cooked food offers. It is the willingness to allow individuals to act anomalously, and to seek to explain their odd behavior, that would offer an evolutionary advantage. Note that many new technologies can be learned through the explanatory drive, without active teaching or apprenticeship, or the ability to explain behavior. Opportunity learning coupled with the drive to explain behavior could have led a juvenile hominin observing the construction of a hand axe to pick up a flint and examine its affordances. The hand axe becomes functionally transparent to him as he learns that it is sharp and can cut his flesh, without anyone needing to teach him. As improvements in tools are made, these new affordances can be discovered by examining the tools, coupled with observational learning when the tools are in use. Thus, cumulative culture can also occur without mindreading.

It is this drive to understand things in one's environment that likely led to the development of mindreading in humans. When the drive to explain is directed at other beings, mindreading is the natural next step. But there is an important move in this story that I have so far not emphasized, and it is this point that will be the focus of the remainder of the chapter. In both examples, someone was acting oddly. It is odd behavior that we seek to explain; there is no need to explain normal behavior. Here comes the important step: to realize some behavior is worth explaining is to realize that it is not normal, and this requires an understanding of what is normal, or what individuals should be doing. It requires some understanding of social norms. Now of course not all statistical regularities are social norms; using your left hand to eat will not lead others in the West to seek an explanation, because there are no social norms against eating with the left hand for Westerners. However, in other cultures where there is such a prohibition, an explanation would likely be sought (and offered). Those violations of statistical norms that are the typical subject of explanatory interest are the ones that indicate the existence of social norms. Among our hominin ancestors, there were likely social norms about the proper treatment of fire. The first fire-cooker had to violate a norm to innovate a new beneficial technology. And the others in the community had to recognize that he had violated a norm, and ask themselves why he did it, for the practice to catch on.

The upshot is that a society in which mindreading might develop is a society that already has social norms. But this claim might make some uneasy, for how could there be moral norms without moral, and hence autonomous, agents? And how could there be autonomous agents who don't mindread – or more specifically, who lack the ability to think about beliefs? I will turn to these questions now.

Agency and mindreading

An autonomous agent knows that her actions have consequences, and she thinks she can affect change in the world through her actions. While accounts differ, as a foundation we can agree that "to be autonomous is to be one's own person, to be directed by considerations, desires, conditions, and characteristics that are not simply imposed externally upon one, but are part of what can somehow be considered one's authentic self" (Christman, 2009). Put this way, an autonomous agent is contrasted with an individual whose every act is controlled by external forces, so an animal whose behavior was completely controlled by, for example, fixed action patterns (such that all behavioral sequences are inflexible and determined by environmental stimuli) could not be an autonomous agent. However, an animal might be an autonomous agent if his behavior is flexible and the result of internal cognitive processes rather than mere reflex or association with environmental stimuli. For such animals, the question would then be whether the internal processes are of the right sort.

There is a tradition in philosophy that would include mindreading (in terms of understanding belief) among the processes required. Christine Korsgaard illustrates this approach in her argument that animals lack autonomy. Because animals can't mindread, they cannot self-govern, i.e., they cannot decide whether an act is justified and then act from that judgment rather than from one's desire. She writes:

What it [normative self-government] requires is a certain form of self-consciousness: namely, consciousness of the grounds on which you propose to act *as grounds*. What I mean is this: a non-human agent may be conscious of it as *fearful* or *desirable*, and so as something to be avoided or to be sought. This is the ground of his action. But a rational animal is, in addition, conscious *that* she fears or desires the object, and *that* she is inclined to act in a certain way as a result. That's what I mean by being conscious of the ground *as a ground*. She does not just think about the object that she fears or even about its fearfulness but about her fears and desires themselves.

(Korsgaard, 2006: p. 113)

The argument that autonomy requires mindreading could go like this: in order to have moral agency, one needs also to have autonomy. An autonomous agent is able to act for reasons. And since acting for reasons requires the ability to recognize *that* one has reasons for actions, and reasons for actions are sets of beliefs and desires that motivate behavior, it follows that acting for reasons requires mindreading (of one's self). Thus, the worry arises that the moral agent must understand belief, and if non-human animals lack the capacity they cannot be agents.

But there are problems by setting the requirement so high. First, note that this requirement excludes human children and many adult humans as autonomous agents. On this view, a moral agent has explicit knowledge of her reasons for action and the ability to analyze them. It would follow that children, who do not seem to understand belief until around four years old (Wellman *et al.*, 2001), and who don't recognize some of the limitations of belief attributions until mid-childhood (Apperly and Robinson, 2001, 2002, 2003) are not autonomous agents. Worse yet, until adolescence humans are unable to evaluate the reasons they may know they have (Pillow, 1999; Morris, 2000; Moshman, 2004). And it is evaluating one's reasons that is the hallmark of agency in Korsgaard's view.

The idea that normal adolescent humans are not autonomous agents flies in the face of the typical approach to developmental moral psychology – and common sense – according to which children have a very early entry into the domain of agency, even though the range of intentional actions is more limited than that of adult humans. By examining moral development starting with ten-year-old boys, Kohlberg presumed that there was moral reasoning of some sort at this age, and that assumption mirrors the commonsense conception that children are moral agents, even though they cannot be held responsible for all their actions (Kohlberg, 1981). That we recognize children as not *fully* responsible for their actions is reflected in their special legal status. Since children are still developing their cognitive capacities and their ability to control their impulses and emotions, children are limited in what they can do. Given the acceptance of the "ought implies can" principle, children enjoy this special status. But this doesn't mean that children are not agents, and that their behavior cannot be categorized as good or bad.

Korsgaard's requirement that agents have explicit knowledge of their reasons and are able to analyze those reasons also runs into problems when we consider adult human cognition. Research on adult moral reasoning suggests that adults do not generally consider their reasons when making moral judgments (Haidt, 2001). Korsgaard may respond that the requirement is only that an agent *can* consider her grounds for action, which is consistent with the findings that humans very often do not consider their reasons for their moral judgments. However, to take this position as a response to Haidt's findings would be to accept that humans are not living up to their moral obligations most of the time, and that most of our actions cannot be understood as autonomous.

If the standard is too high for many humans, it shouldn't be surprising that the standard is also too high for apes. Rather than taking this as an admission that apes aren't agents because they don't mindread, we can take it as evidence that the requirements for autonomy stated by Korsgaard are too high for the type of agency we are concerned with; we want to find an account of agency that can be fulfilled by very young children, since such children already have some understanding of social norms (which, as you may recall, is what we're really interested in here). Instead we should look at an account of agency that is inclusive of these young children.

Consider again Christman's account of autonomy as "to be autonomous is to be one's own person, to be directed by considerations, desires, conditions, and characteristics that are not simply imposed externally upon one, but are part of what can somehow be considered one's authentic self" (Christman, 2009). We can appeal to this definition in order to ask about the cognitive capacities required, and whether children have them, and if so, whether apes do as well. Let's focus on two aspects of this definition in turn; first, we'll look at being directed by internal considerations, and then we'll turn to the issue of being one's own person.

What does it mean to be directed by internal considerations rather than those that are externally imposed? Let me suggest a sufficient condition: the ability to distinguish intentional from unintentional action. An intentional action is done purposefully, and would often be described as being done for reasons. If someone can sort intentional actions from other kinds of actions, then there is at least an implicit recognition that these two kinds of actions are different, and that some actions are the responsibility of the agent, and others are not. If one can distinguish internal considerations from external ones in others, there is every reason to expect that one could do so for one's self as well.

There is evidence that apes have the cognitive capacities for identifying intentional behavior. For example, there is evidence that chimpanzees understand that others have mental states such as seeing, and that seeing motivates individuals to act (Plooij, 1978; Goodall, 1986; Hare *et al.*, 2000, 2001). There is also evidence that chimpanzees understand goals and intentionality (Uller, 2004; Tomasello and Carpenter, 2005; Warneken and Tomasello, 2006). For example, Claudia Uller found that chimpanzees, like human children (Gergely *et al.*, 1995), seem to perceive the behavior of geometric shapes moving in the right way as intentional (Uller, 2004). For both humans and chimpanzees, a violation of expectation paradigm was used to measure the subjects' responses. While Gergely concluded that the infants' surprise response to "irrational" behavior suggests that they attribute goals and rationality, Uller's conclusion was more circumspect, even though her infant chimpanzee subjects responded in the same way as the human infants. She concludes that chimpanzees, at least, have an understanding of agency.

This finding concerning infant chimpanzees is consistent with other experimental studies of intentionality attribution. Call and colleagues found that chimpanzees are more impatient with humans who are unwilling to give them food compared with humans who are unable to give them food; they beg more from the capable person who is unwilling than they beg from the person who is unable to access the visible food (Call *et al.*, 2004). Warneken and Tomasello (2006) found that chimpanzees respond appropriately to the communicative gestures of human caregivers. While engaged in what appeared to be informal social interactions with the experimenter, the chimpanzees would be tested on their ability to respond to a non-verbal request for help. In one condition, the experimenter dropped an object and requested that a chimpanzee pick it up, which she readily did.

Apes' understanding of intentionality has also been investigated by looking at contingent responsivity. For example, a chimpanzee named Cassie responded differently when being imitated by his caregiver than he did when his caregiver engaged in non-imitative behavior (Nielsen *et al.*, 2005). Like human infants, Cassie would systematically vary his behavior while closely watching the imitator. Nielsen and colleagues describe one bout of behavior while Cassie was being imitated: "Cassie poked his finger out of the cage, wiped the ground in front of him, picked up a piece of straw and placed it in his mouth, pressed his mouth to the cage, then poked his finger out of the cage again" (Nielsen *et al.*, 2005: p. 34). Such repetitive sequences were the norm when Cassie was being imitated, but not when the caregiver engaged in non-imitative behavior or no behavior at all. Cassie's response demonstrates that he was aware that his caregiver was acting purposefully, further evidence that the chimpanzee has a notion of agency.

Also, research on natural pedagogy and ostensive communicative cues could offer additional information about apes' ability to recognize intentional agency in others. Csibra suggests that in humans, ostensive cues that lead infants to come to understand intentional communication are direct gaze, infant-directed speech, and contingent responsivity or turn-taking (Csibra, 2010). We know that direct gaze does occur in chimpanzee mother–infant dyads (Bard *et al.*, 2005), and as reviewed above there is evidence that apes test contingencies in order to determine whether actions are intentional or not. Mother–infant ape research can investigate parallels of these ostensive cues in the development of ape communication.

Now, on to the second aspect of Christman's definition. What does it mean to be one's own person, and act from one's authentic self? One interpretation is that an authentic self is one that is self-created, rather than given to the agent fully formed. To create one's self requires the ability to deliberately change oneself. There are various ways in which one can act to change or control one's behavior. One way of changing oneself is to work things out in the space of reasons. But other ways would be to develop habits of behavior and to self-modulate emotional responses.

As far as I know, the ability to self-create by purposefully changing oneself has not been given direct attention by great ape researchers. However, there are some findings that suggest great apes do act to improve themselves in various ways. For example, the social learning literature indicates that great ape species do learn from observing others' behavior (Tomasello *et al.*, 1987; Call and Tomasello, 1994; Whiten, 2000). Orangutans will position themselves so that they are only a few inches away from the behavior that they are observing, and will subsequently attempt the behavior themselves (Call and Tomasello, 1994; see Figure 9.1). Some scientists think that great apes practice behaviors in order to develop competences (Anne Russon, personal communication). We have clear evidence that apes spend years engaged in complicated tasks like nut-cracking and cooperative hunting before they become proficient, and there is certainly valuable

Figure 9.1 Human-reared orangutan, Princess, trying to write with a lit mosquito coil, with her daughter looking on and becoming acquainted with a novel technology. Photo courtesy of Anne Russon.

information about the role of practice in the development of such skills. More research on the question of practice in the learning research can help to determine whether and to what extent the great apes act to purposefully change themselves.

Two other areas of research are ripe for exploring the question of great ape self-improvement: personality and teaching.

Personality

The field of cross-species personality research has identified six personality factors in chimpanzees (King and Figueredo, 1997) and five factors for orangutans (Weiss *et al.*, 2006). While the initial goal of this research was to determine whether there are personality differences in other species, and what the personality factors consist of, future work could be on changes in personality and the events and behaviors that drive personality changes. Given the research of Francys Subiaul and colleagues, we know that chimpanzees can learn some of the traits of unfamiliar humans by watching them observe the human interacting with another chimpanzee (Subiaul *et al.*, 2008). The question is whether individuals can categorize themselves in the same way, and modify their behavior based on that knowledge. This sort of investigation may further our understanding of whether apes or other animals take steps toward something like self-improvement.

Teaching

Self-improvement is related to helping others improve, and any evidence of pedagogy in apes can offer evidence that the apes also act to self-improve. Teaching has been operationally defined by Caro and Hauser (1992) as requiring the teacher to modify her behavior only in the presence of a learner, such that the teacher gains no immediate benefit from the behavior, and the learner acquires the skill being demonstrated. There is some evidence for teaching under Caro and Hauser's criteria for three species: the ant species *Temnothorax albipennis* (Franks and Richardson, 2006), meerkats (Thornton and McAuliffe, 2006), and pied babblers (Raihani and Ridley, 2008). While there are occasional anecdotes about teaching in apes, these are often very controversial (e.g., Boesch, 1991, 1993), and there have been no published systematic studies of pedagogy in apes. This may be due to the difficulty of observing that an instance of teaching fulfills all three of Caro and Hauser's criteria. Another way of investigating teaching in apes would be to look more toward the mechanisms presumed to be at work, and look for evidence of such mechanisms. For example, one could follow Csibra and Gergely's (2006, 2011) recent suggestions about natural pedagogy (which they think is unique to humans) and look for the building blocks of teaching in attention monitoring, understanding of reference, and acquisition of general knowledge in ape species.

But what about prediction?

One might worry that none of the capacities discussed above can be sufficient for being a moral agent, which requires agency that is socially directed. And, the objection might go, to be a social agent, one has to predict behavior, and one needs to mindread to be a good behavior predictor.

I hinted at my answer to this objection earlier. While I think it is true that on most theories in normative ethics a moral agent has to be able to predict behavior, I don't think mindreading is necessary for predicting the majority of human behaviors – and in fact, when we do mindread we are much less likely to make accurate predictions. I have argued that some of the presuppositions about human cognition that have been made in the ape theory of mind research are unwarranted (Andrews, 2005, 2012). The assumption that humans attribute beliefs and desires in order to predict simple behaviors has had a huge influence on subsequent research, though it is by no means clear that humans do need to attribute beliefs and desires to predict behavior. I have argued that even in false-belief cases and in deception cases, individuals can predict behavior using a number of different heuristics that social psychologists have shown are present in humans, including generalizations over past behavior, social roles, personality traits, generalization from self, and stereotypes. Given that predicting behavior is usually seen as being an important part of being a social agent, the ability to predict behavior without attributing belief allows us to set aside this objection to seeing apes as intentional agents.

Conclusion

Whether or not apes or any other non-human species have an understanding of others' beliefs is an open question, but it is one that need not be answered before answering the question of whether other species have autonomy, social norms, pedagogy, or cumulative culture. We already know that other species are able to discriminate intentional from non-intentional agents, and there is evidence that apes may actively seek to self-improve or to create themselves. Such striving may sound rather high and mighty, but it need not be. Choosing a mate, engaging in a dominance battle, increasing hunting territory (as described by Watts *et al.*, 2006), and exposing one's offspring to the technologies and practices of the community may all be seen as part of self-creation. When humans engage in such things, we often do think of them in rich or evocative terms such as *building a family*, *improving one's lot in life*, *seizing power*, *mothering*, and so forth. But the descriptions are less important than the acts themselves, and those we share with other species. If we can come to see other animals as agents who engage in self-creation, and who live according to community-wide social norms, we will be less likely to exclude them from the moral and legal standing we grant human agents, and this change in attitude can help with the move from speciesism toward respectful co-living. Apes that express grief when they lose their loved ones (see King, this volume), and ants that share food when their resource stocks are high (see Kirksey, this volume) are showing behaviors that can be investigated as non-human social norms. The status of agent doesn't require sophisticated cognitive capacities such as mindreading or language capacities, but rather suggests a more robust role for social relationships in our understanding of agency and social norms (see Gruen and Hutto, both in this volume).

I've suggested that a society in which mindreading might develop is a society that already has social norms. We need not worry about moral norms existing without mindreading, given an account of agency that consists of (a) recognizing intentional agency, and (b) active self-improvement. However, evidence that apes mindread will also serve as evidence that they have moral norms, on the account I offer. Thus not only the evidence reviewed above but also evidence that apes understand others' perceptual states, informational states, and goals would serve to defend the claim that they are agents with social norms. Even understood as an ongoing research program, the current body of evidence should make us think more than twice about continuing practices that treat other apes as objects, rather than as subjects with rich individual and social lives. Using apes as actors, as subjects in medical tests, or otherwise as objects for human enjoyment does not respect their probable nature as agents with social norms. As scientists continue examining ape behavior, the teachers, judges, and builders of our society should seek to protect other apes like the orphan Cecep from our overwhelming interest in them and their natural environment, by taking the default position that non-human apes, along with many other species (including dolphins, as per Marino, this volume), are autonomous agents with their own projects that need to be respected. This suggests to me that we should grant these other species moral and legal standing, thereby demonstrating what humans can do with our ability to consider our own reasons for action.

10 Intersubjective engagements without theory of mind

A cross-species comparison

Daniel Hutto

Although I feel for the plight of the great apes and other animals, my interest in the topic of this chapter is an outcome of curiosity-driven research that began with a quite different focus. My work on the nature of basic cognition and social cognition has its historical roots in questions and puzzles about beliefs, and related phenomena, such as whether involuntary believing is possible, and how self-deception is possible. Puzzles of this kind were the topic of my MPhil thesis. I was happily working on such questions in St Andrews, UK when I happened to attend a talk by Professor George Graham, a visiting fellow from the USA. He brought news of a new trend in academic thinking from across the water that questioned the very existence of beliefs – and threatened the very future of our everyday ways of making sense of another. Suddenly, I felt pressured to move from working on puzzles about belief to defending the very existence of beliefs. From that moment on, my research agenda was set. It inspired empirically informed philosophical investigations into the nature of animal minds (including ours) and how they interact with one another. This chapter is one of the fruits of those investigations.

In naturalistic settings, great apes exhibit impressive social intelligence. Despite this, experimental findings are equivocal about the extent to which they are aware of other minds. At the high level, there is only negative evidence that chimpanzees and orangutans understand the concept of belief, even when simplified non-verbal versions of the "location change" false-belief test are used (Call and Tomasello, 1999). More remarkably, even the evidence that they are aware of simpler mental states – such as seeing – is equivocal and "decidedly mixed" (Call and Tomasello, 2008: p. 61). Putting all of this together suggests that simian capacities for engaging with other minds have a particular, signature profile. In addition, there exists a range of proposals about what lies behind their particular form of social intelligence. Within the cognitivist camp – which assumes that social cognition essentially involves the use of mental representations of the mental states of others – these range from positing a naïve, weak, or minimal theory of mind (Tomasello *et al.*, 2003; Apperly and Butterfill, 2009; Bogdan, 2009); perceptual mindreading (Bermúdez, 2009); an early mindreading system (Nichols and Stich, 2003); or a theory of behavior (Povinelli and Vonk, 2004).

The Politics of Species: Reshaping our Relationships with Other Animals, eds R. Corbey and A. Lanjouw. Published by Cambridge University Press. © Cambridge University Press 2013.

Respecting the experimental evidence, this chapter motivates a different explanatory possibility. It defends the view that the basic ways that apes and humans (both children and adults) engage with other minds is emotionally charged, enactive, and non-representational. If true, this would be important for thinking about the moral status of animals capable of such mind minding. This is not only because it would identify an important cross-species commonality, but also especially because it would promote the importance of non-cognitive aspects of such engagements. For example, it is at odds with the idea that having a "theory of mind" is required for being phenomenally conscious (Carruthers, 2000).

Two types of human intersubjective engagement

Normally developing humans can relate to others by attending to mental states in basic and more sophisticated ways. Put another way, they have more than one way of minding minds. Very often, in line with the demands of particular encounters, we are required to bring these different capacities into play, operating at more than one level at the same time. This requires bringing an array of cognitive (broadly conceived), emotional, and behavioral resources to bear flexibly and in concert.

Sometimes our intersubjective dealings call on distinctively high-level, cognitively demanding and conceptually based capacities. These are needed, for example, when making sense of others by means of attributing sophisticated psychological attitudes. For example, to understand why this or that individual might accept to speak at a closed roundtable in New York City requires appreciating the particularities of their circumstances. This will include, among other things, getting a handle on their unique mindset – the constellation of psychological attitudes – hopes, beliefs, desires, and so on – that enable us to make sense of the choice. Doing so necessarily calls on one's (a) explicit mastery of a wide range of psychological attitude concepts; (b) ability to represent the contents of such attitudes; and (3) grasp of how such attitudes interrelate.

Arguably making sense of ourselves and others in this sort of sophisticated and articulate way is a uniquely human capacity, one that emerges late in ontogeny. Notably, even in the industrialized Western populations, where such abilities are most reliably and robustly evident in the form described, they aren't securely in place in children's performance until after the age of five or six. Emerging evidence suggests that:

Five- and six-year-old children (who are old enough to pass false-belief tasks) still have problems understanding: how beliefs are acquired (Carpendale and Chandler, 1996; Apperly and Robinson, 2001), how beliefs interact with desires (Leslie and Polizzi, 1998; Leslie et al., 2005), and the emotional consequences of false beliefs (e.g., Harris et al., 1989; Ruffman and Keenan, 1996).

(Apperly and Butterfill, 2009: p. 957)

There are various proposals afoot about what lies at the basis of these capacities. I defend the view that the normal route for acquiring our folk psychological capacities may well depend upon engaging in special kinds of linguistic, specifically, narrative practices (Hutto, 2008). The narrative practice hypothesis (NPH) conjectures that children acquire the component capacities for full folk psychological competence slowly, over time in partial and piecemeal fashion. Moreover, it holds that they only manage to do this by engaging in shared, discursive practices. In line with this, only having some capacities for explicit mental state attribution – including the attribution of beliefs – where these capacities are not fully integrated turns out to be necessary but not sufficient for being fully folk psychologically competent.

Apart from the sorts of developmental evidence gestured at above, why should we believe this sort of proposal? On the empirical side, there is a wealth of evidence that strongly links language capacities with so-called "theory of mind" abilities (de Villers, 2007). On the philosophical side, there are arguments for thinking that the capacity to (a) "represent" someone "as" possessing this or that propositional attitude, and to (b) attribute requisite contents to such attitudes is decidedly dependent on linguistically based capacities.

While not obviously or indisputably true, I believe in the final analysis we have every reason to believe that folk psychological or "theory of mind" capacities – when understood in the demanding way described above – are linguistically based capacities. But it is equally true that we do not call on such high-level capacities for the great bulk of our intersubjective engagements. More often than not we relate to and engage with others in face-to-face encounters in entirely more basic, visceral, and low-level ways. For example, when one responds immediately to another's expressed emotion it is possible to pick up on and react to the other's psychological situation without ascribing any mental states. Such responding involves being aware of another's expressed attitudes, emotions, and moods and, if all goes well, in making timely, affectively appropriate, and well-managed responses to such attitudes. Doing this successfully is partly a matter of picking up on subtle cues as well as controlling and regulating one's own emotions and expressions, keeping them within an acceptable range and register.

That humans operate on more than one level in their intersubjective engagements is backed by empirical findings from several disciplines. There is strong evidential support for the claim that adult humans do in fact have more than one way of minding minds. There is compelling evidence for the existence of at least two, functionally isolated and anatomically quite separate, networks – one is implicated in our more basic intersubjective dealings and the second in our more sophisticated ways of making sense of other minds (Gallese, 2007, 2010; Saxe, 2009a,b; Fonagy and Luyten, 2010). These distinct modes of mind minding are called into service for and enable very different kinds of interpersonal relating.

Preliminary findings from cognitive psychology suggest that low-level capacities for responding to other minds are apparently enlisted during fast and fluid non-verbal intersubjective engagements. Such capacities are in place early and appear to be preserved throughout development. Adults continue to call upon them in a default manner even after more sophisticated capacities for mental attribution and understanding others are fully in place (for a review see Apperly and Butterfill, 2009). The existence of these parallel but distinct systems is supported by analyses of interference patterns that are evident in experiments that tax subjects' capacities for making mental state attributions while they are simultaneously engaged in tasks that require the low-level monitoring of others' perspectives.

Low-level mind minding is assumed to be relatively automatic; it seems that a degree of inflexibility is the price of a speedy response. Hence our more basic ways of minding other minds has signature limitations. In contrast, reaction time data suggest that actively and explicitly monitoring and attributing contentful mental states, such as false beliefs, in a controlled manner calls on quite different cognitive resources. Moreover adults only do this sparingly and selectively when it is demanded by, and specifically relevant to, the interpretative task in hand (Saxe, 2009b).

Against the idea that such low-level capacities are nothing but high-level capacities that have become automatic through practice, it appears that they are part of a phylogenetically quite ancient system for dealing with others. Fonagy and Luyten (2010) identify the brain regions associated with it as including the amygdala, basal ganglia, ventromedial prefrontal cortex (VMPFC), lateral temporal cortex (LTC), and the dorsal anterior cingulated cortex (dACC).

An exciting possibility, recently much touted, is that at least some low-level mind minding in humans might involve mirroring mechanisms. In such systems sets of cortical neurons tied to specific types of goal-related actions fire when actions of that type are perceived (visually or audibly) and also when observers execute actions of that type. Such systems are akin in functionality to those discovered in the premotor cortices of macaque monkeys (Gallese *et al.*, 1996, Gallese and Goldman, 1998). It seems that humans have several different neural systems that function in this way. Some, located in the premotor and posterior partial cortices, are implicated in the imitation of simple movements, imitative learning, and the detection of action intentions. Others involve the activation of brain regions associated with specific emotions, such as disgust, and sensations, such as pain, which are set off by observing expressions of these emotions and sensations in others (see Gallese, 2010 for discussion).

Consistent with the two-systems hypothesis, mirroring activity in humans is regarded as "immediate, automatic, and almost reflexlike" (Gallese, 2005: p. 101). Some hold that mirroring activity is sufficient to enable a special variety of non-conceptual and prelinguistic form of understanding of actions and intentions (Rizzolatti and Sinigaglia, 2006, 2010; Sinigaglia, 2009). Although the claim that mirror systems yield anything that might be properly called an "understanding" of minds is disputed, it is agreed that whatever precise role they play in interpersonal relating, mirror systems come before and below, and remain quite distinct from high-level folk psychological abilities of the sort that involve the ascription and interpretation of mental states that have content – such as beliefs that represent how things stand with the world – in systematic ways.

A different set of brain regions is implicated in the reflexive, controlled, and cognitively demanding tasks of explicit mental state attribution. For example, successfully attributing beliefs that diverge from one's own requires simultaneously inhibiting one's own current view of how things stand with the world. This entails keeping track of the inferential connections that hold between the attributed attitudes, professed or inferred, and checking these for coherence. For this reason, sophisticated attribution and interpretation is slower than low-level forms of engaging with another's attitudes and emotions. Apart from calling on specialized capacities for representing and understanding mental states it also involves deploying domain general, cognitively costly resources for inhibitory control, working memory, and language.

The high-level network for making sense of others in ways that involve the attribution of contentful mental states is thought to be formed by an alliance of dissociable brain regions including the right and left temporo-parietal junction (TPJ), medial parietal cortex (including posterior cingulate and precuneus), and medial prefrontal cortex (MPFC). These areas exhibit significantly greater hemodynamic activation, for example, when subjects read about beliefs as opposed to other purely non-mentalist topics (Saxe, 2009b).

Functional magnetic resonance imaging (fMRI) experiments reveal that the right temporo-parietal junction (RTPJ), in particular, is selectively enlisted for tasks requiring the interpretatively complex attribution of mental states. Saxe and Wexler (2005) discovered that RTPJ activity is enhanced when the professed beliefs or desires of story protagonists conflicted with subjects' expectations about what such characters ought to believe or desire, based on background knowledge about them. Moreover, this region is not similarly recruited for other tasks that involve assessing other, more general, socially relevant facts about persons. Saxe and Wexler also reported that none of the other brain regions in the wider network for controlled mental state attribution – i.e., the left temporo-parietal junction (LTPJ), posterior cingulate (PC) and MPFC – exhibited equally selective activity. Summarizing, Saxe (2009b) reports that the "fMRI literature suggests a division in the neural system involved in making social judgments about others, with one component (the RTPJ) specifically recruited for the attribution of mental states, while a second component (the MPFC) is involved more generally in the consideration of the other person" (p. 405).

In line with the Gogtay *et al.* (2004) neurodevelopmental data about the stages of cortical maturation of gray matter, Saxe *et al.* (2009) found "that selectivity for thinking about people's thoughts emerges in the RTPJ between ages 6 and 11 years" (p. 1206). In attempting to understand what might drive this process they consider the "intriguing recent hint that middle childhood is a critical time for interactions between language and theory of mind" (p. 1207). The authors recognize that these new findings present a challenge "for theories of cognitive development that emphasize an innate and early-maturing domain-specific module for theory of mind" (p. 1207).

Not only are the two ways of minding minds functionally and neurally separate, they do not appear to directly communicate. Saxe (2009a) observes "there is substantial evidence for co-opted mechanisms, leading from one individual's mental state to a matching state in an observer, but there is no evidence that the output of these co-opted mechanisms serve as the basis for mental state attributions. There is also substantial evidence for attribution mechanisms that serves as the basis for mental state attributions, but there is no evidence that these mechanisms receive their input from co-opted mechanisms" (p. 447). This raises important questions about the extent to which and precisely how the different mind minding systems interact.

Figure 10.1 Emotional engagement and meeting the gaze of the other without theory of mind. Photo courtesy of Jamie Lantzy.

Do apes mind minds? And, if so, how?

Premack and Woodruff's (1978) landmark paper, which asked if the chimpanzee has a theory of mind, launched a thousand ships. Since its appearance there has been enormous interest in the question of whether our closest living relatives, the great apes, are mentalizers or mindreaders. However, framing the question with reference to these labels has unfortunate consequences. When focusing on high-end capacities, the question is understood as asking to what extent apes are like us in having folk psychological abilities, as strictly understood. This is to ask to what extent they are capable of explicitly representing and attributing a wide range of sophisticated mental states and their contents in complex, conceptually based, and articulate ways. Those who reject the view that this is necessarily a language-dependent capacity will suppose that this is at least possible since contentful mental representations might play the roles earmarked for the sentences of natural language described in the previous section. Either way, without refinement, to ask if apes operate with a theory of mind, is to ask if they are somehow capable of a very high-level capacity that is only fully and reliably evident in the performance of older human children.

Yet, as the considerations above reveal, our capacities for high-level attribution and interpretation of psychological attitudes are only a small part of the story about how we conduct ourselves in our intersubjective encounters. This suggests that making comparisons between humans and other non-human social animals at the high level is a mistake. It is much more plausible that both humans and apes are alike in being able to attend and respond appropriately to the mental states of others – such as another's anger or fear – in ways that do not involve making any mental state attributions at any level at all. This is to assume it is possible, for example, to respond fearfully to another's anger without invoking or even having a concept of fear. All that this requires is the exercise of low-level capacities for non-conceptual responding and attending to mental states of different kinds.

It is worth reviewing the data on great ape mind minding capacities in the light of this distinction. It is widely assumed that being able to ascribe false beliefs in an explicit way is the best and most secure evidence for thinking that someone positively has an explicit if minimal mastery of the concept of belief. And, since explicit mastery of that concept is also assumed to be a pre-requisite for being in the folk psychological club – it is one necessary but not sufficient condition – false-belief tests are often used as the litmus test for determining if one has the right credentials for joining that club.

Yet, as is well known, the great apes perform dismally even on specially modified non-verbal versions of standard false-belief tests (Call and Tomasello, 1999). Defenders of the idea that apes might nevertheless have a concept of belief are quick to point out that it is not safe to infer that apes necessarily lack that concept based on negative performances on these tests alone. Other factors might explain these poor results. Certainly it would be a mistake to infer a complete lack of mind minding capacity – i.e., a blanket mind blindness – from these findings.

Yet, even taking the full set of existing evidence into account, there is uncertainty about whether apes are mind minders and exactly what their abilities are in this regard. Call and Tomasello (2008) highlight this in revisiting Premack and Woodruff's big question, 30 years later. In replying to it, they tells us that:

recent evidence... suggests in many respects they do, whereas in other respects they might not. Specifically, there is solid evidence from several different experimental paradigms that chimpanzees *understand* the goals and intentions of others, as well as the perception and knowledge of others. Nevertheless, despite several seemingly valid attempts, there is currently no evidence that chimpanzees understand false beliefs. Our conclusion for the moment is, thus, that chimpanzees understand others in terms of a perception–goal psychology, as opposed to a full-fledged, human-like belief–desire psychology   (Call and Tomasello, 2008: p. 187; emphasis added)

Summarizing the performance data from controlled experiments and more ecologically valid studies the general consensus is that however apes conduct their intersubjective engagements they do so using means that decidedly do not equate to the use of full-fledged folk psychology (Suddendorf and Whiten, 2003).

Despite acknowledging this most theorists are nevertheless inclined to talk of their having some "understanding" of some mental states, as Call and Tomasello do in the quotation above. This goes hand in hand with the idea that if apes are not using all of the folk psychological concepts then they must be using some of them or low-level variants.

I return shortly to the question of whether it is appropriate to talk of apes "having an understanding of mental states," but before doing so it is necessary to deal with an otherwise potentially confounding issue.

Why think apes mind minds *per se* and not just behavior? Like most others, Call and Tomasello (2008) reject the idea that apes get by in their social engagements by using a set of behavioral rules and representations. They hold that given the variety of disparate situations with which apes deal this strategy would be unnecessarily cumbersome given that there are common psychological factors at play in all such cases – i.e., another's goals, perceptions, and intentions. Hence they take it to be more parsimonious, more elegant, to suppose that apes have some understanding of these psychological states.[1]

Even more positively, Krachun *et al.* (2009) recently devised a competitive version of a non-verbal false-belief task precisely because apes generally perform poorly in cooperative contexts. In at least one version of the false-belief trials apes looked more often at an unchosen container. The experimenters interpreted this as possibly indicating some degree of implicit understanding of belief. If true then ape mind minding capacities look to be on a par with those of very young human infants. A range of experiments using violation of expectation and anticipatory looking paradigms show that children as young as 25 months (Southgate *et al.*, 2007), 15 months (Onishi and Baillargeon, 2005), and even 13 months (Surian *et al.*, 2007) can pass language-free versions of false-belief tasks. This is the strongest evidence that younger children must have command of a concept of belief in place much earlier in development than their capacity to pass standard, verbally based false-belief tests suggests. And, crucially, this apparently shows that infantile command of the concept belief is not dependent on mastery of the syntax, semantics, or pragmatics of language.

What do these experiments involve? Infants are familiarized with an actor putting a toy in a green box, as opposed to a yellow box. They are also familiarized with an actor searching for the toy in the green box but without his or her actually revealing it. In the experimental trials the actor is then caused to hold either a true belief (TB) or a false belief (FB) about the location of the toy. In TB trials infants watch as the actor witnesses the object being moved from the green box to the yellow box. In FB trials, while the actor is absent from the scene, infants watch as the object is moved from the green box (where the actor had placed it) to the yellow box. The striking result is that infants expect the actor to search according to his or her belief, whether true or false, and looked reliably longer when he or she did not do so.

Now, *if* the best explanation of this evidence is that these infants are able to attribute beliefs, and *if* this implies having command of the concept of belief (where that implies that at least conditions 1 and 2 are met) *then* we have strong evidence of pre-linguistic mindreading. Indeed some think the mindreading hypothesis is not only the best explanation but that there is no conceivable alternative: that there is "simply no other way of explaining our competence in this domain" (Carruthers, 2009: p. 167).

---

[1] The key premise in the argument is this: if apes are mentalists and not behaviorists then they operate with a smaller and more manageable set of rules. But see Perner (2010) for an objection to this idea.

But the mindreading explanation of the non-verbal false-belief experiments is only attractive if we assume such things as that: "25-month-olds do rely on the content of another individual's belief" (Southgate *et al.*, 2007: p. 591); or that "the child must imagine a thought bubble in [the experimenter's] head that has the actual cognitive content driving his behavior" (Buttelmann *et al.*, 2009: p. 341).

If infants are pre-verbally representing the "actual" content of another's mental state then their elementary mind minding involves more than attending to the other's directedness at the worldly items and anticipating what might be done with them. Note, however, that representing the actual cognitive content of another's belief would require representing finely individuated intensional (with an 's') content. It requires representing the mode of presentation under which the other is thinking about the state of affairs in question.

Those who subscribe to the truth of the representational theory of mind (RTM) may be encouraged to think that infants could and would need to do this because they assume that any intelligent engagement with worldly offerings requires acting on contentful beliefs that might be true or false. Many cognitivists model the content of even the most basic mental states directly on the (apparent) semantic properties of language. Despite providing accounts of (mental) representational content that get their sense by making direct comparisons with the properties of linguistic utterances, all of these theorists reject that such representational content is linguistic. This is the classical rendering of the representational theory of mind. An advantage of operating with this linguistic model is that it is quite clear which sorts of properties are picked out by talk of representational contents of mental states.[2]

The fly in the ointment is that there is, as yet, no credible naturalized theory of content that explains how it is possible that mental representations could have semantic properties of just the same sort (apparently) possessed by linguistic representations. So far all such theories have consistently failed to deliver the promised goods. Moreover, we live in interesting times, it is now widely questioned whether representational theories of mind are the right way to think about the nature of basic mentality. As Ramsey (2007) notes "something very interesting is taking place in cognitive science... cognitive science has taken a dramatic anti-representational turn" (p. xiv–xv).

It is beyond the scope of this chapter to spell out and motivate a non-representational approach to basic mentality (but see Hutto and Myin, 2013). But it is worth bearing this possibility in mind when reviewing the credibility of the mindreading characterization of what infants are doing in these tasks, given that it is not evidently supported by the empirical data.

In terms of content... no study has yet suggested that infants track beliefs involving both the features and location of an object (e.g., "The red ball is in the cupboard"); or that they track beliefs whose contents can be represented only using quantifiers (e.g., "There is no red ball in the cupboard"); or that, in tracking beliefs, they are sensitive to modes of presentation. (Apperly and Butterfill, 2009: p. 957)

---

[2] This is true not only of those who believe in the language of thought, such as Fodor (1975, 2008) – it is equally true of anyone who thinks that mental representations, whatever their vehicular format, have properly semantic content (see, e.g., Searle, 1983; Millikan, 2005).

Based on these observations Apperly and Butterfill (2009) conclude that "whatever [infants] represent, it is not a state with propositional content" (p. 957). With this in mind, they propose that younger infants are really only able to attend to belief-like mental states. It is their capacity to do so that sponsors the illusion that they have the ability to represent genuinely propositional attitudes, such as beliefs, because "reasoning about [belief-like states] would enable someone to track beliefs, true and false, in a limited range of situations. At the same time, reasoning about [belief-like states] imposes signature limits" (Apperly and Butterfill, 2009: p. 963).

The basic idea is that by tracking lesser states of mind infants can, incidentally, have expectations about more complex propositional attitudes, including contentful beliefs. Tracking belief-like states that happen to correlate with beliefs would be a crude, but reliable enough, way of developing expectations about another's propositional attitudes without being able to represent beliefs *as such* or being able to represent their contents.[3] It is surely possible to explain how young children can anticipate what another might do, even where this involves true or false beliefs, without their having the concept of belief or any kind of mindreading abilities, strictly conceived (for a fuller review of the interpretative and explanatory possibilities see Hutto *et al.*, 2011).

Where does this leave us? In short, it raises the possibility that apes or other species (such as dolphins – see Marino, this volume) might mind minds even if they lack even weak theory of mind abilities; even if they are not mindreaders of any kind at all.[4] Certainly this proposal about the nature of their abilities is consistent with the available evidence about the level of their abilities.

Yet if when engaged in basic-level mind minding apes and humans are not employing mental state concepts then there is reason to question that their engaging successfully with others constitutes any kind of understanding on their part (see e.g., Andrews, this volume). For it is one thing to say that a creature has a low level of understanding when it is operating with a minimal grasp of a concept, but quite another to say this when it operates with no concept at all – that is, it literally has "no idea" about what it is dealing with.

In considering this issue it is therefore crucially important to note that: "The best evidence that an observer understands goals... is when she *reacts specifically to the*

---

[3] It might be objected that children must be able to attend to more than evident relations between individuals, the objects of their concern, and the locations of these objects since even registrations must be "goal directed" in some minimal sense. This is true. Attending to another's registration requires attending to the foci of the other's psychological attitudes on which their possible actions concentrate. However, in a paper that adds further detail to their account, Butterfill and Apperly stress that "representing goals in this crude sense of 'goal' does not require representing representations because such goals are non-representational" (Butterfill and Apperly, unpublished).

[4] Capacities for mind minding may be quite widespread in the animal kingdom, with different species exhibiting them to different degrees and operating with quite different styles. I have focused solely on the capacities of apes in this chapter since they are among the most studied. Also, given that apes enjoy the status of being the closest living relatives to humans, there is great controversy and special interest in knowing exactly what they can do when engaging with others and how they achieve it. However, if it can be established that such intersubjective abilities are best characterized in terms of mind minding and not mind reading, then this conclusion will generalize to other species that exhibit social cognitive abilities as well.

actor's goal and not to his overt behavior when he is trying unsuccessfully or having an accident" (Call and Tomasello, 2008: p. 188). Andrews speaks not of reactions but of appropriate responses.

But, as above, quite generally, being able to attend to, track, and reliably react and respond to some X does not imply one has the capacity to represent Xs *as such*; it does not entail that one has conceptual understanding of the things with which one reliably deals. Put simply, low-level mind minding capacities need not be conceptually based (indeed depending on the outcome of certain live debates about the foundations of cognitive science such capacities might not involve representations of any kind at all, conceptual or otherwise). A compelling reason to think elementary mind minding capacities are not conceptual is that there doesn't appear to be any need to make mental state attributions in order to attend to and keep track of mental states at this level.

Perhaps it might be thought that not all understanding needs to be conceptually based, or that there is a reason to adopt a very weak, purely ability-based concept of concept. But this is to take the first step that leads to a fast slide down a slippery slope. As a defender of anti-intellectualist, anti-representationalist enactivism it may seem strange that I should raise this concern. Radical enactivists hold that systematic, sustained interactions with an environment suffice for the most basic kinds of cognition. Accordingly, *not all* cognition requires individuals to construct representations of their worlds. This offers an antidote to those approaches to mind that "take representation as their central notion" (Varela *et al.*, 1991: p. 172).

Enactivists are known for offering a framework for thinking about basic mentality that promises to close the imagined gap or at least blur the lines between cognition proper, on the one hand, and mere bodily and affective activity, on the other. This is achieved by adopting a quite liberal understanding of the nature of cognition. For example, Thompson proposes that "cognitive interactions are those in which sensory responses guide action and actions have consequences for subsequent sensory stimulation, subject to the constraint that the system maintains its viability. 'Sensory response' and 'action' are taken broadly to include, for example, a bacterium's ability to sense the concentration of sucrose in its immediate environment and to move itself accordingly" (Thompson, 2007: p. 125).

I agree that this is the correct thing to say about basic forms of cognition – including those of amoeba and ants (see, e.g., Kirksey, this volume). Nevertheless we should be wary of talk of "interpretation," "sense-making," "understanding," and even "emoting" in describing much of the responding of simple living systems. Such talk is misplaced and misleading. To some extent enactivists admit this. Indeed, on this score, when they giveth they then quickly taketh away. Thus Varela *et al.* (1991) note that they use the terms "significance" and "relevance" advisedly in speaking of creatures enactively bringing forth their worlds. Consequently, even though these authors assume that some kind of interpretation takes place in such cases, they confess that "this interpretation is a far cry from the kinds of interpretation that depend on experience" (p. 156). I am inclined to take a harder line: this is no kind of interpretation. The simplest life forms are capable of basic cognition that takes the form of intentionally directed responding, whether this is of the fairly fixed or more flexible dispositional variety. But basic cognition of this sort does not equate to contentful forms of cognition, interpretation, understanding, or sense-making. Rather it provides a necessary platform for these forms of mentality, when suitably augmented and scaffolded.

Still, there seems no reason to rule out that the activities of even the simplest living beings constitute a basic kind of intentional directedness – intentional directedness of a sort that is shared with more sophisticated forms of life that possess more elaborate states of mind. The task is to explain how such basic states of mind become possible when other proper supports are in place in the form of the right kinds of interactive engagements and practices.

In this light a more modest claim becomes attractive – it is that basic interest-driven ways of responding provide the right platform for understanding how certain kinds of intersubjective engagement can involve intentionally directed attitudes toward the intentionally directed attitudes of others in ways that are essentially emotionally charged. Certain organisms are not only set up so that they are intentionally directed at the attitudes of others, their ways of responding constitute phenomenally charged ways of responding that are properly emotional, despite being wholly non-representational.

To respond emotionally is to respond in ways that have a distinctive phenomenal character. But it also assumes that to respond in such ways is identical with, and to be understood in terms of, concrete patterns of environmentally situated organismic activity; nothing more, nothing less. Phenomenally charged experiencing simply equates to the way in which certain creatures are disposed to respond to a range of worldly offerings – this includes, for social creatures, what is afforded by attending to the psychological attitudes of others. That is an important difference between the humble amoeba and the great apes, and also us.

Moral agents? Conscious beings?

Where does this leave us with respect to Andrews' proposal (this volume) that the great apes, despite lacking folk psychological abilities proper, may nevertheless be capable of moral agency? Her tactic is to raise doubts about the kind of agency that is typically assumed to be required for *bona fide* moral agency.

Certainly, we do much more than we choose to do. Only a very little of what we do is freely chosen or autonomous. This surely includes most of our attitudes and ways of responding toward others, including those that are of fundamental moral significance. We form relationships with and make commitments to others. We engage others in sharing ways that affect and shape both them and us. This need not involve explicit representation of, or reflection on, first-order psychological attitudes *as such*.

Of course, acknowledging this is wholly consistent with insisting that genuine moral agency and thinking requires more still – specifically, many hold that it requires being able to reflect on and assess someone's reasons for acting. Velleman (2000) is an appropriate spokesperson for this sort of view:

Purposeful activity is motivated by desire and belief, but it may or may not be regulated by the subject's grasp of what he is doing. Autonomous action is activity regulated by that reflective understanding which constitutes the agent's rationale, or reason – the reason for which the action is performed.   (Velleman, 2000: pp. 29–30)

Accordingly, being able to make sense of reasons for acting is necessary for deliberative reflection and evaluation about which reasons are good reasons for acting. If evaluating which reasons are good ones is required for moral evaluation generally, then it follows that having a capacity to represent and reflect on reasons is required for moral evaluation. As Velleman (2000) unashamedly admits "It's an intellectualist view, all right. But we are intellectual creatures, and our autonomy may well be a function of our intellect" (p. 31).

I am sympathetic to Andrews' suggestions and there are good reasons for thinking that the basic intersubjective engagements of apes and ourselves are potentially morally significant. However, more would need to be done to convince skeptics that, if they truly lack a capacity for understanding and evaluating reasons, apes qualify as moral agents worthy of praise and blame, of moral commendation and condemnation, in their own right. But even if they should fail to qualify as moral agents in this sense this would not threaten their status as morally significant beings, i.e., as moral patients.

Hence, whatever the outcome of this debate, those concerned with our attitudes toward and treatment of non-human animals should be much more concerned with the claim that having high-level "theory of mind" capacities is required for being phenomenally conscious. For if apes and other non-human animals lack phenomenal consciousness this would potentially threaten their status as morally significant beings – or at least be an undisputable reason for thinking them to be such. Although it is suggested that there may be grounds for special pleading in certain cases, the bottom line is that to think great apes and other non-human animals genuinely lack phenomenal consciousness is surely to remove the most compelling reason for treating them with the care and concern that any genuinely feeling creatures – those that are loci of experience – deserve.

Higher order theories of consciousness maintain that being phenomenally conscious requires attending to, or noticing one's own, mental states in ways that necessarily involve making reference to them in higher order acts of perception or thought. Accordingly, phenomenal consciousness requires higher order perceptions or thoughts (Lycan, 1996; Carruthers, 2000; Rosenthal, 2005). If such higher order operations are, in fact, partly constitutive of phenomenal consciousness then the neural basis of experience must include machinery for "inner sensing" or for making "theory of mind" ascriptions. These are the standard proposals for making sense of these meta-representational abilities (see Carruthers, 2006 for an overview).

For those who doubt that such mechanisms form part of our basic biological equipment, Dennett's account has an advantage. He regards such capacities to be the result of cultural design, a kind of Joycean software. He tells us that consciousness is, "largely a product of cultural evolution that gets imparted to brains in early training" (Dennett, 1991: p. 219). But critics regard this as an admission that non-verbals – such as animals and younger infants – are incapable of having phenomenal experiences. Dennett's response to this worry is that our folksy intuitions regarding animal and infant consciousness are not sacrosanct. Jettisoning some of our most deeply held intuitions concerning the nature of experience may be a price for a neater criterion of consciousness.

With direct reference to non-human apes, Carruthers (2005) – a foremost defender of higher order thought theories – tells us that:

> If possession of a conscious mental state M requires a creature to conceptualize (and entertain a [higher order] thought about) M as M, then probably very few creatures besides human beings will count as having conscious states... There is intense debate about whether even chimpanzees have a conception of perceptual states as such... in which case it seems very unlikely that non-apes will have one. So the upshot might be that phenomenal consciousness is restricted to apes, if not exclusively to human beings. (Carruthers, 2005: p. 49)

He goes on to add, "I conclude that higher order representation theories will entail (when supplemented by plausible empirical claims about the representational powers of non-human animals) that very few animals besides ourselves are subject to phenomenally conscious mental states" (Carruthers, 2005: p. 51). While many find this conclusion incredible, Carruthers, like Dennett, happily bites the bullet, questioning the legitimacy of deeply rooted intuitions that lead us to think the opposite verdict is beyond question.

The root problem for these sorts of proposal is that they allegedly place too much emphasis on sophisticated add-on extras or supplements and confuse these with necessary conditions for being conscious. Being able to reflect on one's own mental states may be plausibly required for having certain kinds of conscious experience but should not be confused with the essential ingredients of a more basic, biologically grounded capacity for phenomenal experiencing itself.

Both strong and weak versions of representationalism about consciousness face a number of serious objections (for details see Hutto, 2009). Arguably, a major problem with all such accounts is that they attempt to understand basic perceptual activity by illicitly importing features that in fact necessarily depend on being a participant in sophisticated, linguistically based practices – e.g., having mental states with the kind of semantic content that requires assessment by appeal to public norms and concepts, as in the attribution of "blueness" to aspects of the environment. If so, in imagining basic experiences to have more properties than is necessary or possible for them to have, such accounts make the opposite mistake to Dennett and higher order thought theorists.

Plausibly, having a capacity for phenomenal experiencing is more rudimentary and fundamental than the capacity to represent the world as being a certain truth-evaluable way. Consequently, experiencing aspects of the world might be thoroughly non-contentful (and not just non-conceptual). Experiencing might not be intrinsically content involving even though there is something-it-is-like to experience worldly offerings in phenomenologically salient ways. This non-representationalist view of experience features as the central plank of a radically enactivist approach to phenomenality; one that seeks to understand phenomenal experience by focusing on the ways in which creatures actively sense, perceive, and engage with their environments (see Hutto and Myin, 2013). Enactivists propose that the core features of experiential properties are best explained by appeal to specific patterns of sensorimotor activity, through which complex self-organising systems interact with aspects of their environment. If true, this non-intellectualist account would secure the status of non-human animals as properly sentient beings. On this view, even if non-human animals lack theory of mind capacities, they are still beings clearly deserving of moral care and concern.

11 "Unnatural behavior"

Obstacle or insight at the species interface?

Lucy Birkett and William McGrew

One of us (WM) first met chimpanzees in 1972, and in that calendar year spent time with them housed individually in a biomedical (virology) facility, living in a semi-free-ranging group in a 1.5-acre outdoor enclosure, and in the wild in western Tanzania. The marked contrast in settings and the apes' consequent behavior made a lasting impression. For the last 40 years, William (Bill) has mostly carried out field studies of chimpanzees in nature, but he has never lost sight of the plight of the captives. Over that period he has supervised students doing projects on chimpanzees in zoos and other confined settings. No matter how good the setting, it always paled by comparison with the natural context (or rather, contexts, as nature is not uniform either!).

At various times, Bill sought to influence captive facilities either unofficially or officially, hoping to improve the lot of the chimpanzees. He served on the Board of Directors of Chimp Haven, Inc. (Louisiana) for five years, accordingly. Perhaps some of these efforts bore positive fruit, but sadly his overall conclusion is negative. If a single event was needed to nail that down, it was seeing the first author's (LB's) carefully collected dataset, which has both breadth and depth. It showed that captive chimpanzees are inevitably behaviorally disordered, and so they suffer accordingly.

As an anthropologist specializing in primatology, and in particular on the basis of the research discussed in this chapter, Lucy, just like Bill, is convinced that the difference between humans and non-human animals is greater in general human opinion than the scientific results and scientific theory suggest. In this she follows the arguments laid out by Charles Darwin, Donald Griffin, and Bernard Rollin, and is optimistic that current and future students of animal welfare law find and will continue to find these authors to be of influence.

In the following, we both question the practice of keeping apes in captivity. Both the physical and behavioral evidence suggests that chimpanzees and many other animal species are conscious, emotional beings, with the capacity to suffer. When non-human animals are kept in human-created environments it is our responsibility to monitor their welfare, and to question our institutions and practices if they are unable to meet the needs of animals' physical and mental health. Scientific studies such as these discussed below are important for informing welfare law and policy. We encourage future discussion and debate on the arguments for and against keeping apes in human-created

The Politics of Species: Reshaping our Relationships with Other Animals, eds R. Corbey and A. Lanjouw. Published by Cambridge University Press. © Cambridge University Press 2013.

environments, although we believe that the only defensible reason for doing so is to offer lifelong sanctuary to those who cannot be returned to nature. Following this logic, we argue that, while there may be no alternative to the plight of apes already living in human-created environments, it is unnecessary to breed apes into captivity. To do so perpetuates a situation of abnormal behavioral patterns and suffering. This chapter shows some of the evidence that has informed this stance.

Organisms survive and thrive best in the total environment to which they are optimally adapted, all other things being equal. That is, for every living species, natural selection has shaped its structure and function to work best in the environment of evolutionary adaptedness (EEA). The great apes (bonobo, chimpanzee, gorilla, orangutan) are no exception. Just as the brain is a physical phenomenon, so is its functioning, including consciousness, which also has been shaped by past evolutionary processes. Thus, the conscious experiences of both human and non-human animals are consequences of selective forces that acted in the EEA. For optimal functioning and well-being in the present, an organism fares best in environments that come closest to its EEA. Therefore, when creatures such as chimpanzees are confined in captivity rather than roaming free in nature, they are vulnerable to psychological abnormality and suffering.

Captivity is problematic for the welfare of non-human apes (and for other non-human animals), and in trying to show this, we suggest that the species interface between human and non-human ape is best understood by using natural behavior as the baseline against which to compare unnatural behavior. More precisely, we seek to scrutinize the chimpanzee (*Pan troglodytes*) in nature versus various types of captivity, in the hope of shedding light on how these two counterparts relate to one another. When contrasts emerge from this comparison, we strive to infer their significance, especially with regard to thorny issues of what constitutes suffering and how it might be detected.

Wild chimpanzees

Of the living apes, the chimpanzee is best known by far, both in nature and in captivity. Field studies to modern scientific standards have been underway for more than 50 years, led by the pioneering research of Jane Goodall (1986). In captivity, only the chimpanzee has become a laboratory species, bred accordingly in massive numbers, and is found worldwide in zoological gardens. Despite this ubiquity, it is difficult to generalize about "the" chimpanzee, as will become clear below. Variation is the problem: chimpanzee behavior varies greatly across a wide range of environments, from rain forest to savanna, and from laboratory to range country sanctuary.

Methods of studying chimpanzees vary from natural history notes to tightly controlled experiments, with various intermediate states, such as quasi-experimentation in nature and systematic observation in captivity. Large brains and intelligence results in complexity, and chimpanzee groups vary surprisingly often in their cultural norms and practices. Finally, ape–human relations range across those of prey, predator, competitor, and that of mutual dependence and benefit. We cannot hope to do justice to this diversity here, but will attempt a précis, as background related to specific topics covered in this analysis.

In the wild, chimpanzees inhabit a range of vegetation types, from evergreen tropical rain forest, through deciduous woodland, to grass-dominated open savannas. They range from lowland swamps to montane meadows. The most productive field studies are of habituated groups, that is, chimpanzees who have come to tolerate human observers at close range from dawn to dusk. However, this state of tameness has sometimes been achieved by human intervention, usually in the form of artificial feeding, which carries distorting effects on natural behavior. One study-site, Bossou in Guinea, maintains a clearing in the forest where items (such as raw materials for tools, or novel foods) are available to test wild chimpanzees who choose to drop in. This arrangement allows for non-invasive, near-experimental data collection (Matsuzawa et al., 2011).

Chimpanzees show cross-cultural variation on various fronts (McGrew, 2004). Groups that use elementary technology to fish for termites do not crack nuts, and vice versa, although both are feasible. Groups that use wands of vegetation to dip for army ants use different techniques and different types of tools, even when the prey species of ant is the same. A food-item, oil palm fruits, which is most important at one site (Gombe) is ignored at another one of similar ecological make up (Mahale). The standard long-distance vocalization of chimpanzees, the pant hoot, shows subtle variation from population to population, reminiscent of dialects in human language. Finally, some aspects of the "glue" of chimpanzee society, social grooming, vary in non-functional ways across groups, even in the same population.

On the other hand, there are universals in chimpanzee behavior, just as there are in humans. In all known chimpanzee groups (or communities), males are philopatric, that is, they are born, live, and die in one group, while females typically disperse, that is, at sexual maturity they leave their natal group and migrate to join another, in which they stay and reproduce. Thus lifelong male bonding is a key part of group continuity, and they are the more sociable sex. Within a group, all adult males, no matter how doddering, dominate all females, no matter how robust. Females are more solitary and one-parent-family oriented. All groups are territorial and show aggression, sometimes fatal, to their neighbors. Chimpanzee communities rarely congregate at one time in one place, but instead practice fission–fusion, whereby subgroups (or parties) constantly form and divide, on a daily basis. Every great ape over the age of weaning constructs at least one bed (or nest) per day for sleeping, usually up in the canopy.

Relations with humans living in the same local area vary immensely too. In some cases, humans hunt apes for meat, so that the local villager is the predator, and the ape the prey. In other, rare cases, chimpanzees may kill and eat human infants or toddlers, thus reversing the roles. Depending on local cultural mores, especially religious beliefs, human and ape may live in peaceful coexistence, or, in other cases, especially when there is ape crop-raiding on cash crops, they may be bitter enemies, to the point of extermination. Sometimes apes and humans are competitors for the same forest produce, such as prized seasonal fruits. Humans may pose hazards to apes, even unintentionally, when they set snares for ungulates that maim or kill the apes. Pathogens may be transmitted accidentally either way, from human to ape (respiratory ailments) or from ape to human (ebola).

Captive chimpanzees

Marked differences exist between the social and physical environments in which the chimpanzee mind evolved compared to those in which captive apes are confined. Captivity presents both costs and benefits. The costs vary but are unacceptable when mental and physical health are compromised (McGrew, 1981). Wild chimpanzees have the freedom to choose their companions, mates, and home ranges. They choose when, where, and what to eat; their natural diets include many species of flora and fauna, with a large variety of foraging, extracting, and processing methods, including hunting. Their daily activity budgets vary accordingly, and they range widely over varied landscapes and habitat types.

In comparison to this, zoo chimpanzees face the chronic presence of human observers and restrictive human management of most aspects of their lives (Hosey, 2005). The constant composition of zoo groups means that there are few if any opportunities for the species-typical fission–fusion group structure of the wild. In the interests of economy and management, males often have no same-sex companions, and females lack matrilineal kin, if their offspring are removed. Captive chimpanzees are usually prevented from reproducing, either by sexual segregation or contraception. Thus they may be denied copulation, consortship, parenthood, and even inbreeding avoidance. Captive environments are usually cramped, sparse, and simple compared with the complexity of the natural environment, and controlled diets and feeding times contrast radically with the ever-changing foraging and decision-making processes of daily life in the forest.

Diminished stimulation (deprivation) takes many forms: sensory (sunshine), motor (climbing opportunities), perceptual (interactions with other species), physical (varied substrates), nutritional (natural foods), social (agemates, newcomers), intellectual (novelty). Captive apes must cope with boredom and the inability to escape the attentions of other chimpanzees or humans (Hosey, 2005). Behavioral strategies that evolved to solve social and environmental problems across a variety of natural habitats may founder in the face of novel environmental challenges (or a lack of such challenges) presented by captivity, and so normal development patterns may fail to occur. In the face of such paucity, the benefits of captivity, such as abundant food, lack of predators, and veterinary care pale by comparison.

The study of animal behavior takes into account both the life of the individual and the evolutionary history of the species. For the individual, maturation, development, and its present-day responses to stimuli, such as the biotic and abiotic environment (proximate mechanisms) are examined. For the evolutionary history of the species, the responses and strategies that have evolved in adaptation to past environments (ultimate mechanisms) are considered (Tinbergen, 1963). In applying these concepts to chimpanzees, we see that many pressures are exerted because of mismatches between the EEA and the captive environment. Reaction to these pressures may be expressed through the repression of normal behavior, and the occurrence of abnormal behavioral patterns.

Measuring effects

So, how does this natural behavior compare with captivity? Of what are confined chimpanzees deprived, either by elimination or reduction? This unnatural existence brings both costs and benefits, which means that any analysis requires a careful calculation of trade-offs. For example, captive chimpanzees in "good" conditions are freed from disease (by prevention or cure), parasites (either endo- or ecto-), predators (carnivores, poachers), and other hazards (snares, venomous snakes). Many live to ripe old ages, in apparently good physical health.

The costs come in various forms of loss of freedom: movement, association, demography, procreation, stimulation, or even activity. Captive chimpanzees are unable to range freely, on a daily, seasonal, or annual basis; they have no territory to patrol and no neighbors to seek out or to avoid. Captive chimpanzees cannot choose with whom they associate, or to be solitary; they have no chance to seek quiet, one-to-one times with friends, or to avoid bullies. They are trapped in constant, unavoidable, total sociality. Most captive chimpanzees live in unnatural groups, in terms of numbers and age- and sex-classes, if they are lucky enough to live in groups at all. Even the most basic activities with which wild chimpanzees fill their waking hours may be unavailable, such as foraging, tool making and use, hunting, and nest building.

Confusingly, not all captivity is the same. Laboratories range from biomedical centers that are sterile and secure but offer minimal amenities and solitary housing, to those emphasizing behavioral and cognitive research by non-invasive experimentation. Some laboratory housing is for breeding alone, or even for long-term "warehousing" of subjects until they are needed in research protocols. Zoos, wildlife parks, safari parks, etc., typically house apes in more spacious outdoor settings, but with the added source of stress from public exhibition. Sanctuaries, refuges, and retirement homes offer respite from invasive research and intensive scrutiny, but most are limited by over-crowding and under-resourcing. Many are charitable, volunteer operations operating on limited budgets.

How then to detect the net effects of captivity? An obvious indicator is the normality of the behavior of the apes, when "normality" equals the performance of the full behavioral repertoire of their wild counterparts. This repertoire (ethogram) is well known, at least for some populations of chimpanzees, and numbers hundreds of behavioral patterns (Nishida *et al.*, 2010). Given this, "abnormality" can be defined as behavioral patterns rarely or never seen in nature. "Abnormality" has various connotations, ranging from neutral, statistical rarity to the pejorative sense of aberrancy; here we opt for the former. However, such abnormality is not so easy to establish: coprophagy (eating one's or other's feces) by apes is rare in nature, but can be functional, if what is re-ingested is undigested content that remain nutritious. This can make sense, and is sometimes done even by foraging peoples, but it may differ from the compulsive, non-nutritional consumption of fecal matrix, especially when it is regurgitated and re-ingested repeatedly (Bertolani and Pruetz, 2011). This non-functional form of coprophagy is what is most commonly observed in captive chimpanzees and was the form observed in LB's study. In contrast, some patterns of abnormal behavior are clearly maladaptive, such as self-mutilation.

The simplest way to measure behavioral abnormality is *prevalence*, that is, the number of individuals in a group that show it. This is done by recording whether or not an individual shows an abnormal behavior during the study period. However common is this measure, it is crude, as it lumps together habitual and occasional performers. Better is *diversity*, which is the number of types of abnormal behavior shown by an individual. More discriminating are quantitative measures of frequency or *rate* (number of events per unit-time) and *duration* (amount of time spent in a state, per unit-time). Such fine-tuned measures are possible only with focal-subject sampling, in which all acts performed by a subject are recorded, during a standard sample period of focused observation.

So, how to measure the impact of various captive contexts on various kinds of abnormal behavior? Hypothetically, in descending order of "merit," one might expect more abnormality in laboratories > zoos > sanctuaries > released into wild > nature. In reality, there is much published data from labs, little from zoos, and virtually none from the other categories. In folklore terms, there may be a general perception that relief from the constraints of the lab may be found in re-homing to a "good" zoo, which can be defined as one meeting the standards of professional organizations, such as AZA (Association of Zoos and Aquariums) or BIAZA (British and Irish Association of Zoos and Aquariums).

Because key evolutionary processes occurred in the past, we can only *infer* these events, rather than show them conclusively. Thus if we wish to ask questions about the relationship between abnormal behavior and evolutionary processes, we must devise careful experiments, with as much appreciation as possible for the complexity of the species and its natural environment. For example, asking if an individual's unusual behavior sets her apart from the rest of a captive group (and therefore potentially reduces her survival chances) may not yield reliable results compared to the wild. The responses of captive chimpanzees, who have probably spent most of their lives in the company of others who show abnormal behavior, may well be a different response from those who have never or rarely seen such behaviors. In order to ask appropriate questions, and to set up successful studies, an understanding and appreciation of both captive and wild situations is needed. Despite these methodological challenges, the considerable advances made by human evolutionary psychopathology (Baron-Cohen, 1997; Brüne, 2008) illustrate that inferring the evolutionary roots of abnormal behavior is highly beneficial in seeking to understand these phenomena.

Empirical tests

Lucy Birkett sought to test this by comparing the occurrence of abnormal behavior across six accredited zoos in the UK and USA (see details in Birkett and Newton-Fisher, 2011). She did 30 hours of observation on each of 40 chimpanzees, using focal-subject sampling; she recorded 37 behavioral patterns that are rare or absent in wild chimpanzees. The range

Table 11.1 Summary definitions of 16 most common abnormal behavioral patterns recorded for chimpanzees in six zoos (from Birkett and Newton-Fisher, 2011).

Bite self:	bite own body part
Clap:	slap palm of hand or sole of foot, making noise
Display to human:	stylized agonistic display and build-up directed to humans
Eat feces:	ingest own feces, both matrix and undigested food
Fumble nipple:	manipulate own nipple by hand or sucking
Groom stereotypically:	repetitively groom specific body part of self, without goal
Hit self:	hit own body part with hand
Jerk:	spontaneous body tic, without apparent external stimulus
Manipulate feces:	hold, carry, or spread on surface own or other's feces
Pat genitals:	repetitively touch own genitals, then often lick hand
Pluck hair:	pull out own hair
Pluck socially:	pull out other's hair
Poke anus:	insert finger into own anus
Regurgitate:	vomit voluntarily, then usually re-ingest vomitus
Rock:	sway repetitively and rhythmically, without bristling
Toss head:	circular movement of head

Other less common patterns include: Bang self against surface, Clasp self, Drink urine, Float limb, Incest, Pace, Pinch self, Poke eye, etc.

of behavioral patterns included "eat feces," "drink urine," "pluck hair," "regurgitate," "bite self," "rock," "pace," "poke eye," etc. (see Table 11.1). The results are sobering.

- Each chimpanzee in the study showed abnormal behavior, yielding a prevalence of 100%.
- Every chimpanzee showed 2 or more abnormal behaviors. The average number shown by each individual was 5. The greatest number was 14.
- The frequency (number of occurrences per unit-time) of abnormal behavior was 1.5 per hour (range: 0.1 to 13.5).
- In the 30 hours of observation for each chimpanzee, the average duration was 79 minutes of abnormal behavior (range: 1 to 141).

Such abnormal behavior was common in all the zoos, and its variation could not be explained by standard independent variables such as age, sex, rearing history, or life history background. In other words, abnormal behavior is endemic, even in "good zoos."

While LB's study explored the breadth of abnormal behavior in zoos, McComb (2009) went into depth: she concentrated on one behavioral pattern, hair-plucking (also called hair-pulling or depilation), in a population of 24 chimpanzees living in a single group of naturalistic age-sex composition. They inhabited one of the UK's most prestigious zoos. She recorded extent of abnormal hair loss on 25 areas of the chimpanzee body surface, using a 7-point rating scale of hairlessness, from 0% to 100%. She distinguished between self-depilation and social-depilation, and sought to relate frequency and duration of hair loss to other measures of behavior, especially to its close relation, grooming. Seventeen of the 24 (71%) individuals plucked out their hair (self-directed) or that of others (socially directed) or both.

In a follow-up study, Woolloff (2010) sought to determine if the performance of hair-plucking could be ameliorated by short-term intervention. More specifically, she asked if two types of environmental enrichment would affect the frequency of hair-plucking, hypothesizing that such enrichment might lower stress levels, and therefore reduce depilation. Her study design was A1–B1–B2–A2, in which A1 = baseline before enrichment, B1 = provision of tool-use opportunity (enrichment type 1), B2 = scatter feeding to encourage foraging (enrichment type 2), and A2 = baseline after (same as A1). The results were disappointing: although enrichment affected other behavioral patterns, such as foraging and social grooming, it did not change levels of hair-plucking. Remediation of habitual abnormal behavior via distraction from environmental enrichment appeared to be ineffective, at least in the short-term timescale of weeks.

Abnormal behavior

Abnormal behavior has been recorded many times in captive chimpanzees, and most studies report a wide range of aberrant behavioral patterns and a high prevalance rate (Walsh *et al.*, 1982; Fritz *et al.*, 1992; Bloomsmith and Lambeth, 1995; Nash *et al.*, 1999; Martin, 2002; Bradshaw *et al.*, 2008). For example, in one group of zoo-housed

Figure 11.1 (a) Two grooming chimpanzees with normal hair. Photo W. C. McGrew.

Figure 11.1 (b) A zoo-housed chimpanzee with depilated state, that is, having plucked out its hair – this is the most common abnormal behavioral pattern. Photo W. C. McGrew.

chimpanzees, four out of six individuals showed abnormal behavior, with one individual having an average frequency of 15% of total activity dedicated to abnormality (Celli *et al.*, 2003). A group of 13 laboratory-housed chimpanzees showed 11 types of abnormal behavior, and 85% of the individuals were regurgitating (Baker and Easley, 1996). One laboratory individual averaged two self-injurious episodes per day (range: 1–4), a third of which caused tissue damage (Bourgeois *et al.*, 2007). Although Hosey and Skyner (2007) found a low rate of self-injurious behavior in zoo-housed chimpanzees, their findings are inconsistent with most of the published scientific literature, including the new results discussed in this chapter (McComb, 2009; Woolloff, 2010; Birkett and Newton-Fisher, 2011). Although abnormal behavior can be reduced through intervention practices, at least in the short term (Clarke *et al.*, 1982; Bourgeois *et al.*, 2007), elimination of these behaviors almost never occurs (Baker and Easley, 1996; Woolloff, 2010).

Opinions vary on how seriously these anomalous behaviors should be taken – ranging from reluctance to associate abnormal behavior with poor welfare (Nash *et al.*, 1999) to seeing it as a major welfare problem (Bradshaw *et al.*, 2009). Ferdowsian *et al.* (2011) used psychiatric methods to study unusual activity budgets and behavioral patterns in captive chimpanzees. Their findings indicate symptoms correlated with forms of post-traumatic stress disorder and depression, clearly signaling welfare problems. When viewed from an evolutionary perspective, as signs of potential mental health problems, these abnormal behaviors are indicators of poor welfare (Broom, 1998).

Evolutionary psychopathology

Psychiatry entails the assessment, description, prevention, and treatment of mental disorders and mental health problems. The discipline has an organic basis, in which psychological problems are seen to result from activity in the central nervous system. These biological roots mean that it is necessary to consider evolutionary processes if we seek a cohesive view on mental illness, because all living organisms are subject to evolutionary processes (Brüne, 2008). Evolutionary psychopathology attempts to do this, as the study of mental illness within the framework of modern evolutionary theory (Baron-Cohen, 1997).

Evolutionary psychopathology follows from evolutionary psychology, which is the study of the evolutionary factors that have shaped and affected the normal mind. As the brain is a physical organ, like any other part of the body, we need to appreciate the impact of evolution on the development of this organ, just as evolutionary theory is key to understanding any other part of the body. In the same way that evolutionary theory is important for understanding maladaptations such as physical disease (e.g., sickle cell anemia), it is also important for understanding maladaptations in mental health.

Evolutionary psychology supposes that our minds – including our thoughts and feelings – and behavior are shaped and affected by natural and sexual selection. This overarching concept has been important in helping us to understand human behavior

(Buss, 2004; Confer *et al.*, 2010). When looking at normal psychology, it is only a small further step to ask questions about abnormal psychology. Evolutionary psychopathology, which looks at evolutionary effects on abnormal psychological states, is not a new field, although it has gained momentum only recently (Kennair, 2003). Probably the most researched area in terms of evolutionary explanation is depressive disorders, and evolutionary theories have done much to increase understanding and knowledge in this area (Gilbert, 1992; Nesse, 2000; Kennair, 2001; Wolpert, 2001). Evolutionary analyses of anxiety (Poulton and Menzies, 2002; Rachman, 2002), autism (Baron-Cohen, 1995), and schizophrenia (Polimeni and Reiss, 2003) also yield important results and insights.

The most comprehensive way to investigate behavior (given that behavior is an expression of the mind) is to consider both the present life of the individual and the evolutionary processes that shaped it. The Nobel Prize-winning ethologist, Niko Tinbergen (1963), set this out clearly with his four 'why' questions: "Why" can refer to issues of (1) proximate causation or control; (2) development or ontogeny; (3) function or survival value; or (4) evolution or phylogeny. His framework forms the basis of animal behavioral study, and has been used effectively in trying to understand animals in the natural world. Most psychopathology focuses on only one of his questions, that of proximate causation. Ultimate mechanisms are less frequently discussed or are combined with proximate mechanisms, but recent work shows how useful the inclusion of ultimate mechanisms can be (Nesse and Williams, 1994; Gaulin and McBurney, 2004; Brüne, 2008).

Psychopathology is the study of brain and behavioral irregularities. When looking at psychopathology in chimpanzees, we can learn from authors such as Brüne and colleagues (2004, 2006), who argue that both proximate and ultimate mechanisms are important for understanding abnormal behavior in this and other species of apes. As chimpanzees show similarities to humans in both their brain structure and behavior, it is logical to hypothesize that the basic proximate and ultimate mechanisms that cause psychopathological states in humans and chimpanzees are similar.

What do we perceive to be the root cause of the abnormalities observed in captive chimpanzees? Is their abnormal behavior triggered by poor treatment or by captive conditions per se? These questions can be re-phrased as: is it solely because of proximate mechanisms that chimpanzees act abnormally, which, if altered, could allow them to lead a life free of abnormal behavior? Or does the problem lie at the ultimate level, such that the conditions of captivity are so different from the wild environments in which the chimpanzee evolved that abnormal behavior is inevitable?

From the above discussion of chimpanzee life in nature, it is easy to see how ultimate mechanisms of environmental mismatch between the EEA and the captive environment may cause enough stress to trigger abnormal behavior. From the above discussion of chimpanzee life in captivity, it is also easy to see that there are enough immediate stressful factors to affect negatively the life of a chimpanzee. Therefore, chimpanzees in captivity face a two-fold challenge – to be confined in an environment to which they were not evolved to respond, that a priori places strain upon them, and to cope with stressful events in daily life that are serious enough to trigger behavioral and mental abnormalities. So, there are likely to be both proximate and ultimate deleterious effects on the behavior of chimpanzees. The results presented above suggest that abnormal behavior is widespread in captive chimpanzees, regardless of background and regardless of type of captive environment. Thus it seems that captivity per se is the problem, and it is likely that proximate stressors negatively impact on the individual on top of this.

There is much variation both in the behavioral patterns displayed and across individual chimpanzees. Some behavioral patterns clearly are more serious than others, i.e., self-mutilation, and some individuals are more affected by mental health problems than others, i.e., those that engage in abnormal behavior for long periods, even incessantly. How can we disentangle some of these differences? Psychiatry has no hard and fast rules, and we always are likely to find explanations to which there are exceptions. Then, we can begin to make useful inroads into non-human abnormality by borrowing some of the principles and methods of human psychiatry.

Evolutionary psychiatry suggests that maladaptive behaviors often are a sign of psychopathology (Brüne, 2008). Behavioral patterns such as self-injury and hair-plucking are clearly maladaptive: a wounded chimpanzee is more open to the risk of infection or may move at a slower pace than the rest of the group. Similarly, a chimpanzee with reduced hair cover may fare poorly in the downpours of the rainy season, during which a coat of hair provides important thermoregulatory protection. Other behaviors are more subtle – a chimpanzee that repetitively blows air through its lips may not cause itself damage – but if this behavior distracts it from other behaviors needed to survive, or sets that individual's behavior apart as different or strange compared to the rest of the group, this could reduce its chance of survival. When sensitively evaluated within the context of the natural environment, abnormal behaviors seen in captivity may be better understood. However, it is essential to have comprehensive knowledge of the natural behavioral repertoire and natural environmental diversity of the species in question, in order to fully appreciate not only the extent of the gulf between captivity and nature, but also in *what ways* this mismatch affects the captives (McGrew, 1981; Pruetz and McGrew, 2001).

Is it suffering?

Huge numbers of species show abnormal behavior in captivity (Mason and Rushen, 2006) including domesticated birds (Garner, 2005), laboratory rodents (Würbel et al., 1996; Garner et al., 2003, 2004), and cetaceans (Marino, this volume). Within the order primates, chimpanzees are not alone in displaying aberrant behavior – abnormality has been recorded in gorillas (Gould and Bress, 1986; Hill, 2009), macaques (Hook et al., 2002; Lutz et al., 2003), baboons (Brent et al., 2002), and others.

Viewed from the perspective of human psychiatry, many anomalous behaviors recorded in chimpanzees (and many other species) are signs of mental health problems. Although the published literature from human psychiatry shows that not all signs of mental health problems are associated with reduced quality of life, suffering is often manifest in these disorders (Brüne, 2008). Debates about suffering in animals have produced volumes of literature, with opinions ranging to both extremes, but we argue that non-human animals can, and do, suffer, and that most behavioral abnormalities are a welfare concern (see Broom, 1998). Given that chimpanzees are intelligent, conscious, self-aware, and emotional beings, who doubts that behavioral patterns such as self-mutilation, regurgitation, and hair-plucking are serious problems?

In considering the animal mind, we follow authors such as Donald Griffin (1976, 1984; Griffin and Speck, 2003), Bernard Rollin (1989, 2006), and Marc Bekoff (2007a), who argue that the evolutionary continuity displayed across animal species also indicates continuity in conscious experience (Baars, 2005). Therefore as conscious beings, animals suffer accordingly (Rollin, 1989; Bekoff, 2007a; see also King, Marino, this volume). Conscious suffering has major serious ethical considerations. This assertion is easier to argue for mammalian species, because there is clear evolutionary continuity across all mammalian brains. For birds, reptiles, and fishes, convergent structures are required to make a plausible case of suffering in these species. Ample biological and behavioral evidence exists to suggest that such convergent structures occur (Pepperburg, 2002; Garner et al., 2003; Chandroo et al., 2004; see also Marino, this volume), and that there are more similarities than differences in the experience of pain (Allen, 2004).

What often keeps scientists from ascribing consciousness and suffering to non-human animals is that it is thought to be impossible to prove suffering in species other than humans (Keeley, 2004; Andrews, 2009b). This is mistaken logical thought, because it is impossible to prove consciousness and suffering in any organism, human or otherwise. In philosophy, this is known as the "other minds" problem (Nagel, 1999; Hyslop, 2010; Hutto, this volume; Andrews, this volume). Other humans, such as our friends and family, may tell us that they are unhappy, but because it is impossible to have direct access to the experience of another being, we can never confirm that they are having the experience they tell us they are having. We therefore have to infer that based on our conscious experiences, other humans who act and behave similarly to us must be conscious, and in general we believe them when their behavior indicates suffering. The conscious experiences of animal species are likely to be *comparable* rather than identical (as after all, no two humans have identical conscious experiences either). Thus suffering in another species is not identical to "human" suffering, because it is "chimpanzee" suffering, or "rat" suffering, but still it is real and unpleasant to the being who is experiencing it.

Just like human suffering, non-human animal suffering comes in different forms and degrees (Ferdowsian et al., 2011). Some signs of suffering should raise more concerns than others. Likewise, the abnormalities of captive animal behavior should not be viewed in a single rubric. Instead, different behaviors should be treated differently, while being careful not to isolate behaviors that may be related, with consideration always being given to the cause. This may sound complex, but it is the same way that we perceive indicators of stress in humans – some indicators are more serious than others – but indicators are rarely considered in isolation, and the underlying root of the suffering is considered.

The range and prevalence of abnormal behaviors in chimpanzees indicate that captive apes suffer greatly. While particular behavioral abnormalities in chimpanzees (Walsh *et al.*, 1982; Bradshaw *et al.*, 2009; Birkett and Newton-Fisher, 2011) are of varying concern, with some patterns being more serious than others, overall these aberrancies indicate a problem. Because the underlying cause of these behaviors is their confinement in unnatural environments, there is a welfare issue at a root level.

Future directions

We need to work toward a mutually beneficial future for chimpanzees and humans. Interventions (such as environmental enrichment) in order to improve the welfare of chimpanzees, either by enhancing conditions for individuals (Bloomsmith and Lambeth, 1995) or by improving housing and surroundings (Baker, 1997) are bound to be useful, and we applaud the efforts already made by many zoos and other captive facilities. We also acknowledge the valuable role that some zoos provide as sanctuaries for otherwise homeless chimpanzees. We appreciate that reducing abnormal behavior (let alone eliminating it) is not a straightforward process. This is partly because abnormal behaviors easily become habitual and resistant to reduction, and also because despite best efforts to improve the accommodation, the problematic condition of captivity remains. Re-introduction to the wild is not a sensible option for American and European zoo-housed chimpanzees, who are at the bottom of the list of potential candidates for re-introduction due to their lack of necessary skills for survival in the wild. Many wild-born chimpanzees housed in African sanctuaries would be much better suited for release, if well planned and executed re-introduction programs were undertaken (e.g., Humle *et al.*, 2011).

Conclusion

The results of the studies cited in this chapter, plus many others (Walsh *et al.*, 1982; Nash et *al.*, 1999; Bourgeois *et al.*, 2007; Bradshaw *et al.*, 2008), indicate that chimpanzees show high levels of abnormal behavior in captivity. Here we show that zoos are not exempt, and that all captive environments are problematic, whether laboratories, breeding facilities, wildlife parks, or even sanctuaries. Likewise, these issues are not restricted to chimpanzees. In regard of other species, the widespread occurrence of abnormal behavior in captive non-human animals (see Mason and Rushen, 2006) indicates we have a long way to go for conviviality to become a reality. When viewed from the broad perspective of evolutionary adaptedness, this is not surprising. The chimpanzee mind is conscious, self-aware, emotive, and smart. It is a mind that is in many respects similar to the human mind. As our understanding grows, we see scientific evidence of the consciousness, emotionality, and intelligence of numerous other species – including non-mammalians. The chimpanzee mind – like the minds of all species, including humans – has been shaped over millions of years.

Thus when chimpanzees and other animal species are placed in environments that are so very different from their EEA, their minds are vulnerable to distortion, and the visible effect of this is the abnormal behavior demonstrated by the chimpanzees and many other captive species. As abnormality is likely to be an indicator of suffering, the abnormal behavior displayed highlights a welfare problem with unavoidable ethical implications. While we recognize the admirable efforts of many captive institutions, and we commend those institutions that act as sanctuaries for otherwise homeless chimpanzees, we argue that the problem is captivity itself, which causes abnormality in ape behavior. If every ape born into captivity is condemned to suffer, then why are we breeding more apes?

Acknowledgments

We thank participating zoos for help and support, especially during data collection; BIAZA for approving LB's study and providing letters of support; S. McComb and A. Woolloff for allowing use of unpublished data; N. Newton-Fisher for supervision of LB; S. Hill and E. Thetford for various essential contributions.

12 Animals as persons in Sumatra

Jet Bakels

> "Everything has a soul. There is nothing that does not have a soul."
> Mentawaian shaman (Schefold, 2012: p. 150)

While most of the other contributions to this book focus on issues in the Western world, this chapter explores human–animal relations in two traditional, less modernized societies in Sumatra, Indonesia. It focuses on what connects humans with and separates them from (other) animals, according to local views and as acted out in both daily life and ritual. Special attention will be paid to hunting practices and to indigenous attitudes toward dangerous animals, which differ dramatically from Western mores.

In past centuries, for example, Europeans tried to exterminate the wolf, a dangerous animal often associated with the devil, in particular in werewolf mythology. In the Sumatran societies I will present, however, Mentawai tribesmen and Kerinci farmers do not believe they have a right to kill, consume, or use any animals at will. The position of animals in their traditional worldview can inspire and inform the Western struggle with moral inconsistencies with respect to animals (cf. this volume, *passim*; Herzog, 2010).

In the interior of the isolated Mentawaian archipelago, off the west coast of Sumatra, live some tribal groups that still are animists. They honor their deified ancestors, hunt with bow and arrow, grow sago, and keep pigs while continuing to live in traditional longhouses in the rain forest. The Kerinci people from the highlands of central Sumatra, on the other hand, are village-based farmers who grow rice. They have adopted Islam as their official religion, but a layer of traditional animism lurks beneath. Both societies keep domesticated animals. Pigs and chickens are raised for meat in Mentawai – cows, sheep, and chickens in Kerinci. Monkeys, wild pigs, and deer are hunted for meat in Mentawai – deer in Kerinci. While the Mentawai fear the crocodile as a threat to the lives of domestic animals and humans, the Kerinci fear the tiger.

I have conducted fieldwork in both societies as a cultural anthropologist specializing in traditional Indonesian cultures and in modern and non-modern attitudes toward animals. For my PhD I studied the relationship of the Kerinci with wild nature, in particular the tiger. Much of my interest in and knowledge of the Mentawai is due to my husband, anthropologist Reimar Schefold, who has researched Mentawaian culture in

The Politics of Species: Reshaping our Relationships with Other Animals, eds R. Corbey and A. Lanjouw. Published by Cambridge University Press. © Cambridge University Press 2013.

great detail. As a curator in various museums I focus on the role and meaning of animals in various cultural settings, with a special interest in how respect, fear, and admiration for nature come to the fore in human dealings with birds, whales, big cats, bovines, apes, and wolves. I find the often religiously inspired traditional bond between animals and humans moving and interesting, and a source of inspiration in rethinking the species interface.

What's in a soul?

A mingling of Greek and Judeo–Christian traditions underpin Western attitudes by postulating that man is the crowning achievement of creation, and the only living creature possessing a rational soul and moral personhood. Fed from this source, the perception that a deep gap separates humans from the rest of the animal kingdom persists to this day, despite Darwinist and genetic arguments to the contrary.

The Mentawaians are more generous toward other creatures. They attribute a personal soul and moral personhood not only to animals and plants, but to virtually all aspects of their surroundings: stones, houses, ropes, canoes, trees, and so on (Schefold, 1988). The soul, *simagere* in Mentawaian, can be described as a spiritual equivalent of a human person, animal, or thing. It is often depicted as a small figure that generally stays in or close to the entity it belongs to, but can also leave it and make its own decisions. When it stays away too long a living being may die.

Does the attribution of a soul to an animal guarantee it respect and better treatment? To answer this question we need to investigate the complexities of the Mentawaian way of live and worldview.

In daily life tame pigs and chickens roam around the longhouse and are regularly fed. The eggs are used for rituals and sometimes eaten, and the animals are slaughtered for festive occasion and rituals. The soul of the chicken that is slaughtered as an offering to the ancestors has an important mission: it has to beg the ancestors for protection and convince them to bring good luck to the family group that is making the offering. In the accompanying ritual speech a clear relation is established between the way the animal is treated during its life and its responsibilities after its death.

Before killing the chicken the shaman explains this to the animal in elaborate terms. He holds the chicken to the light of the sun (see the photograph on the front cover of this book) and says:

"[Our] faces are lit by the rising sun every day... until we have white hair, until we walk hunched over... I give you oil, oh chicken, so that your heart will be soft and sweet for all of us. Long, long have I taken care of you. The egg from which you come I have sheltered. Then, when you were there, every evening, and every morning, under thunder and with rain... I have taken good care of you, oh chicken. And now I have come to fetch you, so that you will help me to keep all sickness far away."   (Schefold, 1988: p. 355, my translation)

Pigs, when sacrificed, are addressed in similar terms: "Dear pig, do not feel angry because we now have to kill you... please remember how well we fed you all your

Figure 12.1 Before it is slaughtered a Mentawaian shaman addresses a pig with calming words and touches it with a flower to appease its soul. Photo Reimar Schefold, 1974.

life!" The animal is then caressed with a bundle of magical leaves, to "cool off" and appease its soul, which is presumed to be excited and "hot" by the capture of the animal and the fact that the pig is going to lose its life (Figure 12.1).

Subsequently the pig is killed. Its skull is cleaned, elegantly decorated with plant material, and hung on display in the longhouse. The skulls of game, deer, and monkeys, receive an even more elaborate treatment. They are painted and adorned with geometrically decorated wooden birds, called "toys for the souls." These toys have to please the souls of (living and deceased) animals and humans alike, and figure in many rituals. Positioned near the skulls of game animals in the longhouse, they will cause their souls to stay close and keep them feeling amused and happy.

Although these animals are killed for human consumption, this is not a thoughtless act. Their death is carefully surrounded by rituals to placate their souls. Their treatment springs from the Mentawaian worldview, in which everything has a soul and intricate relations exist between humans and their surroundings, including, for example, utensils and the trees one picks fruits from. These relations have to be balanced at all times.

Animals have to be treated with respect, in accordance with their worthiness. They may never be ridiculed and laughed at – an ancient and pan-Indonesian theme (Forth, 1989) – and will take offence at impolite behaviors that disturb humans, like farting or belching. The same respect is extended to objects that in Western eyes are inanimate, like a coil of rope or a canoe. When carelessly treated, a rope may take offence, and not be helpful the next time it is used, or even cause misfortune.

Are the souls of killed pigs and monkeys honored purely out of respect, or primarily for selfish reasons, to ensure the well-being of the Mentawaians themselves? This is complex. Both at once, it seems: there is deep respect, and there is also a practical and utilitarian motivation. In the worldview of the Mentawaians, the toys and decorations for the skulls are a prerequisite for the success of future hunting parties. When the souls of the dead animals are happy, they will call their living relatives in the jungle and explain to them how good their existence is in the longhouse, pressing them to "give" themselves to the hunter. The Mentawaians are concerned that no great harm, if any, is done. They want meat, and kill animals for it, but since the animals are respected during their lifetime and their souls are treated well after their death, they are presumed not to blame humans for their behavior. The souls of the killed game will even help the hunters with their hunt. The Mentawaians thus live in a universe where harmony is sought and constantly negotiated. The Mentawaian concept of *mateu*, which translates as "fitting," refers to exactly this type of respectful relation with everything in one's surroundings.

The village and the wild

While the Mentawaians live in longhouses in the rain forest, the Kerinci are farmers who for centuries have occupied the fertile valleys of the interior of Sumatra growing their crops, primarily rice. They enter the forest frequently, to collect products like rattan, fruits, and herbs, and to hunt. In the traditional, animistic worldview of both Mentawai and Kerinci – in Kerinci covered by a layer of Islamic beliefs – the souls of the ancestors are important religious forces. Animals are but rarely animals in the modern Western sense. In both societies the tropical rain forest is a mythical domain, the realm of the spirits of the forest, upon which the villagers depend spiritually as well as economically. The Kerinci people see spirits as the rulers of the forest and owners of the forest animals. In the forest, their abode, humans are their guests and must behave accordingly. When entering the forest, a token of respect is always offered. When hunting, elaborate precautions are taken and substantial offerings made. There is a strong sense that villagers can take something only when they offer correct ritual compensation. Certain elements of present-day Kerinci rituals still betray an old hunting tradition in which the hunter returns the bones of the hunted animal to the forest to ensure its resurrection and thus compensate for its loss.

Various myths in both Kerinci and Mentawai articulate this relation between humans and their surroundings and stipulate behavioral standards. They mention a pact that the first human inhabitants of the area, the ancestors and village founders, established with the animals and spiritual beings of the forest that had lived there since time immemorial. In this mythical pact there is a clear notion of equilibrium, a sort of status quo between the world of humans and that of the forest (Bakels, 2000). The arrangement implies that the creatures of the forest allow human activities in their territory, as long as the humans behave correctly. Both parties are subject to the same ethical code and moral order. Hunting could arguably be considered a dubious pursuit, a rather aggressive intrusion upon another's domain. Therefore it is necessary that the ideal of peaceful coexistence and harmony is reaffirmed during hunting rituals. At the beginning of Kerinci deer hunts utmost care is taken to assure the spirits of the forest, in ritual speech, that the hunters will only take deer that are presumed to have damaged village crops. The spirits are mollified by telling them that the hunt is conducted only to set the balance between man and animal straight (see also Bakels, 2004: p. 153).

Many mythical narratives in Mentawai and Kerinci, as in other traditional societies in Indonesia, dwell on the origin of animals, often from humans, and define how animals and humans should interact. One Mentawaian story, for example, presents an unhappy family that, in a conflict over food, changed into animals – the father into a crocodile, the mother into a deer, the son into a monkey, and the daughter into a wild pig (Schefold, 1988). Some stories proclaim specific taboos before the final transformation of humans into animals. In this case the boy announces that his meat should only be eaten by males, not by women, while the father is very specific about how he is to be treated once turned into a mighty crocodile, and advises humans how to avoid attacks by such a dangerous animal. Thus these animals are encapsulated in culture, drawn into the human domain. As thinking agents, they can be addressed and placated, influenced and appeased.

Animals can also be an outward appearance, a materialization, or a companion, of a spirit of the forest – for example in the case of the gibbon. Whereas in Mentawai animals are associated with the spirit world, in Kerinci, animals can contain the soul of a deceased human, and thus become ancestral figures. This is often the case with the most feared and respected animals of the jungle: man-eaters such as tigers and crocodiles.

A balance of power

Mythical stories in these two societies are most elaborate when it comes to the two animals that pose a real threat to human life and are regarded with awe: the tiger in Kerinci and the crocodile in Mentawai. This makes good sense, for it is important to understand these animals to avoid confrontations and casualties. In Kerinci and, even more strongly, in Mentawai, people depend directly and economically on nature. This circumstance in itself already inspires respect toward nature's forces, explained, as we have seen, in mythical terms. In Kerinci, a modest form of ancestor worship is omnipresent notwithstanding Islam, and tigers are generally considered to be ancestors that return in this particular guise (Figure 12.2). Long-deceased cultural heroes, like the founders of a village, are thought to appear as tigers, along with more recently deceased ancestors, such as one's grandmother. These ancestors can be called upon by their relatives and can appear in dreams and in trances. They are thought to protect and advise their descendants, and believed to live together in a village somewhere deep in the forest, as a tiger people. There they have a human form, while elsewhere they assume a

Figure 12.2 A dead tiger is respectfully transported, Padang Highlands, Sumatra, around 1890. Photo by C. Nieuwenhuis, courtesy of KITLV/Royal Netherlands Institute of Southeast Asian and Caribbean Studies.

tiger's appearance. At some point in a mythical past, humans and ancestral tiger-people struck a pact, in which they promised to respect each other's territory: humans rule in the villages, tigers in the forests. As long as they respect each other, there will be no harm.

The fact that tigers are ancestors does not guarantee peaceful coexistence. Many Kerinci villages have lost people to tiger attacks, but ascribe meaning to such deaths in terms of this ancestor-tiger lore. They believe that the ancestor-tiger will only attack those who have trespassed the rules of *adat*, traditional law, for example in cases of adultery, misuse of power, or unfair division of an inheritance. As such, the villagers explained to me, the tiger acts as a veritable *polisi hutan*, forest police. By doing this, however, the tiger/spirit violently enters the human domain, breaking an essential rule of the pact, and will therefore be killed by the villagers. The dead body is not abused or ridiculed, but treated with respect. It is resurrected and the tiger honored with ritual chants and dances, and subsequently buried just outside the village (Bakels, 2000: p. 280). However, tigers are killed in such cases exclusively, never randomly, as has been confirmed repeatedly in colonial sources. In the nineteenth century, the Dutch colonial administration offered a bounty for every dead tiger, but no Sumatrans responded to the lucrative offer.

In the more isolated and traditional areas of Sumatra, like Kerinci, beliefs blending ancestors and tigers have persisted into modern times. It was still very much alive, although under pressure, in the 1990s when I conducted fieldwork in various parts of the island. A man from a Kerinci village, for example, found a dead tiger caught in a rope trap he had set up for deer. He so feared the wrath from the tiger's family that he never dared to enter the forest again and soon afterwards left Sumatra to sells potatoes in Jakarta.

In Mentawai there are similar attitudes with regard to crocodiles, although the latter are seen as powerful spirits of nature, not as direct ancestors like the tigers in Kerinci. Crocodiles, in the Mentawaian view, only attack people who have disobeyed traditional rules, and got away with it. These people may not have been very cooperative in rituals, or might be suspected to have taken more than their fair share of meat. Crocodiles, like the tigers in Kerinci, will only be killed if they have attacked humans. This view has advantages for both sides: the animals are spared, and humans live with the comforting idea that although misfortune can not be neutralized, it can be avoided. As long as one respects the social code there is no need to be afraid.

Mary Douglas (1970b) has drawn attention to the fact that misfortunes are often interpreted in such a way that they underline social codes. The meanings lent to tiger and crocodile attacks make sense in this light: offences that are hard to punish in these rather egalitarian societies are often dealt with in terms of interventions by these animal/spirits. As "moral persons" they thus do not disrupt but rather underline the social order.

A pact for the future?

The pacts of respect between villagers and these dangerous animals of the surrounding rain forest have long worked well. Retaliation entails big risks for both parties and must be avoided at all costs. Pragmatism is thus combined with respect for what is spiritual in nature in both traditional worldviews. Having a soul and, to a certain extent, being seen as a moral and rational agent, does not put animals on exactly the same level as humans – tribal peoples, too, seem generally to be inclined to speciesism – but it does bring them respect. And they are clearly seen as belonging to the same social and moral universe as humans. Commodification of animals such as is common in modern societies, in particular agribusiness (see Twine, this volume), is a far cry from the way animals are seen and treated in Kerinci and Mentawai.

We should not, however, be too optimistic. These non-modern Sumatran worldviews have proved to be rather vulnerable, and are under threat from processes of modernization and globalization, as well as the strong presence of Islam. The relatively isolated villages of Kerinci are being incorporated into the larger world and into the market economy. Money, not least for the education of children, has become very important, and since the 1970s some of it is, regretfully, generated by selling tiger skins and bones.

The tiger's habitat, the rain forest, is cut down for timber and to make way for oil palm plantations. The tiger's traditional mythical connotation is waning, although not gone.

It is clear that the old concepts that long brought protection and respect to wild animals are eroding. Nevertheless, they can be useful in fuelling the cooperation of local peoples with the protection of animals in forests and in nature reserves. The traditional myths, taboos, and other oral traditions that in many areas are half forgotten, or looked upon as "primitive," must be documented and made available as important cultural heritage of, and for, these areas. They can serve as a source of inspiration for the local people and the modern world alike, and fuel the implementation of new laws from "outside," necessary to protect the forest and its inhabitants (see Fuentes, this volume; Riley, this volume). It is clear from the ethnographic literature that, roughly speaking, Kerenci and Mentawai attitudes toward animals are quite typical for small-scale, non-modern societies at large. As such this case study provides fascinating coverage of intermediate space between exploitation of other animals and a shared, morally equivalent life with them.

I would like to end this chapter with a Mentawaian song on a gibbon. While other non-human primates are hunted for food the gibbon is not because it is so humanlike and has a strange, strong call. It is seen as a companion of the spirits. This was sung by Koraibikerei and recorded by Reimar Schefold: "The gibbon of the mountain ridge / With sorrow in his heart / When the sun declines / Climbs into the *eilagat* tree / In the fogs of the forest / I hear him: *Ko-a-aii*."[1]

---

[1] Mentawaian song by Koraibikerei, from the CD "Songs from the Uma: Music from Siberut Island, Indonesia," edited by Reimar Schefold and Gerard Persoon, Pan Records (Leiden, the Netherlands) 2011/12.

13 Interspecies love

Being and becoming with a common ant, *Ectatomma ruidum* (Roger)

Eben Kirksey

Insect love has lately become the subject of much attention from anthropologists.[1] In confessing my own affections for *Ectatomma ruidum* – an ant species that is flourishing in the forested landscapes, agricultural fields, and suburban lawns of Central and South America – I must be clear that our feelings are not at all mutual. At best, *Ectatomma* ants remain indifferent to human beings. When an *Ectatomma* forager sees a large vertebrate, a potential predator like me, she will often turn her whole body to face-off – jaws open, legs firmly planted, stinger ready. If these persistent threats are empty (any *Ectatomma* aficionado knows that the ant has difficulty stinging humans and will scurry away, and try to hide, upon serious molestation) they still serve as reminders of the unease generated by my fondness for their kind. Threats also became evidence that these ants were capable of returning my gaze (Haraway, 2008: p. 21). Recognizing gaps in our gaze, and disjunctures in our interests, offers a point of entry to rethinking respectful coexistence across the species interface.

Ectatomma ants are flighty nomads – ever moving among worlds. Nomadic subjects, such as these agile insects, can be dangerous, irredeemably destructive, or tolerant, in the words of Isabelle Stengers (2011: p. 373). The challenge, for Stengers, is to trap nomads, to enfold them in production of what she calls cosmopolitical worlds. Cosmopolitics offers an idiom for considering the diverging values and obligations that structure possible non-hierarchical modes of coexistence. "The cosmos refers to the unknown constituted by multiple divergent worlds," Stengers writes, "and to the articulations of which they could eventually be capable" (Stengers, 2005: p. 995). These common worlds involve contingent "political" articulations. We have to build them together, tooth and nail, in concert with other agents (Latour, 2004: p. 455). Cosmopolitical worlds are structured by relations of *reciprocal capture*, a dual process of identity

---

[1] The definitive work on insect love, *The Illustrated Insectopedia* by Hugh Raffles (2010), chronicles diverse ethnographic adventures – among "squish freaks" who obtain pleasure by stepping on bugs, artists who care for mutant insects living in radioactive zones, as well as other queer entanglements connecting humans with arthropods. Lovers of bees and people who eat insects have also featured in recent ethnographies (Morris, 2004; Moore and Kosut, 2013). Insects have also featured prominently in anthropological accounts of animal becomings in the realm of warfare (Kosek, 2010).

The Politics of Species: Reshaping our Relationships with Other Animals, eds R. Corbey and A. Lanjouw. Published by Cambridge University Press. © Cambridge University Press 2013.

construction where each agent has an interest in seeing the other maintain its existence (Haraway, 2008: pp. 35, 42; Stengers, 2011: pp. 35–6).

My tale of unrequited insect love explores the conditions of capture where relations are contingent and not always reciprocal. Tracing actions oriented to the care of beings and things, sometimes across species lines, I consider how agents come to be enlisted in the production of common worlds, and how they escape. I regard *Ectatomma ruidum* ants as agents of cosmopolitical assembly, conscious beings who become involved with other creatures through relations of reciprocity, kinship, and accountability (see Kockelman, 2011). Drawing on my own bio-behavioral experiments and ethnographic observations I will explore theoretical, normative, and ethical proposals for being and becoming with others.

First contacts

In 1997 I volunteered on a study of ant community ecology on Barro Colorado Island, Panama, an "open air biological laboratory," which was created in the 1920s. This man-made island emerged when a small hill-top was surrounded with water during the

Figure 13.1 These two *Ectatomma* foragers have been captivated by a plant. While waiting for nectar – a sugary and nutritious liquid – the ants help protect the plant from leaf-eating insects. This plant, a species of *Inga*, has captivated some humans too. The flesh of its fruit tastes like vanilla ice cream. In other words, these plants (known in English as "ice-cream-beans") have enfolded ants and humans in common cosmopolitical worlds. Gamboa, Panama. Photo courtesy of Alex Wild.

damming of the Chagres River by US engineers who created the Panama Canal (Lindsay-Poland, 2003). This island became imagined as a place that contained the mysterious secrets of nature's past, an exotic field site for adventures in the present, and a place where new discoveries might unlock future possibilities (Strain, 1996/1997). The field station (which came to be known by the island's initials, BCI) quickly became a site of pilgrimage for aspiring scientists who wished to become tropical biologists (Henson, 2002).

While some projects on BCI were imagined as "pure research," my own work in Panama had a clear relation to US geostrategic interests. Laboring as a quasi-insider in the shadows of US military installations, in the midst of failing imperial ambitions, I began to understand how oblique powers and unexpected contingent events were mediating research agendas. The project that brought me to Panama was indirectly in the service of the citrus industry. *Wasmania auropunctata*, an "invasive species" from Central America, had become a common agricultural pest in the southern United States. In Florida and other southern states, these tiny ants were taking up residence on the leaves and fruit of citrus trees. Fruit pickers were demanding premium wages to work in infested groves, because the ants can deliver a painful sting – especially after getting inside of the workers' clothes.

My own work involved setting tuna fish baits on the forest floor to lure *Wasmania*, and other ants, out of the leaf litter. *Ectatomma,* one of the largest ants in this ecological community, fancied tuna too. I became familiar with the habits of this charismatic ant and came to easily recognize it with my naked eye. One day, while walking the trails of BCI, an unusual sight arrested my attention. I watched two *Ectatomma* workers, one carrying another, exit out of a colony entrance and make a bee-line toward the entrance of another colony several meters away. When the pair reached the other entrance, they disappeared inside.[2]

In the era when I made this observation, the late 1990s, the genetic determinism of E. O. Wilson's sociobiology held sway among ant experts. In the ideal ant colony (at least according to the ideals of Wilson and his followers) there is a single queen and all of the workers are sisters: non-reproductively viable females. There is considerable deviation from this ideal type. In many species, *Ectatomma* included, colonies can have multiple queens. Workers can also lay eggs – some of which are eaten by other adults and others which develop into larvae. Male ants – with wings, small heads, and a waspy look – take little part in colony life other than mating.

Sociobiologists were asserting in the 1990s that the ant colony "is a superorganism." Nests of ants were "analyzed as a coherent unit and compared with the organism in the design of experiments, with individuals treated as the rough analogues of cells." In an encyclopedic tome published in 1990, simply titled *The Ants*, Bert Hölldobler and E. O. Wilson speculated that "natural selection can produce selfish genes that prescribe unselfishness" (1990: pp. 2, 179). As an undergraduate, majoring in anthropology and biology, I became fascinated by behaviors of *Ectatomma ruidum* that did not fit with the

---

[2] Later I found an article by Stephen Pratt (1989) describing the "kidnapping" of young workers by *Ectatomma* ants from neighboring colonies.

prevailing consensus of the 1990s. Carefully observing ants in the field, I speculated that they were embedded in endlessly expansive networks. If ant colonies were to be understood as superorganisms, my observation of workers moving among colonies suggested that the cells were running wild.

My love of *Ectatomma* developed from these initial surprising observations. Later, while watching different colonies on separate occasions, I observed the transfer of food, larvae, and even winged queens among distinct nests. Putting up a barrier around one focal colony, I let the ants collect all the tuna fish they wanted for an hour. After removing the barrier, and the bait, I watched as tuna fish was redistributed. Ants exited the focal colony and carried it into the nests of neighbors. Minutes after watching tuna entering one neighboring nest, I watched as it was carried out again to an even more distant nest.

Cutting the network

Human social worlds, according to a classic definition from sociology, involve collaborating and doing things together. They are communities of practice and discourse engaged in collective action. Fluid exchanges of material and semiotic elements, a discourse of sorts, structures the social worlds of ants (see Haraway, 2008; Hayward, 2010). While much of the literature about humans is preoccupied with the roles of entrepreneurs, agents that are viewed as being central in the construction of common worlds, it is clear that a multitude is involved in the coproduction of ant worlds.

Insects are generally thought to be incapable of recognizing each other as individuals. With upwards of 300 ants in an *Ectatomma* colony, it is highly unlikely that each colony member recognizes one another. A colony scent, "a complex Gestalt of hydrocarbons" on the cuticle of their exoskeleton, is instead learned by ants. This odor is largely independent of genetic factors and is instead thought to be spread through shared food exchange and grooming (Reznikova, 2007: p. 365). Most ant species vigorously defend the boundaries of their colony – killing intruders from different colonies of the same species on contact. For most ant species the stranger is the enemy "with whom there is the real possibility of a violent struggle to the death" (Balakrishnan and Schmitt, 2000: p. 108).

Ectatomma ruidum is different than most ants – in a certain sense this species is exceptional, in fact. Workers will sometimes stand in their nest entrance, and occasionally bite or drag away other *Ectatomma* ants that are trying to get inside. But often the nest entrances stand empty. "Guard" ants also sometimes stand aside, letting members of neighboring nests, or even ants from colonies several hundred yards away, pass unmolested. Once inside, these neighbors have access to caches of food.

While volunteering on BCI in 1997 I began excavating *Ectatomma* colonies and keeping them in transparent test tubes in the Smithsonian labs. Inside of these nests adults spent much of their time grooming themselves and others. Introducing ants from other colonies, I found that they were often bitten at first, and pulled around the chamber by resident ants. With time, I found that the strangers were sometimes adopted – enlisted into the social world of the colony. They began doing things together with the other ants – grooming the adults and caring for the larvae.[3]

In the field I found that *Ectatomma* ants sometimes become captured by multiple social worlds. Marking individual adults with paint, and gripping their hind leg with a pair of steel forceps, I positioned them at the entrance of colonies that were not their own. Almost unfailingly, when released, the ants went inside. On follow-up visits to these same nests, I found marked ants foraging for food and bringing it back to their new homes.

Adult ants are only able to eat solid food in concert with their anatomically flexible youngsters. With ultra-thin waists, called petioles, adults cannot move solid foods into the digestive organs of their own abdomens. The larvae of ant colonies are thus agents of *interessement* – to deploy a keyword from actor–network theory. "*Inter-esse*" means being in between or interposed (Latour, 1987). The larvae are obligatory points of passage for solid food that stabilize networks of adults living together in the same nest or colony. The embodied differences of adults and the larvae thus keep them interested in one another.

With a conjoining of diverse body parts, with an intermingling of mutual utility and perhaps pleasure, adult workers and larval ants often eat solid food together. Chopping up the food with their mandibles, adults position manageable tidbits within reach of larvae. Ingesting bits of food, and excreting enzymes to predigest other solids, the larvae break the food down into chemical components. Larvae of many ant species generate nutritious liquids that adults, in turn, drink (Hölldobler and Wilson, 1990: p. 348; Cassil *et al.*, 2005).

Marilyn Strathern astutely observes that the power of actor–network theory (ANT) also presents a foundational problem: "theoretically networks are without limit." Cutting the network, using one phenomenon to stop the flow of others, is what makes this analytic useful in the eyes of Strathern (1996). My study of *Ectatomma ruidum* found that individual ants in colonies are always cutting the network, making high-stakes and potentially arbitrary distinctions between who is enemy and who is ally (see Kirksey, 2012: p. 177). Rather than a categorical rejection of all non-kin, I found a nuanced pattern of graded recognition, where the frequency of hostility increased over topographic distance.

During experimental trials I spent close to 150 hours in the field – staring at small holes in the ground, squatting on my knees, waiting for something to happen. In short, during all this waiting and watching I found that *Ectatomma* ants regularly enter the nests of their neighbors. I also discovered that ants from distant nests – from more than 300 meters away – can readily enter experimental colonies. If conventional models of the ant colony resemble "a hub, or star, network in which all lines... radiate from a central point along fixed lines," I began to understand the social world of *Ectatomma* ants as something like a "distributed, or full-matrix, network in which there is no center and all nodes can communicate directly with all others" (Hardt and Negri, 2004: pp. 56–7).

---

[3] "Social worlds," according to a classic definition from the realm of humans, involve "doing things together" (Becker, 1986).

If Hölldobler and Wilson speculated in 1990 that "natural selection can produce selfish genes that prescribe unselfishness," after more than two decades of searching, with genomic technologies of ever-increasing sophistication, a gene for altruism has yet to be found. Departing from the notion of superorganism, I suggest that *Ectatomma* colonies might be understood as ensembles of individuals – associations composed of conscious agents who are entangled with other beings through relations of reciprocity, accountability, as well as kinship.

The notion of *ensemble* is borrowed from Paul Kockelman, who in turn, has purloined William James' ideas about the self – the sum total of things we call our own. Selfhood involves what constitutes part of the ensemble. In human realms the self-as-ensemble includes one's clothes and house, one's ancestors and friends, one's nail clippings and excretions, one's body, soul, thoughts, and ways of being in the world. Actions oriented to the care of beings and things enlists them in the ensemble (Kockelman, 2011: p. 13). "To care for others is to care for one's self," write Deborah Bird Rose and Thom van Dooren in a related vein. "There is no way to disentangle self and other, and therefore there is no self-interest that concerns only the self" (Rose and van Dooren, 2011: p. 27).

Fluid exchanges

Feeding nestmates, with fluid exchanges of material and semiotic elements, enfolds individual adult ants into ensembles. Stephen Pratt, who studied communication behavior in *Ectatomma* in the 1980s, described the sharing of liquid food in this species with loving attention to detail:

Droplet-laden foragers returned immediately to the nest tube and, after a few seconds of excitation behavior, either stood still or walked slowly about the nest with [their] mandibles open and mouthparts usually retracted. They were generally approached within a few seconds by unladen workers who gently antennated the clypeus, mandibles, and labium of the drop-carrier, using the tips of their antennae. The carrier then opened its mandibles wide and pulled back its antennae, while the solicitor opened its mandibles, extruded its mouthparts and began to drink. During feeding, the solicitor continued to antennate the donor, who remained motionless. Usually the solicitor also rested one or both front legs on the head or the mandibles of the donor.   (Pratt, 1989: p. 327)

William Morton Wheeler, who was perhaps the most prominent early twentieth-century ant biologist, developed an elaborate model to explain the origin and continued functioning of insect societies based on his observations of exchanges of liquid food. He coined the term *trophallaxis* – deriving from the Greek words for "nourishment" and "interchange" – to describe this behavior in 1918.[4] Assuming that the proximate cause of certain behaviors was genetic, Wheeler argued that "the origin of the behavior of

---

[4] *Six Legs Better*, a cultural history of myrmecology (the scientific study of ants) by Charlotte Sleigh, offers a nuanced account of Wheeler's intellectual formation and his later battles with E. O. Wilson (Sleigh, 2007: p. 248, n. 4).

individual ants *within the context of the colony* could not be explained in terms of individual inheritance. Mutual feeding relations were the true and necessary cause of social forms of life" (Sleigh, 2007: p. 79).

At least since the time of Wheeler's writings about *trophallaxis*, biologists have drawn analogies between the productive capacities of human societies and those of social insects – comparing the ability of human workers to earn wages to the ability of ant workers to collect food; comparing the collective wealth of a nation to the amount of energy stored in nests with caches of food or in the bodies of workers; comparing systems for producing commodities to systems for reproducing new ant queens (Sleigh, 2007: p. 169). These comparisons have been grounded in economic models of rationality and scarcity.

Wheeler based his model of ant society on the work of Vilfredo Pareto, an Italian economist from the early twentieth century, who in his early writing argued that human beings act rationally in pursuing their economic ends. Later in life Pareto studied celebrations of great occasions, jubilees, graduation ceremonies, religious ecstasies, and excesses of all kinds (Millikan, 1936: p. 327). Pareto suggested that human proclivities for these excesses were evidence of what he called "residues," forces which were distinct from instincts or biological drives. But, in Wheeler's hands, Pareto's work on "residues" was inflected with functional evolutionary explanations. Wheeler suggested: "The residues of the common man condemned him to a life that was functionally similar to the ant's" (quoted in Sleigh, 2007: p. 86).

Figure 13.2 Treehopper nymphs feed on sap from plants by piercing the stems with their beaks. Excess sap, concentrated in a honeydew, is exuded out of the nymphs' anus and this sugary liquid often attracts ants. In this picture *Ectatomma tuberculatum*, a closely related species to *E. ruidum*, is tending a treehopper nymph in the Jatun Sacha Reserve in Ecuador. Photo courtesy of Alex Wild.

Paging back and forth again, past nearly 100 years of intellectual history, produces a parallax effect that brings new dimensions of insect sociality into focus (see Strain, 1996/1997). While holding on to Wheeler's insights about nourishment/interchange, I returned to Panama in 2008 and began to rekindle my collaborations with Bill Wcislo, a Smithsonian staff scientist who supported my undergraduate studies of *Ectatomma*. Fujimura (1998: p. 347) suggests that the Science Wars were "not about science versus antiscience, not about objectivity versus subjectivity, but about authority in science: what kind of science should be practiced, and who gets to define it?" Keeping Fujimura's words in mind, I began to design an experiment with Bill that would speak to timely concerns in both of our disciplines.

Bill brought me up to speed about developments in research on *Ectatomma* including a new study of "thievery" by Michael Breed. I had read Breed's earlier papers more than a decade earlier. Breed suggested that individual "thief ants" use chemical camouflage to gain access to neighboring colonies. "Thief ants have reduced quantities of cuticular hydrocarbons on their surface," Breed reported, "and their cuticular hydrocarbon profile is intermediate between the hydrocarbon profile of their own colony and the colony from which they are stealing" (Breed *et al.*, 1999: p. 327; see also Breed *et al.*, 1990, 1992).

My own observations of *Ectatomma* colonies as an undergraduate had already led me to suspect that there was more to the story than "stealing." At the time I speculated that they might be engaged in "trading" rather than thievery. While talking with Bill in 2008, while working with him to design an experiment, I said: "Breed's characterization of these exchanges as thieving has always seemed hasty to me, perhaps neighboring colonies can become allies." Making a quick interdisciplinary translation and conceptual imposition, Bill said: "Nobody has ever demonstrated reciprocal altruism among distinct ant colonies. Let's see if you can." And after thinking a moment, he added: "I would never suggest this as a project to a biology postdoc. It won't involve any new techniques or fancy toys."

Breed's study of thievery was restricted to watching solid food move among nests above ground. Bill and I decided that further studies should focus on the exchange of liquid food, trophallaxis, in laboratory colonies. This would enable us to know if thievery was taking place or if gifts were involved, what Bill glossed as reciprocal altruism.

As I began collecting *Ectatomma* colonies for this experiment I visited a festive space, a place where the value-added excess of late capitalism is routinely consumed. I found a lively patch of ant nests in the leaf litter of a huge *Pseudobombax* tree and in the plastic litter left behind by human picnickers. In a fragment of forested land next to a waterfall in El Giral, a small farming community about an hour outside of Panama City, I uncovered six *Ectatomma* nests among packaging of two brands of chocolate chip cookies (Choki's and Creamas Cuky), a supersized Cheetos bag, and some discarded wrappers of Papitas, a cheese-flavored snack. Amidst a leftover cardboard case of Miller Genuine Draft, as well as Balboa and Panama brand beer cans, I discovered a red bottle cap, a product of the Coca Cola Company, with a cryptic message printed inside: "*Sigue participando*" – keep participating.

After having a picnic of my own in El Giral with friends – Daniella Marini, an Argentinean ecologist who earned a Masters degree from Yale's Forestry Program, and Jesus Hernandez-Montero, a bat specialist from Mexico – I enlisted their help in observing and recording the transfer of food among *Ectatomma* nests. In the shadows of human surplus, in this place where the excess fructose corn syrup and grain from North America and elsewhere was being expended in celebrating minor occasions and jubilees, we found certain species flourishing. Distinct nests of *Ectatomma* were exchanging small insects, crumbs left by picnickers, as well as small protein-packed snacks from *Cecropia* plants called Müllerian bodies. Worker ants exited the entrance of one colony and marched, usually unmolested, into the entrance of another colony.

After unearthing three colonies in El Giral I transported them back to the Smithsonian laboratories. There I assembled an experimental apparatus out of found objects and specialized equipment – plastic tubs, petri dishes, dental cement, aquarium tubes, a slippery substance called fluon, and a Sony digital video camera. In working to produce an experimental matter of fact, that members of distinct *Ectatomma* colonies exchange liquid food via *trophallaxis*, I embedded certain assumptions in this apparatus – namely that these ants would come to regard my assemblages of plastic and plaster as "a nest" and that their behavior in such a nest, exposed to the light of day, is analogous to what they do underground (see Shapin and Schaffer, 1985: pp. 14, 112). After attaching two nests to a common foraging arena, and giving the ants a week to adjust to their new circumstances, I let the paired colonies interact.

Inside this experimental apparatus I duly observed and recorded *trophallaxis* among the colonies I collected from El Giral – workers holding drops of sugar water opened their mandibles, retracted their mouthparts, and fed workers from another colony who gently antennated the donor's clypeus, mandibles, and labium. When I paired the colony I collected from El Giral, with one from nearly ten miles away in the Canal Zone, I initially observed aggression among the ants – biting and dragging each other around the foraging arena. After growing accustomed to each other, after about a week, these unrelated ants started venturing into each other's colonies, and eventually feeding each other with *trophallaxis*.

These observations do not yet constitute a scientific fact – at this point there is a sample size of two paired colonies. If these observations can be replicated in other colonies, then it will be clear to the peers of Michael Breed that *Ectatomma* workers are not just engaging in thievery, as he suggested. Painting individual ants with a unique color code, and tracking their social interactions over long periods of time, would let us gather data that speaks to Bill Wcislo's hypothesis – that members of distinct *Ectatomma ruidum* colonies engage in reciprocal altruism. Finding that individual ants seem to be rational economic actors, like a long list of other animals – lions, crows, and baboons, for example – would certainly be of interest to many biologists. Perhaps, though, these creatures don't have good economic sense. Further research with *Ectatomma* might reveal that their gifts of liquid food might happen according to fleeting whims, sentiments about the distribution of surplus that escape rational calculus.

Becoming with significant others

If adult ants are part of ensembles with their own kind, if individuals are enfolded into relations of care through fluid exchanges with their peers and with their larvae, perhaps they also care for other species of beings and things. The lives of *Ectatomma* ants are entangled with plants that secrete sugary liquid offerings, phloem-sucking leafhoppers that exude honeydew treats out of their anus, and caterpillars that communicate with the ants in high-pitched stridulatory sounds (DeVries and Baker, 1989; DeVries, 1990). Using a particularly clever trick some *Ectatomma* sniff out the pheromones of other ants, smaller species like *Pheidole*, and follow their chemical trails to sources of food (Perfecto and Vandermeer, 1993). To play with Martin Heidegger's language, *Ectatomma* workers are captivated (*benommen*) by other beings and are open to possible becomings – new kinds of relations emerging from non-hierarchical alliances and symbiotic attachments with other agents (Heidegger, 2010).

Wandering within the *cosmos*, the riotous diversity of the rain forest, individual *Ectatomma* ants form *political* articulations with particular individual plants (Stengers, 2005: p. 995). Building cosmopolitical worlds – together, tooth and nail, with other organisms – ants form stable, but contingent, relations against the backdrop of the unknowable beyond. "The species of *Ectatomma* are widely distributed, enterprising ants," according to an early fellow aficionado, Dr. O. F. Cook. "Instead of being a rare 'archaic' curiosity, [it] is decidedly the... most abundant insect of the Guatemalan cotton fields" (Cook, 1904b: p. 611). Cook's work also offers ample evidence that *Ectatomma* ants are not trapped, as philosophers in Heidegger's tradition might have it, within a particular environmental world (*Umwelt*). In a separate article, he wrote: "the insect is not, like some of the members of its class, confined to a single plant" (Cook, 1904a: p. 864). Since Cook's time, other investigators have found this ant tending the extra-floral nectaries of many other plant species, for example, on woody liana vines (*Dioclea elliptica*) in the canopy of a low-land Amazonian rain forest of the upper Orinoco and on saplings of a tree in the legume family (*Stryphnodendron microstachyum*) on the Caribbean slope of Costa Rica (de la Fuente and Marquis, 1999; Blüthgen *et al.*, 2000).

Diverging values and obligations structure ambivalent relationships between ants and plants – cosmopolitical articulations characterized by mutual utility and mutual exploitation. Douglas Altshuler has found that the presence of my favorite ant species has certain positive effects for *Psychotria limonensis*, a common shrub in the forest understory of Central America. *Ectatomma* foragers increase the rate of pollination for this species – likely because they startle pollinators, like butterflies, making them move to other plants. Ants also serve the interests of *Psychotria* by defending the plant from herbivorous insects and preventing the loss of ripening fruits. The cosmopolitical world of *Psychotria* also includes fruit-eating birds – tanagers, manakins, and neotropical migrants – that eat ripe fruits and disperse the plant's seeds. Even if both *Psychotria* and *Ectatomma* have cause to be interested in the continued existence of each other, the ants do not always act in the best interest of the plant and its avian companions. Ants scare off fruit-eating birds. After fruits ripen, the continued presence of ants thus does not serve the assumed interest of the plants in seeding new territory (Altshuler, 1999).

While jealously guarding their plants from flighty interlopers, *Ectatomma* ants remain open to overtures from other entrepreneurial agents – creatures that work to enlist them in competing cosmopolitical worlds. "Adding insult to herbivory," in the words of Philip J. Devries, *Ectatomma* ants sometimes welcome leaf-eating caterpillars to feast alongside them on plants with extra-floral nectaries. These caterpillars have noise-making organs that attract *Ectatomma* and other sorts of ants. The sounds made by the caterpillars average at 1,877 hertz, which would be audible to human ears if they were not so very faint. Their repertoire ranges from simple "bub... bub..." sounds to fancier noises such as "beep ah ah ah beep" and "biddup... biddup... biddup." Caterpillar calls summon ants to their defense against predatory wasps and parasitic flies. As a reward for responding to the summons, the caterpillars secrete a liquid gift – a nutritious liquid that is significantly higher in amino acid concentrations than the plant nectar. *Ectatomma* ants tend the caterpillars "with greater frequency and fidelity" when compared to the plant (DeVries and Baker, 1989; DeVries, 1990).

Lori Gruen's notion of entangled empathy might help explain why ants have greater fidelity for caterpillars rather than plants. Entangled empathy is not a mere instinctual response, but involves a commitment to the well-being of others – an awareness of others' interests and a motivation to satisfy those interests. Gruen is developing her ideas about empathy to understand multispecies entanglements – specifically her own interactions with chimpanzees (Gruen, this volume). Exporting these ideas beyond our own situated perspectives, the embodied perspective of primate vision, contains the danger of imposing anthropomorphic assumptions on other worlds. Even still, Gruen's work prompts me to ask: do ants perceive the interests of the plants they protect? Do they recognize plants as beings in the world? Quite possibly not. Are ants aware of the caterpillars' interests and are they motivated to fulfill them? Quite possibly yes. With intriguing sounds, and an anatomical structure similar to ant larvae, it seems plausible that these caterpillars appear to *Ectatomma* as beings (cute baby insects) that demand empathetic regard.

Gruen's work also offers a point of entry to what Matthew Chrulew (2011: p. 134) has identified as one of the central ethical questions of our time: how should we love in a time of radical ecological transformations? The agency of *anthropos* – the ethical and reasoning being that Enlightenment Europeans conjured as their inheritance from classical Greece – has recently been scaled up to embrace and endanger the whole planet. In the Anthropocene, the era of excess when humans have become a geomorphic force, our species has been figured as the agent driving climate change and the large-scale destruction of ecological communities (see critical discussion in Kirksey and Helmreich, 2010: p. 549). In this context, Deborah Bird Rose and Tom van Dooren have asked:

Given that creatures who are so vividly present in our imaginative lives are nonetheless on the edge of loss, what hope could there possibly be for the countless other creatures who are less visible, less beautiful, less a part of our cultural lives? What of the unloved others, the ones who are disregarded, or who may be lost through negligence? What of the disliked and actively vilified others, those who may be specifically targeted for death?

(Rose and van Dooren, 2011: p. 50)

Escape

With these questions in mind I ventured beyond the realm of the Smithsonian Tropical Research Institute, a social world of ecological scientists where my own love for *Ectatomma* was unremarkable. I began living as an ethnographer in the City of Knowledge – formerly Clayton Army Base, the one-time command/control/intelligence center of the US Military's Southern Command. My temporary residence was an army barracks that had been converted into a backpacker hostel. The landscape of empire had become a picturesque spot of refuge for road-weary travelers on the gringo trail.

The City of Knowledge is now a suburban enclave populated by middle-class Panamanians, indigenous Kuna, staff of international organizations, and a few remaining white Zonians. Here transnational institutions of governmentality and medicalization have begun to inhabit the infrastructure left behind by the US military: the Red Cross, the Nature Conservancy, the United Nations, and the Organization of American States are among the new resident organizations. On an evening bicycle ride in misty rain, I found many other residents engaged in the pursuit of physical fitness. An aerobics instructor was screaming out chants at the top of his lungs to a group of women doing exercises on big inflatable balls inside a huge Kiwanis Club gymnasium. A pair of men, pitcher and batter, were at work in a nearby cage. Joggers, and many other bikers, hailed me with smiles, nods, and lifted eyebrows – recognizing me as a fellow recreator and a possible neighbor.

I found *Ectatomma* ants foraging in the shadows of abandoned satellite dishes, collecting dead insects under electric lights, and living in an expansive network of nests in neatly manicured lawns. Few of my fellow humans were articulate about the ants living in the grass, all around them. More than one of my interlocutors looked at me as if I were a little off, for initiating a conversation about insects. Only after living in the Reverted Zone for several weeks, did I discover some housewives and grounds keepers periodically going around their lawns with boxes of powdered poison, sprinkling it on nests of *Ectatomma* and multiple other species of ants.

Occasional attempts to senselessly poison them aside, *Ectatomma* is flourishing in the Anthropocene. Quick to exploit emergent opportunities, never just sticking to one world, these ants are constantly moving among different beings and are open to possible becomings. This species is proliferating largely beyond the purview of human dreams and schemes. Perhaps these small animals are comfortable with their status as "unloved others," anxious to escape from fleeting encounters with humans into the cosmos, into the unknown beyond anthropocentric worlds.

While refusing the cosmopolitan illusion of Immanuel Kant that there might ever be a final peace (Stengers, 2005), I suggest that we should learn to better embrace species such as *Ectatomma*, cosmopolitical creatures that are good for humans to live with in common worlds. Being with this species responsibly might involve an openness to possible becomings from a respectful distance. If touching significant others, in Haraway's words, generates lively becomings with certain species of companions, "flesh-to-flesh and face-to-face," then ethical engagements with other sorts of critters demand tactful politeness.[5] Composing common worlds with other species might involve enacting new sorts of loving gestures, making tactful cosmopolitical proposals that leave room for the possibility of escape.

[5] Here I am inspired by the work of Matei Candea (2010, 2011) and Augustin Fuentes (2010; this volume), who both write about "waiting together" with other species. Candea suggests that certain species demand "inter-patience," from humans, rather than straightforward "inter-action" (Candea, 2010: p. 249).

Part III

Toward respectful coexistence

14 Social minds and social selves
Redefining the human–alloprimate interface

Agustín Fuentes

Social and ecological niche construction and hyper-sociality are central aspects of our primate heritage. Social complexity and modern evolutionary theory are core elements in a comprehensive understanding of the interface between humans and alloprimates (the other primates aside from humans). This approach is epitomized by the emerging field of ethnoprimatology: the theoretically and methodologically interdisciplinary study of the multifarious interactions and interfaces between humans and other primates (Fuentes, 2012a). Our perspectives on current relationships and the potential for conviviality with the apes (and many other species), are constrained when they rely only on biological/phylogenetic similarity arguments and simplistic evolutionary understandings, or when they ignore the histories and everyday patterns of the human–alloprimate interface. We have more, and less, in common with apes than many of us think. This matters in talking, and thinking, about our relationships with other primates.

As an anthropologist who studies human and other primates, and who is centered in a modern evolutionary approach, I am forced to consider the ways in which we (humans) think about, interact with, and coexist with other animals. I have lived among humans and other primates in Asia, North Africa, and even the Southern tip of Europe. I have observed the myriad ways in which humans look at, think about, consume, and cohabitate with other animals across the planet. These experiences in various cultures and ecologies have convinced me that the interface between humans and other forms of life is neither uniform, nor simple. I've published elsewhere (Fuentes, 2006) that a core aspect of the overlap between humans and animals is in the shared aspects of personhood. There is humanity in animals and an animality in humans: humans are animals, we are mammals. We are human apes, a particular kind of ape that manipulates ecosystems across this planet and is capable of intense cruelty and amazing compassion via symbol, language, niche construction, and interaction with other animals and ourselves. We are not cruel in our core, but we are quite capable of intense cruelty. Our relationships with other animals are complex and culturally contingent and contextual: no uniform or simple perspectives – ethical, ecological, ethnological, or literary – can effectively categorize them.

In this chapter I would like to make two main points that might facilitate forward movement in thinking about the human–other primate relationships. These are:

The Politics of Species: Reshaping our Relationships with Other Animals, eds R. Corbey and A. Lanjouw. Published by Cambridge University Press. © Cambridge University Press 2013.

1. Evolutionary and ethnoprimatological approaches suggest a reappraisal of main-stream notions regarding the human–alloprimate interface.
2. Considering the concepts of altruism and cooperation, and the social mind–primate mind–cultural mind scenario can help in thinking about human–alloprimate relationships.

Evolutionary and ethnoprimatological approaches suggest a reappraisal of the human–alloprimate interface

Evolutionary approaches are core to understanding long-term biological histories within and between taxa. In the primates, there is a series of adaptive trends and socio-ecological patterns that unite us (continuities), and an array of specializations related to these trends that differentiate elements of behavioral and adaptive trajectories among taxa (discontinuities).

Examining evolutionary continuities reveals specific, undeniable, overlaps and connections between humans and other primates, especially the apes. These connections are not simply in our morphology or general physiology, but rather in the ways in which our social selves interface and engage with the world and with one another. Recent research demonstrates the centrality of complex and long-term social relationships, dynamic social networks, and the origins and patterns of reciprocity, even altruism, in the monkeys and apes (e.g., Silk, 2007; MacKinnon and Fuentes, 2011). These patterns emerge from evolutionary processes and their phylogenetic stamp is felt in humans as well. This means that we must incorporate our anthropoid and hominoid abilities for social connections and complex social networks into our considerations of the human–other primate interface and into our perceptions and behavior toward our primate kin.

At the same time, evolutionary processes also result in divergence via specialization and restructuring in behavioral and physiological processes. We share mirror neurons with our monkey and ape cousins, but their patterning and function varies for humans as do the clusters of von Economo neurons we share with apes and cetaceans (Damasio, 2010; Ramachandran, 2011). Our brains have much in common but also show important deviation in form and function. Behaviorally humans utilize symbolic action and language in the creation and maintenance of our social systems. We employ a diverse array of bodily and external modes to harness the material world in a much more intensive and extensive manner than do any other primates. These discontinuities result in particular and substantial differences in the ways in which humans and apes and monkeys inhabit and engage with the world. In humans, our particularly successful tendencies to alter landscapes and ecologies, combined with our incorporation of symbolic and social processes into evolutionary pathways, sets us somewhat apart from other primates and shapes the ways we see and interact with other species.

The study of the human–other primate interface is an interdisciplinary project that places humans and other primates in integrated and shared ecological and social spaces. This perspective is best seen in the practice of ethnoprimatology (Fuentes, 2012a). Ethnoprimatological approaches force a reappraisal of mainstream notions regarding

Figure 14.1 Across much of the planet humans and other primates are sharing space and place. Increasingly cities and towns are as much home to some monkeys as are forests. In much of Southeast Asia every time you look in the rear-view mirror a monkey is likely to show up. Singapore, photo by Amy Klegarth.

the human–alloprimate interface. We can no longer simply consider that there is a state of "nature" where pristine populations of organisms exist, outside human impact and interface. We reside in the Anthropocene (Rose, 2009; *Economist,* 2011) and anthropogenic influences on the structure and functioning of the planet are ubiquitous; the planet is changing faster than we can study it. To better understand dynamic interspecies relationships we are forced to shift our gaze from the conflict at species' interfaces to encompass the patterns and possibilities of mutual ecologies and histories. These interfaces are interweavings of the social and structural ecologies of the participants into a naturecultural relationship (Fuentes, 2010).

The ethnoprimatological approach, alongside a broader multispecies perspective, is concerned with the integration, engagement, and interface between ourselves and other kinds of life (Kirksey and Helmreich, 2010; Kirksey, this volume; Riley, this volume). Viewing relationships between organisms as shaping, and being shaped by, mutual ecologies in behavioral, ecological, and physiological senses we can make greater inroads into our interdisciplinary conversations about the Anthropocene. Using ethnoprimatological and multispecies perspectives as the frame for studying the human–other primate interface can enhance our abilities to envision emergent relationships in the Anthropocene, wherein human social, political, perceptual, and economic action are entangled with other primates' (and other organisms') behavioral and ecological lives (Fuentes, 2010, 2012a; Kirksey and Helmreich, 2010).

In the case of our interface with the other primates who live alongside us (alloprimates), humans display a myriad relationships, perceptions, and impacts. Ethnoprimatological investigations demonstrate that physical and social ecologies at the human–alloprimate interface are shaped by shared histories, economies, cultures, and landscapes as well as each species' behavior and physiologies (Fuentes and Wolfe, 2002; Fuentes and Hockings, 2010; Fuentes, 2013). These studies hold up hope for some patterns of interaction, such as those of humans and macaques throughout Asia, and at the same time suggest continued destruction and dire futures for the large ape species in their interrelations with humans. Ethnoprimatology shows us that human perceptions matter a great deal in all of these relationships, they structure the human action and the ways in which symbolic worlds interface with the ecological ones. This approach also demonstrates that human perceptions of such relationships can and do shift over time and place; these are seeds of hope.

The human–alloprimate interface

The human–macaque interface is receiving increased attention from policy-makers, academics, and the general public in much of South and South East Asia. Unfortunately, the prominent characterization of this interface is one of conflict and competition. Behavioral, ecological, and pathogen overlaps between humans and macaques are increasingly common and while most studies document the disputes over space and resources between humans and macaques, there are substantive data suggesting a range of sustainable coexistence possibilities (Fuentes, 2012b). For example, nearly all studies indicate that the central focus of conflict arises from the presence, or assumed presence, of potentially contested food sources. Some of these same studies show that even small modifications of human behavior result in substantial changes in macaque responses and reduced conflict. Additionally, overviews of human–macaque interfaces reveal that space conflicts frequently have multiple solutions, few of which are ever implemented.

In Bali there are tens of thousands of macaques and millions of humans. The macaque monkeys thrive on the island, especially in and around villages and temples complexes. At locations where monkeys and humans are co-resident there is substantial overlap in area use, water source use, and potentially shared pathogen environments suggesting that humans and macaques share physiological, spatial, and social connections (Fuentes, 2010). Rather than being only seen as pests or simply parts of the natural landscape, the macaques play important roles in Balinese socio-religious spaces and in local economies. At the same time the Balinese play important roles in the daily lives of the macaques. This interface also plays an important role in the long-term ecology of Bali itself: macaques and humans reside in, and help construct, each other's social and ecological niches. An ethnoprimatological approach, integrating the social, mythical, economic, and historical alongside the ecological and behavioral, enables us to see that development, maintenance, and dynamism of the human–macaque relationships on Bali provide hope for the possibility of generosity between humans and macaques and the potential for a sustainable relationship.

This same kind of ethnoprimatological approach focused on the ape–human interface would bring less optimism. Based on what we know about the ape–human interface, the future of the great apes, gorillas, chimpanzees, and orangutans, is extremely bleak. The basic pattern of this interface is a negative one for the apes. These apes have substantial range needs, a requirement for a diverse fruit representation in their diets, and have slow and easily disrupted reproductive cycles, making them particularly susceptible to radical anthropogenic habitat alteration. Also, these apes are large, their behavioral profiles (and human responses to them) make co-residence with human populations, particularly agricultural ones, extremely problematic (Hockings et al., 2010; Fuentes, 2012a). While there is a small benefit from the presence of taboos on hunting chimpanzees and gorillas in some indigenous peoples in Central Africa, these are fast disappearing with social, economic, and cultural changes. There are no such beliefs about orangutans in South East Asia; however, there is a large and growing orangutan tourism industry. In Africa apes are targeted by bushmeat hunters and until recently orangutans were prized in the Asian pet trade. The bodies (in part or in whole) of all ape species are highly prized by some human cultures for assumed therapeutic benefits, and in Central Africa ape meat can be an economic boon. Across Africa, increased human presence in areas where humans and apes have coexisted without conflict is changing relationships. In areas of increased human influx we see expanded conflict and along with it nutritional stress, disease, and political and economic instability. In the last habitats of the orangutan, the economic drive to convert forest land to economically valuable environments driven by local and global economics is the primary cause of ape population decline.

Serious inquiry into our relationships with other primates must include these both evolutionary and ethnoprimatological perspectives, and their optimistic and pessimistic possibilities. For the theme and goals of this volume it is worthwhile considering, in somewhat greater detail, the core points above: continuity and discontinuity in evolutionary histories and our current understandings about human–other primate interfaces.

Cooperation, reciprocity, complexity, and altruism in primates: continuities and commonalities

Primate species that exhibit complex coalitionary behavior inhabit a broad geographical range, have high levels of behavioral plasticity, an extended period of socially mediated learning, and also display a degree of malleability in their adaptive niches. This malleability can be seen in the formation of complex social networks in many anthropoid primates (the monkeys, apes, and humans). There is an adaptive advantage of social networks in terms of functioning as a niche construction mechanism in many primate societies (MacKinnon and Fuentes, 2011).

Humans are primates, but we appear to display a wider array of social complexity and altruistic behavior (or at least apparently altruistic behavior) than other primates. Can we envision the modern human system in the context of phylogenetic connections to other primates? If so, is this a cause for optimism about the human–alloprimate interface?

We can consider aspects of behavioral flexibility and plasticity as means to an end in hominoid socio-ecological landscapes. There is an emerging recognition in evolutionary theory that phenotypic plasticity – continuous and reversible transformations in behavior, physiology, and morphology in response to rapid environmental fluctuations – is important for many organisms. It is also the case that organisms are not only impacted by their immediate environments but also, in part, shape those environments and thus the selection pressures that they face. In a sense, we see that many organisms are engaged in some level of niche construction (Odling-Smee *et al.*, 2003).

If niche construction is occurring in the hominoids (apes and humans) we can envision the variability at the group and population levels as a major component of the hominoid niche. For example, Flack and colleagues (2006) argue for social niche construction in anthropoid primates where social networks constitute the essential social resources in gregarious primate societies. They posit that the structure of social networks can play a critical role in infant survivorship, cooperative behavior, social learning, and cultural traditions. Hominoids engage in social niche construction and individuals in hominoid groups negotiate social networks, modifying their boundaries and internal landscapes in the context of changing demographics and, potentially, ecological variables. The flexible social and ecological characteristics of the hominoids can act as a niche-constructing mechanism that both maintains behavioral flexibility and modifies selection pressures that then feedback on the system modifying the patterns that the flexibility takes (MacKinnon and Fuentes, 2011; Malone *et al.*, 2012).

Primates and all social mammals share a specific type of social intelligence wherein social relationships are important and kin selection and reciprocal altruism are salient forces. Primates (and a few other hypersocial mammals such as cetaceans and elephants, see Marino, this volume) share a form of primate mind where even more complex social networks and increased reciprocity form a central facet of daily lives and adaptive niches. In the apes we can see an increase in the degree of reliance on, and complexity of, social networks and cooperative alliances and an enhanced degree of reciprocity relative to other primates. In this context, humans are different in the degree and vastness of the amount of information that is disseminated socially through space and time, and the ways in which we do it (e.g., Herrmann *et al.*, 2007). In humans an extreme type of hyper-sociality arises as an emergent property of extensive social niche construction, language, and complex cooperation, with reciprocity becoming the core behavioral pattern (Nowak and Highfield, 2011; Fuentes, 2013).

An extension of the primate social intelligence hypothesis lets us envision a cultural intelligence hypothesis not as some relatively distinct human pattern but rather as one of the possible extensions of the primate mind. All primates have evolved social–cognitive skills for cooperating and competing with group/community members, humans have also evolved skills for establishing distinct, cross-temporal, cultural groups, with different physical and symbolic markers (social institutions, artifacts, language, etc.). Here it is worth noting that while other primates transmit many

Figure 14.2 Home is where the heart is and where the forest was. Other primates are increasingly living in and around humans and land conversion and habitat alteration continue unabated. Some species, like this long-tailed macaque (*Macaca fascicularis*) do well, others do not. As humans we need to work to understand how to construct and maintain sustainable multispecies relationships in and around our homes. Photo Agustín Fuentes.

behaviors socially their societies do not require participation in specific cultural, linguistic, and symbolic interactions to the same extent as in humans. Humans have a *species-specific* set of social–cognitive skills (that emerges early in ontogeny) for participating and exchanging knowledge in cultural groups (see Herrmann *et al.*, 2007; Dunbar *et al.*, 2010).

In each of these "minds" there is a potential for altruism as a spin-off from the increasingly centralized role of social reciprocity. This potential for altruism – or even just the basal proclivity toward an altruistic ability – is probably present in many primate lineages and across the social mammals, and might even potentially reflect a gradient of the "cultural mind as niche construction" concept (e.g., MacKinnon and Fuentes, 2011). Plasticity and resulting adaptability in primates, especially some monkeys and apes, connects well with the patterns we see as central to the human niche. Bonding and cooperation play a significant role in social niche construction among primates and in humans the emergence of regular altruism (or altruistic-like behavior) is likely related to an expansion of this core primate pattern.

Evolutionary discontinuity in humans and alloprimates: two cases to consider

One particular evolutionary discontinuity suggests a kind of optimism for the human–ape relationship. As noted above, bonding and cooperation are evolutionarily important for primates. However, humans are able to regularly engage in physiological, behavioral, and emotional bonding with diverse members of their community, going well beyond the mother–infant or male–female/male–male friendship bonds in many primate societies and more akin to the strong social pair-bond found in a minority of primate societies. In fact, humans have expanded on other primate social bonding behavior and amplified its importance in creating an intricate structure of bonding involving complex social cognition, which makes use of both distinctive and shared neuroendocrinological, symbolic, and behavioral processes. This is, at least in part, proposed to be a major factor in our evolutionary history, and core to the human social niche (Herrmann *et al.*, 2007; Silk, 2007; Fuentes, 2009a). Of relevance to our discussion here, it appears that a distinctive evolutionary discontinuity is that humans can cast this physiological, social, and symbolic bonding "net" beyond biological kin, beyond reciprocal exchange arrangements, beyond mating investment, and, in particular, even beyond our species (with dogs, horses, even other primates) (Mullin, 1999; Smuts, 1999; Haraway, 2003; Fuentes, 2009b, 2010; Olmert, 2009). In short, we have an evolutionary basis for incorporating other species into our network of caring and sharing, our "kin" writ large. This is an important observation as we are not reliant only on economic, phylogenetic, philosophical, or political arguments to justify the inclusion of apes (for example) into our kin networks. We can also point to an evolutionary pattern, a human adaptation, which is ready-made to bring others into moral equivalencies and social kinship with us. This is a pattern that is malleable, and socially and symbolically mediated, thus our efforts to bring apes into closer connections, politically and practically, should exploit it.

A second evolutionary discontinuity leads to a more pessimistic view. There is a distinctly human evolutionary trajectory of niche alteration for the benefit of human populations and the detriment of many other species (Twine, this volume). Although many can argue that the ecological benefits to humans from such large-scale landscape changes are relatively short term (in a global sense), current evolutionary patterns are not based on future success or what is "good" for species, rather they are reflective of populations exploiting their environments effectively in the recent evolutionary past and the evolutionary present.

This pattern of niche construction, and exploitation, is what brought us into the Anthropocene and is how our species has grown from a few tens of millions to over seven billion in an evolutionary second. Our tendency to alter the landscapes around us, our speciesist practices that modify ecologies and exploit other life forms for our benefits have created a diverse array of ecological pressures and inheritances, and stimulated systematic structural changes in trophic relationships. This has, on average, been particularly deleterious for large mammals. As noted above, the apes are a good example. Coexistence with apes is difficult for humans as we overlap in body size, range use, and many social patterns, but have very different landscape and habitat preferences, dietary and spatial needs, and population growth and spread patterns. Free-ranging humans and free-ranging apes conflict when in contact, and the humans seldom lose this type of competition. Population movements, dietary and cultural practices, perceptions of exclusivity of use and ownership of space, as well as construction of new habitats (urban, plantation, etc.) generally affect apes negatively.

Human niche destruction, alteration, and construction are key to our adaptive successes and are a ubiquitous characteristic of our species. As human populations increase, as our ranges expand across more and more segments of the forested areas of Africa and Asia, apes will continue to suffer. We modify our environments to suit us, and we are good at it. Social and legal changes enforced, top–down, via governmental mandate, such as protected areas and illegalization of harming apes, will not be successful in the long term without providing viable and logical alternative options to the current patterns of land and animal use (see also Lanjouw, this volume). For two million years our kind (the genus *Homo*) has been finding more and more effective ways of altering our surroundings to make them work for us, and we are not about to stop that trend. However, niche construction is not always niche destruction and changes to social and symbolic perceptions can affect the ways in which we evaluate changes to the physical and ecological landscapes, so even in the face of convincing pessimism about human evolutionary patterns, there is some small potential for a change in course.

Parting thoughts

This brief overview of core continuities and discontinuities in our adaptive patterns and the insights that we can glean from ethnoprimatological studies is meant as a mini-primer illustrating the amazingly complex, but at least partially hopeful, study of the human–other primate interface. A main point I wish the reader to take home is that our arguments for, and actions related to respecting, protecting, and living with other primates, especially apes, need not be made exclusively on ethical, economic, or phylogenetic grounds. Utilizing the concepts of evolutionary discontinuity and continuity enables us to make phylogenetic similarity arguments while at the same time not ignoring or intentionally diminishing the easily observable and very real differences between humans and other apes. These discontinuities also show us that humans have a capacity for inclusion not typical to other species and that our evolutionary sameness and differences can be used to argue for, and act toward, similar outcomes.

In the same vein, ethnoprimatological investigations can show us the strong and weak ecological and symbolic links between peoples and other primates who share the same landscapes. They can illustrate the ways in which people have or have not coexisted, sustainably, with other primates and potentially provide insight into cultural/symbolic perception shifts that might facilitate improvement of such relationships in diverse places around the planet. However, we have to also be ready and willing to undertake projects that demonstrate the non-sustainability of such relationships. We have to assess, in a data-rich format, scenarios where there is no option of coexistence and learn from them as well.

Humans impact and alter, and are altered by, the world in ways that are more impactful, far reaching, and long term than any other animal on this planet. We also have more, and less, in common with apes than many of us think. These realities and their constituent parts need to be core facets when we think about, talk about, and act on our relationships with other primates.

15 The human–macaque interface in the Sulawesi Highlands

Erin Riley

The field of anthropology is fundamentally concerned with better understanding what it means to be human, across both time and space. An underlying assumption behind this objective is the undeniable existence of uniquely human qualities. Of course, what we consider to be unique is affected by a multitude of factors, including culture, history, and our ignorance of what makes other animals what they are (Asquith, 1997). Nonetheless, to get at this fundamental question, anthropologists (and other scholars) endeavor to address the intrinsic and extrinsic features that define and explain humanity. A notable and increasingly researched area that simultaneously explores both the intrinsic and extrinsic is the interface between humans and other animals with whom they share evolutionary, ecological, or geographical space. Exploring the human–other interface enables learning about ourselves, and why we do what we do, including how we treat our conspecifics. A recent anthropological contribution to this endeavor is "ethnoprimatology" (Sponsel, 1997; Fuentes and Wolfe, 2002; Fuentes, 2012a; see also Fuentes, this volume; Bakels, this volume): the study of the ecological and cultural interconnections between human and non-human primates. The goal of ethnoprimatology is multifaceted; it explores humanity's role in constructing ecological conditions that shape primate behavior; it also examines how humans conceptualize primates and how these conceptualizations in turn shape their behavior toward them (e.g., Cormier, 2002; Fuentes and Wolfe, 2002; Riley, 2010; Riley and Fuentes, 2011).[1]

Given this multifaceted scope, ethnoprimatologists integrate theory and methods from biology, genetics, epidemiology, conservation biology, socio-ecology, cultural ecology, cognitive anthropology, and sociocultural anthropology. Within its home discipline of anthropology, ethnoprimatology occupies a distinct and important place; it bridges the seemingly disparate subfields of biological anthropology (primatology) and cultural anthropology (Riley, 2006). A perennial challenge to the field of anthropology has been the question of whether biological and cultural anthropology can truly coexist given their traditionally (seemingly) disparate epistemologies and methodologies. For instance, it has been argued that primatology and cultural anthropology have

---

[1] See also contributions in the 2010 *American Journal of Primatology* special issue on ethnoprimatology.

The Politics of Species: Reshaping our Relationships with Other Animals, eds R. Corbey and A. Lanjouw. Published by Cambridge University Press. © Cambridge University Press 2013.

Figure 15.1 Two adult male Tonkean macaques (*Macaca tonkeana*) sitting atop a recently felled tree in their home range (village forest in the Lindu enclave, Lore Lindu National Park, Central Sulawesi, Indonesia). This image simultaneously captures the anthropogenic impact on macaque lives and the macaques' ability to cope with such change. Photo Erin P. Riley.

"a lack of common interest" with "a wide epistemological abyss" separating the two (Rodman, 1999: p. 331). I am trained as an anthropological primatologist, meaning that I study non-human primates within an anthropological framework: what can primates tell us about the human condition? But I am also trained in environmental and ecological anthropology, and hence most of my research focuses on how the lives and livelihoods of humans and non-human primates intersect in their increasingly shared environments. Ethnoprimatology therefore focuses less on what non-human primates can tell us *about* people, and more on the ecological, social, and cultural interface *between* non-human primates and people. I would argue that this focus on the human–non-human primate interface is of considerable interest to cultural anthropologists, particularly given a reinvigorated interest in multispecies entanglements in sociocultural anthropology and science and technology studies (e.g., Mullin, 2002; Kirksey and Helmreich, 2010; Lowe, 2010; see Kirksey, this volume).

In this chapter, based on ethnoprimatological research I conducted in a protected area in the highlands of Central Sulawesi, Indonesia, I discuss how people's link to place shapes how they think and interact with the natural world, in particular, their attitudes about conservation and their perceptions and actions toward non-human primates, namely, the Sulawesi Tonkean macaque monkey (*Macaca tonkeana*). I discuss the implications of the research for strategies to alter human actions toward non-human animals and nature in general.

Humans and macaques in Lore Lindu National Park, Central Sulawesi, Indonesia

From 2002 to 2004, as part of a broader study on the ethnoprimatology of *Macaca tonkeana* (Riley, 2005), I examined local knowledge and perceptions of nature and protected area conservation and the cultural interconnections between humans and the Tonkean macaques in the Lake Lindu enclave in Lore Lindu National Park, Central Sulawesi, Indonesia. Lore Lindu National Park, established in 1993 and comprising 217,982 ha, provides habitat for a majority of Sulawesi's endemic mammals, including one endemic macaque taxon, *Macaca tonkeana*, and watershed protection for two major river catchment systems (the Lariang and the Gumbasa-Palu rivers). Situated at approximately 1,000 meters above sea level, the Lindu highland plain is one of two enclaves that are allowed to exist within the park because it is a major rice-growing area and has long-established settlements. The people indigenous to the Lindu plain are members of the Kaili ethnic group, which is further divided into seven distinct groups on the basis of dialect. The Lindu call themselves *To Lindu* (Lindu people) (Davis, 1976). During the designation of the National Park, the granting of enclave status meant that the Lindu people were able to maintain their current agricultural fields and continue to engage in small-scale forest production collection.

Throughout the colonial and post-independence eras the Lindu plain remained a relatively isolated area largely due to official government policy of discouraging immigration to the enclave by restricting education and healthcare facilities (Schweithelm *et al.*, 1992). Over the last 50 years, however, this area has experienced considerable in-migration, including other Kaili from nearby lowland towns who migrated to the area under local resettlement schemes and spontaneously migrating Bugis from South Sulawesi (Acciaioli, 1989). These migrants are attracted to the area for perceived available land for agriculture and, most recently, for opportunities to participate in the lucrative tilapia fishing industry at the 3,000 ha lake situated at the center of the plain (Acciaioli, 2000). Although wet-rice agriculture (*sawah*) predominates in Lindu, and is practiced by both indigenous Lindu and migrants, tree cash crops, such as coffee and cacao, also comprise an important part of the Lindu economy. Cacao (*Theobroma cacao*) cultivation can often be a major cause of deforestation, primarily because new planting along the edges of primary forest is often cheaper than felling and replanting existing gardens when they become exhausted (Donald, 2004). Additionally, the planting of cacao adjacent to forested areas has proven to be disastrous in many areas, as Sulawesi macaques, and other wildlife, are attracted to the crop (Supriatna *et al.*, 1992; Riley, 2007).

Attitudes toward nature and protected area conservation

To understand local perceptions of nature and attitudes toward protected area conservation, I conducted semistructured interviews with community members of the lakeside village of Tomado.[2] When asking respondents whether they perceived negative outcomes associated with the establishment of the National Park (and the concomitant restrictions on forest resource use), I found considerable diversity in perceptions between the Lindu and migrants (Riley, 2005). While the Lindu respondents more frequently responded with "none," migrants envisioned "limited agricultural development" as negatively affecting them. These divergent views reflect differences in the way the forest is conceptualized; the Lindu conceptualize the forest as a source of livelihood and speak to the value of its persistence for future generations, while many migrants envision the forest as an area that should be cleared for additional agricultural development (i.e., rather than a space to be preserved).

In terms of threats to the National Park, both the Lindu and migrants noted rattan collection as the most salient (Riley, 2005). Interestingly, a number of the Lindu respondents envisioned the migrants *themselves* as threats. This perception likely stems from their fear of the overexploitation of forest resources and the disregard of Lindu *adat* by migrants:

They [migrants] go into the forest and make more gardens.   (Lindu villager)

Lindu people like looking at the forest... and rattan that we need isn't too far away here.
(Lindu villager)

Transmigrants don't know the rules of Lindu *adat* about the forest... they only feel that they can benefit from it.   (Lindu man, in his 30s)

Attitudes and actions toward primates

Informal discussions with villagers upon my initial arrival in Lindu revealed that *To Lindu* folk ecology includes stories about human–macaque interactions. Respondents for the formal interviews were therefore asked to recount folklore they knew regarding the relationship between humans and macaques, human–macaque interactions, or human–macaque conflict. Of the 45 people interviewed, 26 were asked to recount folklore they knew regarding the relationship between humans and macaques and human–macaque interactions. The sample of respondents included 17 men and 9 women; 18 were *To Lindu* and 8 were migrants.

A number of themes emerged from their responses (Riley, 2010). For example, one theme involved ideas about human–macaque similarities and the origin of monkeys.

---

[2] Tomado is inhabited by approximately 132 households; 90% of whom are indigenous *To Lindu*. Interviews (n = 45) were conducted with both indigenous *To Lindu* and migrants to ensure adequate sampling of the potential variability in perceptions held by people of different ethnicities and backgrounds. All interviews were conducted in Bahasa Indonesia by the author and then translated into English.

Both Lindu and migrant respondents spoke to the continuity between humans and macaques (Riley, 2010: p. 236):

Two women were collecting *esa*, a man burned the area with the women, and became a monkey.
(Lindu man, in his 60s)

Monkeys come from burned people. The fire created their skin. That is the reason you can't kill them (because they come from people).   (Daughter of a migrant couple, in her 30s, herself born in Lindu)

Monkeys and people are on one line – whereby a human fell into a cooking pot and burned. . . it lost its butt and its hair turned black. I think it makes sense, we have the same hands. (Migrant man, in his 40s)

Monkeys are from people, because they have five fingers too. A human was thrown from the house to the forest and there it became a monkey.   (Migrant woman, in her 40s)

For the Lindu, however, this notion of continuity extends into the cultural realm. Namely, many Lindu envision themselves linked to macaques via folklore that describes a particular tale of a human–macaque interaction that simultaneously imbues respect and disdain (Riley, 2010). The tale tells the story of how a father left behind his young daughter to guard their house and gardens from raiding monkeys while he went to fish. According to the tale, as soon as the father was gone the monkeys began to raid their crops, and ultimately kidnapped the young girl. The young girl screamed, and screamed, and screamed. Upon hearing the commotion from the lake, the father chased after the monkeys, and in his fight to rescue his daughter he left only a male and pregnant female macaque alive. The moral of the story is that the Lindu may not speak harshly or behave negatively toward the monkeys or they will become their enemies and return to wreak havoc in their lives. Interestingly, while younger Lindu respondents were unable to tell the details of the tale, its essence (i.e., human–macaque interconnectedness) remains and continues to guide people's actions toward them (Riley, 2010).

Another cultural interconnection is exemplified in their viewing the macaques as guardians of *adat* (Riley, 2010: p. 237)[3] in a way very similar to traditional Sumatran perceptions of the tiger (Bakels, this volume):

In the past our ancestors made a mistake. A disrespectful courting resulted in a pregnancy, so the buffalo of people here was caught and attacked by all the monkeys of Anca until the buffalo was dead. According to our ancestors' understanding of it, the spirits in the forest were embarrassed because the illegitimate pregnancy was not talked about well but instead the reality was that it was kept quiet. The buffalo was killed by the monkeys as a sign that something bad was committed and that the spirits were angry about it, the land was made filthy. . . this is the story from our ancestors, and almost everybody knows it. The monkeys are really cruel here. . . if there has been a problem they will alert us. . . gardens will be destroyed.   (Lindu woman, in her 70s: Riley, 2010: p. 237)

For some Lindu these cultural interconnections translate into tolerance as they encounter the macaques in their shared ecological space. For example: "If monkeys come to

---

[3] *Adat* is typically defined as the customary laws (e.g., local traditions and natural resource management practices) developed by the members of various sub-ethnic groups.

our gardens we can only shoo them away... you can't be harsh because otherwise they will get angry" (Lindu man, in his 40s: Riley, 2010: p. 238).

So we have already found religion... but don't dare to kill an animal like the monkey. If you find them in the garden tell them to leave. We are vigilant when it comes to the monkeys, we don't just kill them for there will be an effect. If you follow the story that monkeys descended from humans then this means that we are kin.   (Lindu woman, in her 70s: Riley, 2010)

The role of place

Overall, this ethnoprimatological research suggests that respect for nature is linked to place and generated from direct, long-term interactions with it. Encounters with wildlife, such as the macaques, results in an appreciation of the uncanny similarities between ourselves and the rest of the simian world, as well as experiences from which human–primate folklore derives. In Lindu, the outcome of this respect has resulted in local support for protected area conservation and tolerance of macaques, however destructive their behavior may be. However, because cultural norms and actions are not fixed (Sierra, 1999) a reliance on culture for conservation can be problematic (Riley, 2007). Changing socio-ecological and economic realities may result in abandoning the preservationist values that long-held cultural and ecological ties with nature instill. Such a scenario was exemplified in 2007 to 2009 when five prominent men from the Lindu highlands, all indigenous *To Lindu*, were convicted of illegally selling National Park forest to migrants for conversion to agricultural land. While the defense's position was that the defendants did not know it was National Park land (i.e., they saw it as belonging to their ancestors)[4], the defendants also apparently admitted they did it because they needed the money (Niniek Acciaioli, personal communication). Human–macaque folklore held by the *To Lindu* may also eventually be disregarded with changing realities; in fact, there are already some in Lindu who know the folktale but who have chosen to discount it:

He has already heard the story but he is not afraid... but maybe later there will be a problem. He really doesn't want to do it [make traps] again, after I talked with him. [He] said: "as opposed to father tiring out having to go there every day, I want to kill them all. I put traps up along the entire maize crop so that each time they enter they will be finished." I said "don't you do that." After I spoke that way he didn't want to do it again.   (Lindu woman, in her 70s, talking about her son's frustration with crop-raiding macaques: Riley, 2010: p. 238)

Some implications for conservation

How can this case study help alter human behavior toward nature and other primates/beings? It is clear that no one strategy may be enough. The lived experience of the average American or European is drastically different from the community members I describe herein: for the Lindu, their lives are intricately *in* nature, but for the average

---

[4] "Terdakwa Mengaku Tak Tahu Batas TNLL" ("The Defendants did not know the National Park boundaries") by Sahril, 3 February, 2009 in *Media Alkhairaat*.

Figure 15.2 An example of traditional use of forest resources by the Lindu people: the construction of a dugout canoe and the maneuvering of it down the hillside. Photo Rusmanto Lakareba.

Westerner, nature is "out there"; something to be traversed, or even still, feared or conquered. For the latter, encouraging ways for people to reconnect with nature may be an effective strategy to change human behavior (Swaisgood and Sheppard, 2011). The examples I describe from Lindu, however, are sobering; they suggest that one's connection to place may not be enough, at least not anymore or perhaps not for everyone. I suggest two potential strategies for engendering respect and promoting conservation in areas where human–non-human primate overlap or conflict occurs.

The first strategy is using anthropomorphism to garner respect for other non-human beings. In studies of animal behavior, the use of anthropomorphism has long been debated (Asquith, 2011; Fuentes, this volume). While some envision it as an insidious problem, clouding our science (Wynne, 2004), others emphasize the theoretical and practical benefits of anthropomorphism (de Waal, 1999; Daston and Mitman, 2005). De Waal (1999) distinguishes between humanizing or anthropocentric anthropomorphism, which is to be avoided, and animalcentric or heuristic anthropomorphism: a form that emphasizes human–non-human continuity for understanding the function of behavior. It may be time to move beyond the debate of using anthropomorphism in animal studies given the dire state of many primate species. Instead, encouraging forms of anthropomorphism that emphasize human–non-human continuity enable us "to see ourselves with animals as opposed to against them" (Daston and Mitman, 2005), and therefore might be a way to engender more respect for other beings.

My second suggestion has to do with "sharing space" (Lee, 2010). Throughout much of the tropics, primates and humans are increasingly overlapping in their use of space. Lee (2010) rightly notes that "sharing space" requires that people acknowledge the need to do so despite potential negative ramifications (e.g., crop raiding). Encouraging an appreciation of the biological, morphological, and cognitive continuity between humans and other primates might be the key to instilling an appreciation for the need to share space. However, appreciating the need to share and acting on that appreciation are two separate objectives.

A number of the respondents from Lindu noted that the macaques inhabit their land too, with some even recognizing that the forest belonged to the monkeys before people planted their gardens: "Because people cut down fruiting trees, monkeys go to the cacao trees" (woman in her 30s, born in Lindu but whose parents were migrants). Because others may have no qualms about defending their crops from raiding macaques by any means necessary, the ability of macaques and humans to coexist in their shared spaces may ultimately require that villagers perceive a utilitarian basis to macaque preservation. This might include documentation (and subsequent dissemination to local communities) of the role macaques play in forest regeneration, particularly of tree species that are economically and culturally important to people.

This notion of "sharing ecological space" is akin to Rose's (2011) call for pursuing biosynergies (i.e., mutual satisfaction of human, non-human, and ecosystem needs) in order to achieve effective conservation. An example of a way to share space that is a "biosynergy" inspired action would be the planting of tree species that are beneficial to both humans and primates at forest–agriculture ecotones to deter raiding of cash crops.

Acknowledgments

The research reported here was only possible with permission from LIPI and BTNLL, the sponsorship of Dr. Noviar Andayani, funding from the National Science Foundation, Wenner Gren Foundation, Wildlife Conservation Society, and the American Society of Primatologists, as well as the research assistance provided by my dear friends Manto, James, Papa Denis, Pias, and Pak Asdi. I also thank Greg and Niniek Acciaioli for sharing with me the information they gathered on the Lore Lindu National Park court case.

16 The fabric of life
Linking conservation and welfare

Annette Lanjouw

In Western-influenced cultures, and across much of the globe, there is still a dominating paradigm that the natural world (and all its non-human species) is there for human use, consumption, and enjoyment. Although a shift in this paradigm is slowly taking place, it has not led to a reduction in humanity's sense of entitlement and dominion. Rather, it has led to a rising awareness of the fragility of the natural world and the need to preserve the "ecosystem services" that sustain life, in particular human life, in the form of energy, food, land, air, water, raw materials, and other commodities.

The fundamental basis for this sense of entitlement has been documented in previous chapters, in the principles of religion, philosophy (Corbey, this volume), our sense of intellectual superiority over others and an ever-increasing dependence on industrialization and technology for survival (Twine, this volume). A sense of separateness from nature, and the forces that govern in nature, pervades much of our thinking, which has led to the illusion that resources are limitless and that economic growth is a goal that can be achieved in isolation from the conservation of the environment.

In the following I will describe how the desire for respectful coexistence needs to be based on a realistic and pragmatic, yet deep-rooted, motivation to bring together the objective of human economic development with that of ensuring the long-term survival, and well-being, of natural habitats and wildlife.

In an ecosystem, no one species is more important than another, irrespective of its sentience, emotional appeal, or order of classification. Yet from my many years of working on conservation of apes in Central Africa (since 1985), it is clear that for the emotional beings that we are as humans, some species are more effective than others in eliciting concern and moving people to action. My personal motivation for working in conservation was to ensure that the intelligent and sentient species in the wild are protected from human exploitation. Many different objectives, however, can effectively work together to bring about change, and all these objectives are necessary and justified. I have learned, however, that achieving conservation outcomes requires actions and commitment at all levels, and that the people most ready to change their behavior in support of conservation are often those that are the closest to the issues. Distance, and a sense of separateness from the natural world, is one of the biggest barriers we have to overcome.

The Politics of Species: Reshaping our Relationships with Other Animals, eds R. Corbey and A. Lanjouw. Published by Cambridge University Press. © Cambridge University Press 2013.

The importance of conservation

Since the early twenty-first century there has been a fundamental shift in understanding in one significant respect: the vital importance of biodiversity conservation for human survival is no longer in doubt. Numerous studies and reports over the last decade from the world's leading development agencies, financial institutions, and the scientific literature (TEEB, 2010; Turner *et al.*, 2011) have established decisively that biodiversity and intact ecosystems are essential to provide a broad range of ecosystem services – from human health to climate change mitigation and adaptation; from agricultural productivity and livelihoods to cultural integrity (and in some cases survival). Without even considering the ethical imperative to protect all forms of life on Earth, the evidence that biodiversity and intact ecosystems, far from being a luxury, are rather a vital part of human well-being and human development (United Nations Development Programme, 2011) can no longer be refuted (a fact that many non-Western cultures have embraced for many thousands of years). The emphasis of this understanding, it must be noted, is still on the importance of biodiversity to human health and well-being. The ethical, spiritual, cultural, and other arguments for protecting, conserving, and caring for all forms of life on Earth are not universally shared or understood.

Despite the strong consensus regarding the importance of biodiversity, however, most governments, corporations, and other decision-makers have yet to fully integrate the conservation of biodiversity and intact ecosystems into policy and action. Conservation objectives are still often perceived as conflicting with economic development or political goals. The concept of sustainability is still one that seems alien to much of Western thinking, and is rarely effectively put into practice. It is a generalization, but one that holds for the vast majority of people in the world, be it individuals, groups, states, enterprises, multinationals, etc., that we are locked in a tragedy of the commons (Hardin, 1968). In this globalized world that we live in, driven by economic goals, profit margins, and interests of shareholders, the objectives that we focus on are tied to self-interest that is manifested in the short term, this leads to the depletion of limited resources and destruction of the natural environment, despite the harm that it will inevitably inflict on long-term objectives.

Tied to the failure to achieve sustainable development is the failure in protection and maintenance of biodiversity. The decline of species continues at an unprecedented rate. Large areas of wilderness are being converted and degraded for exploitation and short-term economic objectives, with little valuation of the cost and implications of such destruction over the longer term. Approximately 13 million ha of forest were lost annually between 2000 and 2010, and 15 million hectares of forest were lost annually in the preceding decade, and these estimates are likely to be low (Food and Agriculture Organization of the United Nations, 2010). Global forests contain 90% of the world's terrestrial biodiversity. As a result, the numbers of many species have declined drastically and many are close to extinction (Koh *et al.*, 2004), especially in the tropics. It has been estimated that up to 50% of species are predicted to be lost within the next 50 years (Thomas *et al.*, 2004). Although there has been some progress in the establishment of protected areas (IUCN and UNEP-WCMC, 2012) and in community-based

Figure 16.1 Elephant killed in a banana plantation in Rwanda in 2004. Elephants can destroy significant areas of crops, resulting in human–elephant conflict and low tolerance by poor farmers. Photo Annette Lanjouw.

conservation initiatives, this is obviously insufficient as species declines continue. Despite small incremental gains in protection of biodiversity, the massive pressure for economic development will inexorably lead to mass extinctions, which will come at enormous societal costs (TEEB, 2010).

The main cause of all species losses and ecosystem degradation is an increasing and unsustainable level of consumption by burgeoning urban middle classes across the developing world, as well as in richer, developed nations (Pearce, 2012). The African Development Bank reported in 2011 that Africa's middle class expanded by 60% from 2000 to 2010 (Juma, 2011), with a resulting increase in purchasing power and consumption levels. The resources dedicated to conservation, in the face of these enormous challenges, are extremely small, and as a consequence, the impacts of conservation efforts have been limited.

The direct threats that result in species population decline, environmental degradation, and the loss of biodiversity globally are habitat loss, disease, and direct over-exploitation. The key drivers of these threats are poverty and limited livelihood alternatives; unsustainable natural resource uses; lack of information for development planning; inadequate governance combined with key mega-trends such as population growth, and increasing global demand for commodities as a result of increasing wealth and urbanization.

Understanding the importance of biodiversity and functioning ecosystems in sustaining life on this planet, human and non-human, requires understanding that it is precisely the diversity of life and balance in the ecosystems that is critical. Conservation of biodiversity focuses on the relationship between humans and landscapes, ecosystems and species. The emphasis is on the protection of ecosystems and ecosystem functions for the benefit of all species dependent on them. Although the human population is considered by many as the critical beneficiary, if entire ecosystems are protected, not only humans benefit. Conservation can only be effective if the focus is on the ecosystem and the diversity of species in it.

Healthy ecosystems bring critical benefits for the world, by providing clean water, food and energy supplies, acting as buffers against floods and severe climatic events, continually replenishing food and soil nutrients, and preventing soil erosion. Globally they are critical in sequestering enormous stocks of carbon and stabilizing the climate. On average, it is estimated that ecosystems must be about 50% intact in order to maintain their full range of ecosystem services, and some tropical ecosystems require even higher levels of intactness (Schmiegelow et al., 2006; Noss et al., 2012). All ecosystems require a range of fauna and flora to function effectively, and the higher the biodiversity the greater the potential to contribute to ecosystem health. Understanding that species can attract a conservation focus on an entire ecosystem justifies a species-focused conservation approach, and acknowledges the interconnectivity between ecological, social-cultural, economic, and institutional structures.

The case of the great apes

All species are important, regardless of their sentience, their relatedness to humans, or their emotional appeal. Yet it is clear that some taxa are more able than others to draw attention. The so-called charismatic mega-fauna, including the great apes, elephants, pandas, canids, felids, cetaceans, and other intelligent, emotionally appealing animals have frequently been used to foster conservation effort, funding, and public attention. In effective conservation programs, this attention has ensured that large numbers of species, ecosystems, and communities benefit.

I will now focus on the great apes, the taxa that I have worked with most intensively. They are an excellent starting point for species-based conservation for several reasons, not least of which is their ability to attract attention for critical tropical forest ecosystems. Humans have a unique relationship with great apes that is different from that with any other species. Due to our similarity to them, their intelligence, and our ability to empathize with them, great apes have always stimulated our curiosity and drawn our attention. Indeed, studies of chimpanzees in Africa have been able to contribute significantly to our understanding of human evolution and behavior (McGrew *et al.*, 1979; Laden and Wrangham, 2005).

Human behavior affects apes in complex ways in the tropical forest habitat in which they are found. While apes have the ability to adapt to human disturbance to some degree by adapting their behaviors or by existing at lower densities and extending their home-range sizes, the long-term effects of habitat disturbance on factors such as stress level, parasite load, and as a result reproductive fitness, are only just emerging. The tropical forests in which great apes live are among the richest in biodiversity in the world. While a study quantifying the exact number of species harbored within the ranges of great apes remains to be done, a recent analysis (Dinerstein *et al.*, 2010) demonstrated that the range of great apes overlaps extensively with many endemic species, and of course countless more species that range more widely. Chimpanzees overlapped with 144 other species of endemic vertebrates, gorillas with 79, and bonobos overlapped with 9 endemic vertebrate species. On a broader scale, it is clear that the ranges of great apes clearly overlap with internationally recognized priority areas for biodiversity in general, such as "Key Biodiversity Areas" (KBAs), "Important Bird Areas" (IBAs), and "Alliance for Zero Extinction" (AZE) sites. Protection of the different habitats of great ape species in Africa will therefore result in the protection of many other species living within the same habitat. This makes great apes important "umbrella species," which are "species with large area requirements, which if given sufficient protected habitat area, will bring many other species under protection" (Noss, 1990; Caro, 2003).

It is important to note, however, that great apes are no different from many other flagship species that serve as an indicator for an ecosystem. Although one can argue that due to their genetic and evolutionary proximity to humans they are "special," arguments for the uniqueness and exceptionalism of many other species can and have been made equally well. The emphasis on great apes as our "closest relative," however, does help focus on one element that is important to a large majority of the human population in both industrialized and non-industrialized nations.

Great apes have also been called "keystone species" because they play a key role in maintaining the health and diversity of their ecosystems. In order to conserve tropical forest, the presence of frugivores that disperse the seeds of the plants is of critical importance, and apes play a significant role as seed dispersers (Rogers *et al.*, 1998;

Voysey *et al.*, 1999). Chimpanzees and Western gorillas are two of Africa's largest frugivores with fruit making up a large part of their diet. Studies have shown that the passage of seeds through the gut of chimpanzees increases the speed and probability of germination (Wrangham *et al.*, 1994; Lambert and Garber, 1998; Voysey *et al.*, 1999). In addition, the fact that gorillas and chimpanzees travel large distances in a day also contributes to their important role as seed dispersers. African apes, like all large and primarily terrestrial mammals (orangutan are much more arboreal), have a role in shaping the structure of the forest. They trample, bend, and break vegetation through travel, foraging, and nest building (Plumptre, 1995; Rogers *et al.*, 1998), producing light gaps that facilitate the germination of non-shade tolerant plant species in the forest and thus ultimately contribute to forest regeneration. The decline and potential extinction of great apes therefore could precipitate the decline of other culturally, economically, or ecologically important species. Although this has never been documented for apes specifically, studies of countless other species have highlighted the severe consequences of altering species composition and the resulting impact on other species in the ecosystem. A telling example is the role of elephants and moabi (*Baillonella toxisperma*) seed dispersal in Cameroon (Bikié *et al.*, 2000).

More importantly than anything, however, in the context of this book, we have a moral responsibility to protect all non-human species, including great apes, to ensure their survival. The moral imperative to ensure the survival of the great apes is endorsed by all countries where apes are found in the wild, as reflected in their national laws and policies. This moral imperative is also reflected in many legends, proverbs, and cultural taboos throughout Africa and Asia. For example, many people in the Fouta Djallon of Guinea believe that chimpanzees were once human, and will not eat their meat (Ham, 1998). The Mongandu people in Équateur province of the Democratic Republic of the Congo will also not kill and eat bonobos, as they are considered taboo (Walker *et al.*, 2010). Bushmeat forms the bulk of the animal protein eaten by the Mongandu people, but they exclude the bonobo on ethical and cultural grounds. Mountain gorillas are also not hunted or eaten throughout their range in Rwanda, the Democratic Republic of the Congo, and Uganda, where all primate meat is considered inappropriate as food. The rare incidences of mountain gorillas being killed for food were all perpetrated by outsiders. Apes, however, are still killed for their meat and for sale in many parts of their range, and in some places are considered a delicacy or a symbol of status and wealth.

As mentioned, it would be possible to make a similar conservation case for numerous other species, and this has been done on many occasions. Different audiences have different objectives for conservation. Some are purely economic (for example, the conservation of mountain gorillas as a tourism resource), some ethical (based on relatedness to humans, or sentience, for example), and some ecological (as critical for the functioning of the ecosystem). All of these objectives can be justified and are essential to bring together sufficient support for the setting aside or limiting access and use of land and resources for human development and exploitation.

Figure 16.2 Young girl collecting firewood in the Democratic Republic of the Congo in 2000. For rural populations, the natural resources surrounding them directly sustain their livelihoods. Photo Annette Lanjouw.

Linking conservation and welfare

In conservation, the objective is generally a population of a species, rather than individuals. There is a frequently perceived disjoint, or even conflict, between animal welfare and conservation. In animal welfare, the focus is on the individual animal and its well-being; in conservation, on the survival of a population or species, while the individual is subsumed in the interests of the larger group. Although there are cases where such a conflict does indeed exist, it is often more accurate to speak of a continuum, where the welfare of an individual of a certain species is a critical component of the welfare of the population/species.

A concrete example of this are orangutans, who are kept as pets, hunted as noxious animals ("problem animals"), or die as a result of the destruction of their habitat. The number of orangutans in captivity is now a significant proportion of the total global population (Singleton *et al.*, 2004; Wich *et al.*, 2012). The future of orangutans is dependent on the successful release, survival, and breeding of "rehabilitant" and reintroduced animals. Unless individual animals are cared for, the success of the wild population is jeopardized. Another example is the bushmeat trade and hunting of chimpanzees in Central Africa (Ellis, 2000; De Merode and Cowlinshaw, 2006). Orphan chimpanzees are often placed in sanctuaries in Africa, ideally later to be re-introduced to the wild. Sanctuaries are perceived as "animal welfare" organizations, with significant costs dedicated to the survival of a relatively small number of individuals (Meijaard, 2012). At times, these costs are seen as unsustainable, given the enormous need for funding to prevent and combat illegal hunting and management of forests and entire populations. Unless conservation programs focus on the illegal hunting and trade in apes, there will be continuous pressure on wild populations from hunting. Although the laws exist, effective enforcement to prevent illegal hunters from escaping with impunity is absent. For the laws to serve as a deterrent, they must be enforced. Captive animals must be confiscated, hunters punished, and sanctuaries need to take on the confiscated orphan chimpanzees. The latter play a critical role in conservation and in law-enforcement strategies.

The welfare of individuals is a critical part of the welfare and survival of a population. Although the emphasis of conservation is often different than that of welfare, the distinction is often misleading and perceived as more contradictory than it actually is. Where individual animals are respected and valued, there is perhaps less pressure on entire populations.

Conclusion

An acknowledgment of the importance of biodiversity for human survival is not enough. Critical for the realization of respectful coexistence is the will, and the know-how, that links economic and social development with sustainable management of the environment and conservation of ecosystems and species. The shift in understanding must be coupled with a shift in behavior. Human population growth, coupled with growth in consumption levels and ever-increasing hunger for resources, need to become sustainable, not only for powerful nations, but for all people. It is also clear from the preceding chapters that this must be coupled with an understanding that humans are interconnected with non-human species, and that humanity is dependent on other species for its survival. Respect and empathy are essential ingredients of this understanding. Acknowledging the interconnectedness between humans and the natural environment does not automatically lead to respect and empathy, nor does it necessarily lead to behavior change. Self-interest, however, is often a powerful motivator for action. This action then can lead to a gradual appreciation, and foster respect and empathy.

In 1919, at the lecture series that Max Weber presented at the University of Munich (Weber, 1919), this German intellectual and sociologist defined three critical character-istics for political change: passion, a sense of responsibility, and a sense of proportion that balances what must and what realistically can be achieved. All three are required to achieve a more respectful coexistence between humans and non-human species.

17 Home flocks
Deindustrial domestications on the coop tour[1]

Molly Mullin

In 1989, Michael Moore's documentary, *Roger & Me*, depicted the consequences of General Motors' plant closings in Flint, Michigan, as surreal. In one notorious scene, Flint resident Rhonda Britton described her plan to make money raising rabbits in her backyard "for pets or meat." If viewers squirmed, it was only in part because she was shown beating a rabbit over the head with a metal pipe and skinning the carcass. The very idea of trying to make money raising rabbits in a metropolitan area came across as a form of post-industrial craziness.[2]

In the two decades following *Roger & Me*, perceptions of urban and backyard farming have radically changed. In 2009, although Michael Moore's former employer, the magazine *Mother Jones*, acknowledged that "our industrial food system is rotten to the core," it warned that "heirloom arugula won't save us" (Roberts, 2009). Less skeptical publications regularly instruct readers how to, for example, keep a *Barnyard in your Backyard* (Damerow, 2002) or *A Chicken in Every Yard* (Litt and Litt, 2011). By 2011, Backyardchickens.com's membership had reached over 100,000 with a message board averaging five new posts per minute, with about a 1,000 members online at any time. Yet for all the enthusiasm, others raise concerns about disease, odor, noise, property values, lice, rats, and predators. Although less common, concerns have also been voiced about animal welfare (United Poultry Concerns, 2009; Gruen, 2011). Municipal government officials throughout North America have spent considerable time debating the keeping of "farm animals" in small towns and cities. Considering the prolonged and intense deliberations experienced in many locales, officials often fear that even opening up debate on the issue could delay attempts to address more pressing troubles – including how to fix crumbling roads or faltering school systems. In the case

[1] The research for this paper was funded by a grant from the Hewlett-Mellon Fund for Faculty Development at Albion College. For assistance in learning about urban chickens and coop tours, the author is grateful to Rodney Bender, Martha Boyd, Małgosia Cegielski, Sarah Gilbert, Sean and Kath Kelly, Frances Mullin-Saunders, Linda Nellet, Michelle Thoma, and Jeb Saunders. Thanks to Christina Chia, Raymond Corbey, Augustín Fuentes, Eben Kirksey, and Kathy Rudy for comments on an earlier draft.

[2] A follow-up documentary entitled "Pets or Meat: the Return to Flint" aired in PBS's *POV* series on September 28, 1992. Rhonda Britton had moved on to raising rats for sale to snake owners, in addition to working at Kmart and caring for her nine-month-old daughter (Katz, 1992).

of Detroit, officials postponed debates on urban chicken keeping while considering whether to eliminate streetlights and solid waste collection services to the city's less densely populated areas (Hartman, 2012).

As a cultural anthropologist with a longstanding interest in human–animal relationships, I've spent the past two years investigating the urban poultry phenomenon. In addition to attending backyard and urban coop tours in a variety of cities and towns across the United States, I bought my own coop and flock of a half-dozen chicks and made my way – in two states and several cities – through a maze of laws and ordinances regarding where they could be kept. My interest is in understanding diverse "domesticatory practices" in relation to a particular historical context without resorting to simple generalizations. Although domestication was once considered more event than process and a matter of the distant past, we're learning to consider it an ongoing process, one affecting humans and multiple species, involving built environments, technologies of kinship, gender, feeding as well as breeding, production, and reproduction (Cassidy and Mullin, 2007). We have begun to link historical transformations with specific domesticatory practices: the rise of industrial capitalism and production's move away from the "domus;" the shift from "breed type" to "breed lines" (Ritvo, 1995; Franklin, 2007: p. 103); and, as Sarah Franklin observes in her study of the cultural significance of Dolly, the cloned sheep, the rise of "a bio-economy increasingly organized around the bespoke component, the engineered pathway, and the biomimetic assemblage" (Franklin, 2007: p. 195). Where, I wanted to know when I bought my chicks, does the backyard flock stand in relation to this history?

Urban and backyard farming have gained some respectability and popularity as a response to a number of historical developments. Environmental and economic crises have encouraged concerns with sustainability and self-reliance, with survivalism a more extreme expression of such concerns. There are people who begin keeping chickens because they expect "zombies" to arrive and disrupt existing access to food. There are far more people who begin keeping chickens and make jokes about making such preparations. Home food production, including the keeping of egg-laying chickens, has also been fueled by concerns about hazards to human and animal health posed by industrial corporate agriculture, hazards that have been exposed by bestselling books and popular documentary films (e.g., Schlosser, 2005; Pollan, 2007; Kenner, 2009) as well as by scholars, such as Barbara Noske (see Twine, this volume).

Two recent books that address the cultural politics of chickens highlight the efforts of entrepreneurs marketing expensive factory-made mobile coops to wealthy urban consumers (Squier, 2011; Potts, 2012). Such products, in my view, are far less influential than participatory technologies that facilitate information gift exchange. I have only rarely seen factory-made coops on city "coop tours," though in cities and towns throughout North America, people are selling coops they have made themselves, in their backyards and garages. On Backyardchickens.com and other sites, designs for coops are readily shared along with photographs and videos featuring every stage of the assembly process, though there are also many designs offered only for sale. Professional urban poultry consultants have emerged, but "free" information about care and feeding, even slaughtering and "processing," as well as about local ordinances and efforts to change them is also readily available on websites, blogs, listservs, and various face-to-face meetings facilitated by online resources. The fundamental question Squier (2011) and Potts (2012) raised with the topic of expensive mass-produced coops (there are now much more moderately priced ones available) is the extent to which urban flocks have returned to urban and suburban lots only as another extension of consumer capitalism, a system in which even resistance to consumption spawns ever-more commodity consumption, but legitimized by association with caring for animals and the environment. City chickens hardly foretell the demise of the "animal–industrial complex" (Noske, 1989; Twine, this volume). Their current popularity, however, is more complicated than that of the latest mobile phone, with more potential for encouraging worthwhile questions about environments, animal welfare, political geography, relationships among species, and even neoliberal economics.

In early twenty-first century North American cities, along with the perception of home flocks as a kind of ecological good citizenship, a simple novelty factor accompanied the urban chicken's popularity. A man who began keeping chickens a decade ago in Portland, Oregon, told me his friends once considered his hens an extraordinary sight: "'You gotta go over to Rodney's house,'" he recalls friends saying, "'He's got *chickens*!'" But now, he says, "it's Portland, yeah, yeah, of course you have chickens." Or, you no longer have chickens, having decided they tear up your garden too much, require a bit more care than you want to provide, annoyed the neighbors with their cackling, or one of many other reasons people offer for terminating their forays into flock-keeping.

Novelty goes hand in hand with nostalgia. By the 1950s, as Susan Squier observes in her analysis of *Little Red Hen* stories, apron-wearing hens evoked cheerful domestic productivity. At the same time, farms had become "increasingly large-scale, vertically organized and corporate ventures to which the farmer contributed labor, materials, and risk while the agribusiness retained the product and the largest share of the profits" (Squier, 2011: p. 144). In the 1950s, however, consumers still viewed "factory farming" largely as a positive achievement that promised safe and affordable food for all (Mizelle, 2012). Now that industrialized agriculture has lost its shine, urban agrarianism brims with nostalgia. The Urban Farm Store in Portland displays antique feed sacks behind its supplies of live chicks and organic, locally produced feed. Portland artist Joe Wirtheim describes his "Victory Garden of Tomorrow" poster series as "a chance to re-connect with our legacy of overcoming challenges – we've done it before, we can do it again" (http://victorygardenoftomorrow.com/about.html).

Nostalgia is such a pejorative label. Popular magazine articles and guides for backyard and urban chicken keeping often include lists of the many chicken and bird idioms pervading the English language. Acquire your own flock, these lists suggest, and this homespun heritage will once again be yours. Such appeals to nostalgia are easily dismissed from the outside. However, once my family started keeping our own chickens, the idioms on those lists could no longer roll off the tongue without awareness of their connection to actual animals. Many of them seemed as apt as ever: "pecking order," "flighty," "cooped up." Those that suggested chickens as an entire species are stupid or cowardly did not (my daughters were slightly perplexed that their school explicitly forbid calling a person a "chicken"). Our new awareness of language was just one of many ways in which chickens seemed to enrich the lives we shared with one another and with our flock.

Rustbelt chickens

My house in Michigan was built in 1910 and sits on almost an acre of land on a quiet residential street lined with sidewalks and shade trees. I spent a lot of time in that house, especially in the winter, looking out of the windows, imagining what I could do outside. Since my husband and I purchased the house, in 1996, I had gradually expanded the gardens and reduced the size of the lawn. I had several intentions: to provide habitat for wildlife, to create pleasant spaces for spending time outdoors, to reduce the need for lawn-mowing, and, increasingly over the years, to provide food for family and friends. My oldest daughter could eat all the berries and most of the tomatoes I grew. Spending a lot of time at a desk and computer and living in a fairly remote small town, I liked spending time outside accomplishing something: gardening was my entertainment, exercise, and therapy – my preferred "technology of the self," as Foucault described individually chosen programs of preoccupation, self-improvement, coping (Foucault, 1988).

After my husband and I adopted our two daughters, I spent more time at home and more time inside the house. When I looked out of the window, I started wondering what it would be like to see chickens. I had friends living in nearby rural areas with small flocks and my fowl visions were encouraged by articles that had begun appearing in gardening magazines and even in places like the *New Yorker* and the *New York Times*. As with gardening, my interests in acquiring chickens were multiple: I wanted a more convenient supply of eggs from hens kept in decent conditions and fed healthy diets; I wanted fertilizer for my garden; I like being around animals.

There were more complicated emotional aspects also. Chickens at home meant something important to me in the context of a rustbelt town of about 9,000 people that seemed only to become more impoverished and empty. The town's population had peaked in 1960, long before my arrival. Each decade after that, the town lost about a thousand people (Passic, 2001), as once-busy factories gradually closed, in many cases leaving enough toxic waste behind that their locations, designated "Superfund sites," could not be re-used without extensive remediation. During the recession of the late 2000s, many more local businesses closed, along with the hospital and a daycare center. In 2010, while I dreamed of chickens, 33% of the town's population lived below the poverty level (United States Census Bureau, 2010). My husband's law practice in a nearby city, with a focus on employment law and workers' compensation, faced an uncertain future: for years, the state government had been closing courts, making the legal process more difficult and expensive for injured workers. Increasingly, the remaining courts were staffed by judges hostile to labor. The rapid rise in unemployment and election of a Republican majority in the state government were not encouraging.

At Albion College, I felt more powerless and vulnerable as an associate professor than I had as an assistant. I was the primary caretaker of young children in a context where childcare options were limited and comparable in cost to the mortgage on our house. Meanwhile, with the college hit hard by the recession, faculty hours in the classroom and in committee work increased as salaries remained frozen. Staffing was reduced in rounds of budget cuts and I found myself the only woman in a department with five men. The politics of the academic workplace are often described as a minefield in the best of times. When tenure no longer offers any guarantee of survival, tensions are fueled by fear. In such a context, I felt deep longings for abundance, liveliness, and a sense of control over my environment.

My interest in chickens spiked in the summer of 2010 when I followed a link to a page on Katy Skinner's "City Chicken" website. The page features images of "chicken tractors" – mobile house/pen contraptions used to allow chickens to forage in different parts of a garden, field, or yard. Each tractor was unique and handmade. Many were painted cheerful colors and made of repurposed materials. Some looked like play houses for children. As a scholar who had recently co-edited a volume on domestication (Cassidy and Mullin, 2007), I couldn't help but wonder what these structures and their public exhibition represented in terms of the history of boundaries between genders and between human and animal, farm and house, city and country, production and consumption. As someone who felt slightly trapped inside my own house, I wanted one. Virginia Woolf plotted escape through "a room of one's own" – I had two of those, one at home and one at my college, and they offered no particular freedom. Zora Neale Hurston occupied the porch (Friedman, 1998). I had one of those too, but there was nothing interesting going on out there.

I began planning a research project on backyard and urban chickens that would include raising my own flock and working on getting the Albion city government to change chickens' ambiguous legal status. According to the town's ordinances, you could not keep animals "other than those commonly kept as pets" near a human residence or a street. After a month or so of planning to acquire my own chickens, I became discouraged by all the details to be worked out: how to buy or build a coop, where to get the chickens, whether to hatch eggs myself with an incubator or buy chicks, how to keep them safe from predators (which included raccoons, possums, hawks, neighborhood dogs, and our own dogs) and from possible objections from neighbors. Despite these complexities, I wanted to research the urban poultry phenomenon and, as an ethnographer seeking to understand what was going on from something of an "inside" perspective, I felt my own flock was an important component.

Step by step, the details did get worked out, if not always smoothly. On Craigslist in March 2011, my husband and I found a coop and half-dozen young Black Australorps offered by a family in Livonia, a suburb of Detroit where city officials were demanding the hens' removal. The day before we were to collect them, the family called to say they had decided to fight the legal battle to keep them, making use of Michigan's "Right to farm" law, which some lawyers had interpreted as invalidating zoning laws that prohibited chickens. We were disappointed, but a month later received six one-day-old chicks of a variety of carefully selected breeds (with several criteria in mind, including cold hardiness) from a hatchery in Ohio, all vaccinated and sexed, that is, determined to be pullets, or female.[3] Although I knew that animal advocates objected to shipping chicks in the mail, ours arrived in perfect health and quickly settled into the "brooder box" I had set up next to my bed and equipped with a warming gadget. When the kids left the top of the box open, one of the cats tried to eat one of the chicks, but fortunately the chick popped out from his mouth unscathed.

When our pullets were old enough to inhabit it, I purchased a coop for about $500, delivery included, from a retired carpenter outside Detroit. He was doing a lively business selling his coops and chickens to people around the state. "Backyard chickens are the hottest thing around," he told me when he delivered my coop, explaining that he got the idea of raising chickens outside Detroit during visits to Haiti as a missionary. I regretted not having the skills to build a coop myself, but my husband was able to build a pen to surround it using two reassembled large outdoor dog kennels purchased on Craigslist and at a garage sale. Eventually the hens would spend most of their time roaming the yard, but once predator-proofed, the pen provided greater safety when we felt that was necessary. Fortunately our two large dogs (who had killed and eaten many a wild thing) to my surprise did not see the chickens as prey and became their protectors. My idea that chickens would complement my gardening proved true to a point, but was also naïve. I spent far more time than I had imagined erecting fencing to keep chickens away from vulnerable plants. Chickens allowed to forage live up to their origins as "jungle fowl" and can make short work of any cultivated space (Figure 17.1).

When I told a colleague I was studying people raising chickens in urban settings, she responded with disgust. "These people have no idea what it is like to live with farm animals," she said, explaining that she had lived on a farm as a child and knew that chickens were disgusting, smelly creatures, with no business in the city. I would eventually hear others respond similarly, often when arguing against municipal ordinances that would allow chickens in city limits, invoking childhood experience on farms that they felt gave them superior knowledge. But many people raising backyard chickens (or wanting to) in fact do have prior experience with "farm animals." My story is not unusual. I grew up on a farm where we had chickens. I never thought they were disgusting, but certainly it was a very different experience from the way I live with chickens now. The technology is different (for example, coops designed for easy cleaning and integrated with cleansing compost systems). Living with only half a dozen, of different breeds, they are fairly easy to tell apart, with somewhat distinct personalities and different positions in the "pecking order." As I write this paragraph, several of my chickens are sitting outside a glass door a few feet away, occasionally peering inside at me and the dogs. Such intimacy between humans and small household flocks is not unusual.

---

[3] What happens to the "extra males?" According to mypetchicken.com, "The majority of the baby chicks that aren't adopted are offered up for auction near our hatching facility. A portion of the unsold chicks are humanely euthanized and we are currently working on alternatives for the 2012 hatching season." For a critique of the hatchery business, see the Coalition of Animal Sanctuaries' Draft Statement on Urban Chicken-Keeping (United Poultry Concerns, 2009).

Figure 17.1 Visitors inspect a coop's construction on the Windy City Coop Tour, Chicago 2011. According to Mike Davis (2010), "Since most of history's giant trees have already been cut down, a new Ark will have to be constructed out of the materials that a desperate humanity finds at hand in insurgent communities, pirate technologies, bootlegged media, rebel science and forgotten utopias." Photo Molly Mullin.

Many authors (e.g., Squier, 2011; Walker, 2011; Potts, 2012) have observed that humans can find the company of chickens remarkably soothing. Like Alice Walker (2011), who also lived with chickens as a child and was astonished by how differently she felt about their company later in life, my family has been intrigued by our hens' behavior. What, we have wondered, are they eating so eagerly under those leaves? What makes them seek their roosts inside their coop at dusk? Why must they always lay their eggs in the same box? How do they roam as a flock staying within hearing distance of one another? Despite the mystery of some of their behavior, we feel confident we know when our hens are happy and when they are not and we work hard to keep them happy. The most difficult challenge is providing them freedom to forage while protecting them from predators. We are aware we care more about our hens' well-being than many other keepers of backyard flocks. My sense is that we are not unusual, but we are also not typical.

Although initially I planned, as part of my research, to attempt to work on gaining legal status for chickens within city limits, soon after getting my own flock I learned that such an effort would not be welcomed by the City Council – not because anyone was opposed to people keeping chickens, but because members were aware that other municipal governments had devoted many hours to intense, even bitter, arguments. As an Albion resident, I agreed the council had more pressing issues. I was told that the city government had taken the position regarding backyard chickens that "as long as they are pets, they are fine" – a generous interpretation of the ordinance. Along with the rest of my family I worried that this interpretation might not hold if anyone complained, but at least our lot was large and sheltered enough by trees, shrubs, and a fence that made the chickens nearly invisible outside the property.

Chickens' ambiguous legal status hampered open communication about their care. Down the street from us, near the site of what had once been the Union Steel plant, we often saw a bantam hen foraging along the sidewalk. The neighbor who lived there explained she had turned up the previous year when, he surmised, she had been abandoned by college students who often bought chicks and ducklings in the spring and kept them hidden in residence halls. Believing it was illegal to keep her within city limits (and resistant to any information to the contrary) our neighbor told us, as if it was a subversive act, that he let her roost in his garage in the winter. As though he were maintaining his innocence, he insisted that he did not feed her and was not responsible if the neighbors fed her table scraps.

Although attempting to make chickens more explicitly legal in Albion did not appear to be a good idea, I did not have far to go to find other attempts to legalize city chickens. In the town of Chelsea, some 40 minutes east of Albion, a lively campaign was being waged by residents. According to the regulations then in effect, chickens were explicitly forbidden on lots less than five acres. At a public hearing I attended in 2010, it did not seem likely the pro-chicken campaign would be successful. The arguments against legalization included the claims that chickens would hasten the decline of property values and attract lice and "varmints." To my surprise, in March 2012 the Chelsea City Council ruled in favor of residents being allowed to apply for a permit to keep a maximum of four hens. By then, I had already gained firsthand experience with a wider variety of legal situations for backyard poultry than I anticipated when I first planned my research.

Coop tours

Since the early years of the twenty-first century, backyard and urban coop tours, a deindustrial version of the garden tour, have been held annually in many North American cities where residents are allowed to keep flocks. In 2011 and 2012, I attended coop tours in Portland, Chicago, Raleigh, and Durham. I also attended an edible garden tour in a small town in Michigan, where some of the gardens featured chickens, and I attended urban farm tours in Chicago and Detroit. As a public exhibition, coop tours are the products of curatorial strategy that is revealing of cultural context and political geography.

Coop tours exhibit domesticatory human and animal practices very differently than Michael Moore did when he filmed Rhonda Britton and her rabbits in Flint. If industrial domestication, at least in some parts of the world, involved separating urban and rural, city and farm, production and consumption, coop tour hosts enthusiastically display possibilities for transgressing the boundaries, and in ways that seem not surreal or crazy, but as just slightly outside ordinary. Tour hosts are more in control of what they share than Rhonda Britton, and unlike the viewers of Michael Moore's film, tour guests are encouraged to ask questions.

The coop, a shelter and means of containment, and its relationship to the surrounding pen or yard, is meant to be the focus of the coop tour, not the birds that roost inside the coop at night. But the visitors want to see the birds at least as much as their housing. Coop tour organizers know this, and it is the bird totem that adorns posters, t-shirts, and advertising (Figure 17.2). An example of the entertainment value of animals comes from the 14th Annual Detroit Urban Farm tour. "We gotta get us some *animals*!" said a young urban farm manager to his colleague as we strolled the gardens of the Catherine Ferguson Academy, a public high school for pregnant teens and teenage mothers, and watched visitors marveling at the chickens and goats. The lettuce was impressive, but it was the animals that especially delighted and drew visitors. Urban chicken keepers on the tours rave about the entertainment value of chickens: "Some people watch TV, I watch my chickens," one urban chicken keeper told me in Chicago, echoing similar remarks made regularly on internet forums devoted to backyard chickens. The person who told me this, however, did have a large television in his living room. He mentioned that his chicken-raising efforts had been at least in part inspired by watching a DVD of the film *Food, Inc.*

But if backyard coop tour hosts are putting animals on display, it is as part of a system of domestic technology. For contrast, consider fashion photographer Jean Pagliuso's *Poultry Suite* portraits of groomed and elegantly posed solitary chickens standing against blank backdrops as discussed by Susan Squier in her book *Poultry Science, Chicken Culture* (2011). On the coop tour, hosts happily answer questions about their chickens (or ducks, quail, or other animals) – sharing breeds, personalities, histories, names if they have them, or the reasons why they don't have them, but the gift they are sharing, as they see it, is not the spectacle of the animals themselves but knowledge – techniques of protecting, composting, containing, maintaining, reusing, repurposing, producing – a system of domesticatory practices of which the chicken is only a part (and the individual chicken a particularly transient part). And coop visitors, even if drawn in part by the lure of performing or totemic animals, come seeking this knowledge. As coop tour organizers will point out, each chicken-keeping system is unique at least as far as each house and lot is unique and the house-dwellers and coop-dwellers are unique. Even mass-produced coops are situated differently and employed differently in each location. Tour hosts offer knowledge and experiences that can be made use of, if not exactly replicated, by visitors. "I'm always looking for ideas about how I could do things differently," I heard in Portland. Repurposed materials are common in a way that also encourages creativity but not exact replication: a cross-country ski from Portland's Re-building Center is used to frame a coop door; a fence is made with fallen tree branches; a coop made from discarded wood pallets.

Curatorial process is most evident in the large annual city coop tours, such as Portland's Tour de Coops. The tour serves as a fund-raiser for Growing Gardens, an organization

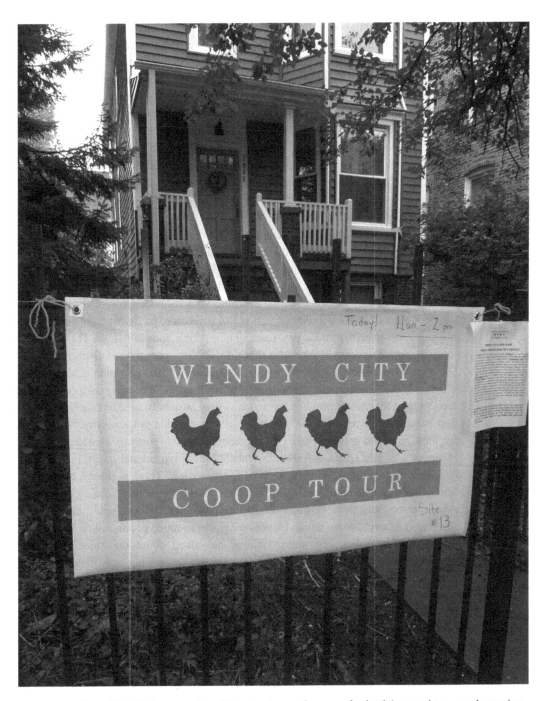

Figure 17.2 The iconography of the coop tour makes use of animals' entertainment and attention-getting value, but from the perspective of tour hosts, it is technology and a set of relationships that is being exhibited more than the animals. Photo Molly Mullin.

that funds education about home food production for low-income Portland residents. The organizers, with a team of interns, work hard to feature a variety of coops: coops in different Portland neighborhoods, made with different budgets, technology, materials, landscapes – as well as being the product of different household organization strategies. Although most belong to single family dwellings some are cooperative coops managed by members of multiple households – so that what is exhibited is not just architecture and animals but also a work schedule and division of labor.

Coop tour organizers are not necessarily promoting urban chicken keeping. In Portland, where "farm animals" have a well-established and relatively uncontested legal place in the city, the current organizers stressed to me that the coop tour is a fund-raiser and not an attempt to encourage people to keep chickens. Tour hosts' motivations are varied. As anthropologists and other scholars have long been aware (e.g., Watson, 2011), hosting and the inviting of strangers inside as well as behaving as a proper guest are actions with philosophical, political, and cultural complexity. The organization member most in charge of running the tour in Portland reported that in recent years people have been so eager to serve as hosts that aspiring hosts besieged him with complaints when only a fraction of the hundreds who volunteered were chosen. To be chosen for the tour is like a merit badge; it is also a means of participating in or joining a community, being expert for the day, meeting new people and old friends. The organizer expressed greater understanding of the appeal of visiting coops than of displaying them: "There's a voyeuristic aspect. Getting to see what you normally wouldn't, how people do things." He seemed concerned to make the tour as interesting as possible for attendees, and that meant providing as much variety as possible and as much creativity. Coop owners not selected for the tour called him up wanting to know why: "they think what they have is so great and how do I tell them, 'nobody's going to want to see *that*'?"

Chicago's Windy City Coop Tour is probably more typical and in 2011 organizers told me that so far there had been no need for selectivity. There is clearly, however, a great deal of self-selection. I attended the Chicago tour with a man keeping 30 hens in his backyard in Bucktown. He hadn't known anything about the tour until I mentioned it and asked if he could come along. He enjoyed it, not so much because of any obviously voyeuristic aspect, but like many visitors, he talked about the value of seeing "how other people do things," getting ideas for the care of his own flock, and also meeting other people nearby with similar interests in sustainability. Living on disability and rental income, he began keeping chickens after reading Michael Pollan and other critics of industrial agriculture; he also mentioned "becoming a steward of the environment." What perplexed him about the tour was the hosts' willingness to admit strangers to ordinarily private spaces. This was a man who lived in the house he had grown up in and inherited from his grandfather. He remembered a time, before the neighborhood became chic, before it had a Marc Jacobs store and fancy espresso shops, before it was known as "Bucktown" and was just "somewhere in Chicago," when people would break into your car to steal the change between your seats. He would worry too much, he told me, to let people walk in and out of his backyard, worry that they could be up to no good. He wondered whether the hosts were reckless or naïve. . . or maybe whether he worried too much.

My Bucktown friend's musings drew my attention to the coop tour's historically specific celebration of conviviality, a conviviality focused on knowledge sharing and the ability to make connections with those with common interests. Urban chicken keeping is not aligned with any particular political position and often involves unusual alliances among libertarians, liberals, and conservatives. But the coop tour's conviviality involves a political geography that is not completely unlike that of the Occupy movement, where public spaces are "occupied" in a way that promotes human interaction and the exchange of ideas and information that would not normally occur in "public" space (Bunting, 2011). The more established institution of the garden tour can do something similar, but has tended to involve less crossing of class boundaries, less intimacy, and a more narrow focus on one aspect of domestic environments.

Many urban coop tours are fund-raisers, some are organized by urban feed stores that have obvious reasons for promoting urban chicken keeping, and others – probably a majority – are organized by groups of volunteers interested not so much in promoting chicken keeping as promoting the freedom to keep chickens (and sometimes other food-producing animals) and to elevate their status in contexts where historically they have been associated with poverty. The annual coop tour is powerful propaganda in cities and suburbs across North America, where household food production, especially that involving poultry, has been tremendously controversial in recent years. In Chicago, poultry have never been banned in residential areas, but the Windy City Chicken Enthusiasts formed in part to fight a 2007 proposal to ban them from one of the city's wards, leading in 2010 to an annual coop tour organized by volunteers. Would-be flock keepers in nearby suburbs with laws against chickens hope the Chicago tour might help them.

In Raleigh, North Carolina, city ordinances were modified in 2004 to allow four backyard hens per household, but many of the surrounding suburbs have maintained bans (Blazich, 2011). Even within Raleigh, despite the lenience of the municipal ordinance, most of the subdivisions and homeowner associations that date from the early 1960s to the 1990s explicitly prohibit the keeping of "domestic fowl" (along with clotheslines, open garage doors, "yard boats," and vegetable gardens). In late 2011, Raleigh residents who learned I had an interest in chicken keeping, mentioned "Henside the Beltline" (Raleigh's annual urban coop tour and a fundraiser for Urban Ministries, a local charity) as "very popular" and suggested that it might be responsible for increasing acceptability of chicken keeping by city and suburb residents.

Wherever tours occur, local journalists tend to provide sympathetic advertising, with appealing depictions of multispecies cohabitation. Of course, coop tours might also inspire opposition from animal rights and welfare organizations, some of which have taken public stances against the use of animals in urban farming. When presenting this material, I was accused of being an enslaver and an abuser, when my intention was perceived as promotion. So far, however, such opposition has not been a common problem for tour organizers. One reacted with surprise at the very possibility, stating that certainly these chickens were far better off in people's yards than in factory farms.

If coop tour organizers sometimes play a curatorial role, the tour hosts also make decisions about what to exhibit. For many hosts, one motivation for displaying a coop on the tour was the pressure to make presentable: to redo, to organize, to tidy. A woman who would later become a friend told me she had often wanted to put her coop on the tour as an incentive to get her coop and yard looking more presentable. She had an MBA but was working as a freelance writer while taking care of her three children. Her husband was in the military and often away. She was amused to find out that her neighbors referred to her family as "the hillbillies."

Even on the tidy properties included on the tour not all is equally shared with visitors. A host on one tour, as we engaged in a lengthy discussion, invited me into her garage to show me an injured hen and her chicks and to tell me the story of her injury, a story that involved unexpected and still unresolved moral dilemmas. Her hen had "gone broody" and wanting to satisfy what seemed the hen's desire to raise chicks (she seemed quite aware her concerns could be seen as example of anthropomorphism), she purchased fertilized eggs to put under the hen to hatch. This is a common practice for flock owners without roosters and wanting chicks. The hatching took place as intended, but soon afterward the hen became injured trying to protect the chicks from other flock members. The tour host, who had been advised too late that she should separate the mother hen and her chicks from the others, was now trying to determine what she should do about the hen, who at that point was unable to walk. Most visitors got stories of the building of the coop, its day-to-day maintenance, a present-tense discussion of the flock, but not the narrative in the garage.

Some scholars have emphasized that in early periods of domestication, humans were focused on taming the wild, domestication's imagined opposite (Hodder, 1990). Although on the coop tour you might hear talk of predators (hawks, raccoons, neighborhood dogs), I would propose that if these domesticatory practices are aimed at taming anything, it is visions of an apocalyptic future where strangers cannot be relied on for assistance. If Michael Moore presented Rhonda Britton and her rabbits as a tidemark of degeneration, on the coop tour, food-producing animals suggest to the human participants not degeneration but ingenuity, at least a minimal form of resistance to consumerism, expertise reclaimed from domestication's long past, forms of community in which human and animal strangers can be invited in, and a modest hope of survival.

Migrant chickens

I'd started imagining chickens outside my window in Michigan during the recession of the late 2000s. By the fall of 2011, when the pullets I ordered from Mypetchicken.com had begun laying eggs, my family's economic situation seemed more precarious than ever. On sabbatical for the academic year, I was making half my usual salary when, along with so many middle-class US families, we were hit by enormous and unexpected medical expenses for family members. Meanwhile, Michigan's Republican-dominated legislature was moving ahead with plans to make the state more "business friendly" by further restricting employees' access to compensation for workplace injuries. With a specialization in workers' compensation and employment law, my husband feared for the survival of his legal practice. When he was offered a position with the Consumer Protection Division of North Carolina's Department of Justice, we faced an agonizing decision. In late December, we rented out our house in Albion and moved to North Carolina. Our coop and flock came with us, the coop pulled in a trailer we bought on Craigslist.

My husband's job was in Raleigh and I would be a visiting scholar at Duke University in Durham, about 40 minutes away, so I looked for rental houses in both cities. Our commitment to bringing the chickens was not absolute. The chickens had never made economic sense. Even the glossy magazine articles that promote urban chicken keeping warn that it is unlikely to be a cost-saving endeavor. We had friends in Michigan who I thought would agree to take them, but we all preferred to keep them if possible. They made us happy and our other animals made no economic sense either.

Although I found several houses to rent in Raleigh where chickens (and vegetable gardens) were allowed, our best choice turned out to be in nearby Durham, where we leased a house a ten-minute drive from family members, surrounded by several acres of woods on the edge of the city, but still within city limits and the city public school district. The landlord said the chickens were fine, as long as I agreed to apply to the city government for a permit. I was told I would need to pay a $25 fee and fill out a lengthy application that included the names and addresses of all the surrounding neighbors, who would be asked to state whether they had any objection. Since the chickens couldn't be seen or heard from outside the property and my family found the neighbors to be friendly, we were optimistic that we would not meet any objection. However, the property was so large I wasn't entirely sure exactly which properties bordered it. In January, when I visited the city planner's office to inquire, the official on duty helped me fill out all the documents, carefully printing by hand the neighbors' names and addresses. After we'd been at this for some 45 minutes he stopped writing and muttered something about what sounded to my more Northern ears like "residential rule." After much confusion, I realized he was saying that our rental house was in an area of the city zoned "residential rural" and that meant we didn't need a permit for our chickens. He directed me to the relevant documents where I read that according to the city's zoning laws, residential rural zones "are designed to discourage the development of urban services and to encourage the maintenance of an open and rural character" (City of Durham, 2013). Although I was not aware of lacking any "urban services," living where I lived, and there was an enormous shopping mall down the street, I did not need a permit for chickens at all. I could even keep horses, he pointed out.

For the next six months, our chickens spent their days roaming those several acres, surrounded by a six-foot fence that had been erected to contain large dogs sometime in the mid-1980s, when the house had been built. The flock seemed especially content foraging under the magnolias and large pines, sometimes finding small rat snakes and worm snakes to gobble nearly whole. When we heard hawks overhead, we raced to corral the flock in the predator-proof pen that surrounded their coop. As the days grew longer, their eggs became more plentiful. We gave many boxes away to friends. My oldest daughter used them to make large batches of noodles and waffles.

But as satisfied as the hens were with their wooded acreage, we humans felt unsettled. We chafed at the amount of money we spent on rent; even more at the amount we spent on utilities. The location was not as desirable as it could be in terms of the public schools and transportation to my husband's office in Raleigh. In June, we bought a small house in Carrboro, where we had access to excellent public schools, parks, swimming pools, and a free-for-state-employees express bus to downtown Raleigh. Nearby there are more manicured subdivisions, some luxurious, others "low income," but our house, built in 1985, lies in a densely wooded neighborhood, filled with small, simple houses on large lots. Ours is next to a set of power lines, a price-suppressing factor, and when we first looked at the property, it did not escape our notice that our neighbors used the treeless land under the power lines for vegetable gardening. When first perusing the property we were also quick to identify the sound of chickens – some ten kept next door by a schoolteacher and her family; fifteen across the street kept by a nurse. Chickens in this neighborhood are not entirely legal: although city ordinances permit chickens, according to a legal covenant applying to properties in our subdivision, they are prohibited by a now defunct neighborhood homeowners' association.

It was only after we moved our coop to Carrboro that my husband observed that it was painted nearly the exact shade of brown as the house, something I had noticed immediately. In Michigan, where our chickens' respectable legal status depended on them being considered "animals commonly kept as pets," our daughters had helped paint the coop a brown that would provide camouflage among a grove of evergreens. After a year of uncertainty and difficult decisions, we cherished this token of suitability.

The *New York Times*, which publishes articles concerning urban chicken keeping in the United States several times a year, recently featured a piece regarding "retirement homes" for urban chickens (Van der Voo, 2012). By the time they are several years old, hens capable of living about a decade present people keeping them on small urban lots with a dilemma. Living in close proximity to the hens, even people who eat poultry regularly may be reluctant to slaughter them or give them to others who might. If they want to continue getting large numbers of eggs from their own hens, they need to replace the older birds with new ones. Hence, according to the *Times*, people in more rural areas were responding by setting up retirement establishments where older birds could be sent, for a fee, to live out their days, for those lacking farm-dwelling friends and relatives willing to take them in for nothing. On Internet forums devoted to urban chicken keeping, the article revived familiar debates. Jennifer Murtoff, who works as an urban chicken consultant in the Chicago area, argued that such practices should be discouraged. "Are we exchanging one unsustainable practice (large-scale egg farming) for another (the potential of filling rural farms with former urban pets)?" Murtoff asked (2012).

Though more clearly and thoughtfully articulated than most, Murtoff's position is not unusual. Memoirist and blogger Novella Carpenter (2009) provides an even more powerful and comprehensive commitment to urban farming and the reasons why the animals she raises for food are not "pets," despite her affection for them. Pet keeping is rife with hypocrisy and obscures so many undesirable realities (Shir-Vertesh, 2012). Behind the cozy image of the family pet lie entire industries devoted to their care and feeding that involve very different sensibilities than the ones we typically associate with companion animals. The pet food industry, for example, promotes what industry people call "humanization" (Mullin, 2007) and along with associated industries including veterinary medicine encourage us to part with enormous amounts of money to keep our companions alive and well. I respect those like Carpenter and Murtoff who manage to avoid the pet trap. If I made decisions about my household's flock on my own, it is possible that I would consider following their lead. My husband and children, however, are adamant that we are not "culling" any of our hens, even if it means we eventually resort to buying expensive eggs at the farmers' market from pastured country hens. Or we could buy the eggs most US Americans buy, the cheap ones at the supermarket that come from hens raised in appalling conditions (for an excellent overview and a critique of the "organic" egg industry, see Cornucopia Institute, 2010). Or we could stop eating eggs. I don't like to make such choices. For now, our flock is still young and produces far more than we can use ourselves.

In his book on the history of humans' relationships with animals, Bulliet (2005) argues that we are living in a period he calls "postdomesticity." Although individuals vary in their choices and positions, the human population overall relies on extensive use of animal products. It is because of our spatial separation from the animals that produce what we consume, Bulliet argues, that we tend to be uncomfortable with animal slaughter and suffering. Periodization always involves simplification. Along with many interested in making agriculture more sustainable, I am more disturbed by suffering than by killing. The notion of "postdomesticity" also obscures the diverse and dynamic nature of domesticatory practices. The direction of change has not been so straight, nor is its future so predictable. But I sympathize with Bulliet's interest in considering a variety of positions and sensibilities in relation to a larger time frame and political and environmental context. It seems appropriately humbling, without discouraging efforts to do things better.

Urban chicken keeping has plenty of contradictions and is hardly a paragon of virtue, in terms of animal welfare or sustainability or any other measure. Along with other "technologies of the self," through which citizen-consumers work to exercise control over their minds and bodies, on its own the practice has little potential for altering existing power relations (Foucault, 1988; Smith, 2008). It is a mistake, however, for animal advocates and others concerned with the ethics of the human–animal interface to dismiss anyone involved in exploiting animals for food as devoid of moral conscience.

An animal rights activist asked me whether keeping backyard and urban chickens altered people's perceptions of animals. I sensed she was hoping I would report that they became more concerned about animal well-being, more likely to become vegan. I could not provide the answer she sought and not only because I did not set out to measure such changes. Ethnographic research on chickens (as well as other animals) often has revealed humans capable of caring for animals and exploiting them at the same time (Geertz, 1973). There is no neat and tidy evolution – from wild to domestic, from pre-modern to modern, or from not caring to caring about animal well-being.

When I first began this research, I frequently doubted the wisdom of acquiring my own flock. Most daunting was the instructions I came across regarding how to perform surgery on a chicken with "bumblefoot," a localized staphylococcus infection and perhaps the most common ailment household flock keepers encounter. Although

I liked the thought of seeing chickens pecking around outside my kitchen window, I had no desire to acquire such know-how. Disgust was not the only reason. I had been studying gender, consumer culture, and expertise for a long time. Didn't I have enough domestic responsibility to manage in the course of a day, without adding bumblefoot? Anyway, I knew how it worked: you depart from your typical purchases, you acquire new knowledge or new sensibilities to go along with this departure, and before you know it, you have invested far more time and money than you ever imagined (McCracken, 1988).

The other day my family and I, with the help of a 13-year-old neighbor (who happened to have some experience with this task), watched my husband perform bumblefoot surgery on our back porch. Princess Karate had been limping for some time and the cause was clear: bumblefoot. I consulted the websites of schools of veterinary medicine and agricultural extension services. We watched a few minutes of instructional videos on YouTube. I came home from the local farm and garden store and the pharmacy with a handful of supplies. The children helped us capture this least docile of our hens. They draped a napkin over her eyes to keep her calm while I held her and while my husband soaked her foot, drained the abscess, cut away the dead tissue, applied antibiotic ointment, and finally wrapped her foot in gauze and pink "vetwrap." One of the videos we watched noted that the surgery tended to be harder on the humans than on the chicken. This appeared to be the case: when we were ready to let her go, she hopped down from my lap and ran across the lawn to compete with her flockmates for the remaining bits of a pile of table scraps. Watching her, I was amused to recall my earlier fear of bumblefoot. I hadn't been wrong about one thing leading to another or about the dangers of taking on more domestic responsibility and expense, but there was so much I never anticipated. I did not anticipate the degree to which members of my own family would be willing to share the joy and burdens of our flock's care. Nor did I anticipate the amount of thinking our chickens would inspire, thinking that led far beyond my own household and flock.

18 Entangled empathy
An alternative approach to animal ethics

Lori Gruen

Ethical arguments for considering the claims of the more than human world have tended to parallel arguments that extend ethical consideration outward from those who occupy the moral center. Historically in the United States and Europe, for example, we have seen white, Christian men extending rights to non-Christian, non-white men, and then women. As Peter Singer has noted, W. H. Lecky described a similar extensionist movement as "an expanding circle which begins with the individual, then embraces the family and soon the circle... includes first a class, then a nation, then a coalition of nations, then all of humanity, and finally, its influence is felt in the dealings of man with the animal world" (Singer, 1997).

Expanding the circle is one way scholars and activists have tried to combat what is alternatively termed "speciesism," "humanormativity," or "human exceptionalism" – terms that have been used to identify a perceived ethical problem with human attitudes toward and treatment of other animals. One of the main strategies for expanding the circle is to turn to empirical work designed to show that other animals are really similar to those at the center of the circle and thus deserve consideration. To be considered consistent and fair, we are implored to treat like cases alike. If those on the margins of the sphere of moral concern can be shown, through ethological and cognitive research, for example, to have some of the relevant qualities that we admire in ourselves and to which we attach value, then we ought to admire and value those qualities in whatever bodies they arise.

I too have drawn on this strategy, highlighting the way many species of non-humans have rich social relationships, are known to sacrifice their own safety by staying with sick or injured family members so that the fatally ill will not die alone, grieve their dead, respond to emotional states of others, engage in norm governed behavior, manipulate and deceive, understand symbolic representations, pass along culture, etc. Indeed many valuable human behaviors have been observed, often in less elaborate form, in the non-human world and these observations help us to see the irrationality of drawing a stark line that separates all and only humans from all other animals.

Some of the work that has drawn on identifying cognitive, behavioral, and social similarities between humans and other animals has helped us rethink how we have conceptualized certain ideas and practices. This work has also helped us to understand

The Politics of Species: Reshaping our Relationships with Other Animals, eds R. Corbey and A. Lanjouw. Published by Cambridge University Press. © Cambridge University Press 2013.

individuals of other species better. And acknowledging similarities can be motivation-ally powerful. Once we see that some other being is like us or our loved ones in ways that matter, we are more prone to consider their interests and be mindful of the ways their lives can go better or worse. But, the sameness view can also lead us astray.

The sameness approach is helpful in drawing our attention to excluded others and getting us to begin discussions about our responsibilities, but arguably, it hasn't led to a fundamental shift in our treatment of other animals. The United States still uses chimpanzees in invasive biomedical research, for example, although there is hope that will change soon. Whales and elephants are kept captive for entertainment, despite weighty evidence that they are similar to us in many ethically relevant respects and that captivity causes them deep distress. Around the globe the numbers of animals used for food is skyrocketing. Sadly, much of the reasoning that has led to the massive increase in the consumption of farmed animals, when reasoning is involved at all, is that people in developing countries want to be viewed as *similar* to those living in affluent nations.

Why entangled empathy?

When what we are looking for is similarities – how we might share the same general type of intelligence or cognitive skills, the same sensitivities and vulnerabilities, the same emotional responses – we tend to obscure or overlook distinctively valuable aspects of the lives of others. We assimilate them into our human-oriented framework; we grant them consideration, when we do, in virtue of what we believe they share with us; we allow them to be seen, perhaps for the first time, through our distinctively human gaze, and we often project our own beliefs and desires onto them. Yet in our magnanimous embrace of the other, we end up reconfiguring a dualism that will inevitably find some "other" to exclude.

Feminist theorists have long been skeptical of the problem of assimilation and have argued that the constructed opposition between "equality" and "difference" itself is problematic.[1] Some feminists have also been skeptical of the individualism that is central to standard approaches to moral considerability, arguing instead that we ought to see ourselves, not just as individual animal beings, but also as inextricably entangled with other animals; as central parts of relationships in which individuals shape each other. This skepticism invites new possibilities for thinking about animal ethics.[2]

We are already in relationships to other animals (and the rest of nature) and being in relationships always has ethical consequences. When we acknowledge this we can begin to reflect on the nature of those relationships in an effort to be responsible to those with whom we are engaged. In being responsive, we exercise our moral agency. Being in ethical relation involves, in part, being able to understand and respond to another's needs, interests, desires, vulnerabilities, hopes, perspectives, etc., not simply by positing, from one's own point of view, what they might or should be but

---

[1] See, for example, Joan Scott (1988) and Iris Young (1990).

[2] Actually, it's not that new as feminist care theorists (Carol Adams, Josephine Donovan, and Marti Kheel, for example) have been arguing in this way since the mid-1980s.

Figure 18.1 Susie Coston playing with a rescued friend, Gabriel, at Farm Sanctuary, Watkins Glen, New York, in 2010. The sanctuary houses chickens, geese, ducks, turkeys, cows, pigs, sheep, rabbits, and goats. Photo Jo-Anne McArthur, We Animals.

by working to try to grasp them from the perspective of the other. This may require learning about species-typical behaviors, personalities, and the ways of being of other animals. It also involves a recognition that one is distinct yet always co-constituted by the relationships one is in and appreciating fallibility and the necessary limits to understanding the experiences of another while nonetheless acknowledging the ethical commitments that relations engender. Though we are distinct from each other we are also shaping and co-constructing each other's needs, interests, desires, and even identities.[3] I suggest that entangled empathy is a way of connecting to specific others in their particular circumstances and thus is a central skill for being in ethical relations.

What is entangled empathy?

As I understand it, entangled empathy is a process whereby individuals who are empathizing with others first respond with a precognitive, empathetic reaction to the interests of another.[4] There are myriad ways such reactions can go wrong, but they are also often right in appropriately directing needed attention. From these reactions, we move to reflectively imagine ourselves in the position of the other, and then make a judgment about how the conditions that she finds herself in may contribute to her perceptions or state of mind and impact her interests. These perceptions will involve assessing the salient features of the situation and require that the empathizer seek to determine what is pertinent to effectively empathize with the being in question. Entangled empathy requires room to correct empathetic responses.

Entangled empathy involves both affect and cognition. The empathizer is also attentive to both similarities and differences between herself and her situation and that of the fellow creature with whom she is empathizing. She must move between her own and the other's point of view. This alternation between the first and third person points of view will minimize narcissistic projections, a worry associated with some forms of empathetic engagement.

Standard accounts of empathy suggest that what one does when one empathizes is put oneself into another's shoes. This does not require that the empathizer accurately characterizes the person or being with whom she is empathizing as the empathizer can maintain her own perspectives, values, beliefs, and attitudes, just from someone else's embodied position, as it were. This is one way to understand problematic anthropomorphizing (not all anthropomorphizing is problematic)[5] with animals and can lead to profound mistakes both in judgments and in practices. But in moving between the first and the third person perspective, one genuinely attempts to understand how the one being empathized with experiences the world and one tries to gain as much knowledge of the ways she lives as is possible. It also usually involves accepting the best explanations, even if they may ultimately be flawed. In the case of other animals, understanding how they experience their worlds, as well as individual personalities, is by no means easy. Indeed, many current discussions of the claims animals make on us are faulty in that the particularity of individual animal lives is not adequately attended to.

Some have argued that empathy with others needn't bring about good results for the individuals who are being empathized with as it doesn't have any motivational pull.

---

[3] See, for example, Karen Barad (2007), Donna Haraway (2007), and Stacey Alaimo and Susan Hekman (2008).

[4] This discussion builds on my earlier work on what I was calling "engaged empathy." For more detailed discussions, see Gruen (2004, 2009, 2012).

[5] See, for example, Brian L. Keeley (2004).

It has even been suggested that "good" torturers are good empathizers, which allows them to more fully access the tortured individual's weak spots. Yet, in the psychological literature, empathy is often coupled with a motivational state that leads to "helping action."[6] If the development and exercise of empathy involves both affective empathy (emotional contagion, imitation, etc.) and cognitive empathy (reflective engagement with the feelings of the other, perspective taking, etc.) then it is likely that empathy is motivational. But the motivations can take different forms. Some people moved to help a distressed individual with whom they are empathizing may be motivated to end the distress because it causes them discomfort; others may be moved because they are unable to imagine themselves in a situation in need in which others do not come to their aid; others may be motivated because their sense of themselves as an empathetic person requires it. Indeed, some combination of motivations may be operating much of the time. Unlike sympathizing with someone in distress in which the sympathizer feels bad or sorry for the person, entangled empathy involves the empathizer directly and thus is motivating.[7]

Correcting empathetic failures

Empathy can be inaccurate and empathetic inaccuracies can take a variety of forms, both epistemic and ethical. Epistemic empathetic failures can involve over-empathizing or incomplete empathizing. In cases in which over-empathizing occurs, the empathizer over-identifies with the emotions of another. This might occur between individuals who already have strong personal bonds – between friends, lovers, or between parents and their children or individuals and the other animals they care for. In cases of this type of empathetic inaccuracy, the empathizer exaggerates the emotional states of the one with whom she is empathizing. The empathizer may need to be less entangled. And over-empathetic inaccuracies are not limited to cases in which the empathizer is entangled with those particularly close to her. Sensitivities to political injustices may also lead people to over-empathize with those who are traditionally the targets of these injustices. Anti-racist white people, or pro-feminist men, or "straight but not narrow" individuals may fail to accurately empathize with people of color, women, or LGBT people when they observe or learn about a racist, sexist, or homophobic insult or injury. In these sorts of cases their empathetic imagination of the others' experiences may be exaggerated based on their identification with the other as a member of a politically oppressed group rather than as the particular individual who is suffering from a particular insult or injury. This is worrisome in the case of other animals who cannot easily correct these inaccuracies.

[6] See, for example, C. Daniel Bateson (1991) and Martin Hoffman (2000), on internalization and guilt.
[7] Because of the variability around the understanding of the terms empathy and sympathy not everyone would agree here. For example, Stephen Darwall (1998) claims that "empathy can be consistent with the indifference of pure observation" whereas sympathy involves concern for the person sympathized with and this concern is motivational. It should be clear from my discussion that empathetic distress is not something one can remain indifferent to and this is the case whether or not the empathizer has concern for the individual with whom she is empathizing. See too Diana Meyers (forthcoming).

Of course, the mistakes of perception that happen by those who are oversensitive to injustice pale in comparison to those who fail to notice. More often than not, injuries to members of historically marginalized groups are overlooked by those who do not see the insults or injuries as injustice, even those who are attempting to empathize. This leads to the opposite kind of empathetic inaccuracy – incomplete empathy. When we empathize with others we are putting ourselves into their position, we respond to the environmental cues that they are responding to, and we attempt to imagine ourselves in their frame of mind. But that frame of mind is often shaped by experiences that we ourselves haven't had, and thus our empathetic engagement will be limited or incomplete. Of course, we are all limited by the resources of our own minds, as Nagel famously put it, resources that are developed in response to experiences we have had or knowledge that is available to us (Nagel, 1974). In cases of incomplete empathy, the empathizer lacks the experiences and information necessary to accurately empathize.

Over-empathizing is a result of heightened sensitivity that blocks one's ability to accurately assess the situation. One remedy for this sort of empathetic failure is greater self-knowledge. When the empathizer becomes aware of her tendency to let her emotional or political dispositions cloud her ability to understand the perspective of another, in principle, she will be able to correct this failure either by keeping the tendencies that contribute to over-empathizing in check (which admittedly may be difficult to accomplish, but worth trying) or by critically reflecting on the judgments she makes paying particular attention to the distortions that are likely to emerge. (This will be especially important in the context of overly-righteous animal advocacy.)

Incomplete empathy results when either the relevant information is unavailable to the empathizer or when the empathizer is unable to grasp the information that is salient in a given situation. This too is correctable with effort. If the situation is one that could be understood, but isn't immediately for the empathizer, then the empathizer can seek out more details in an effort to correct the failure. It may be that the empathizer is unable to fully understand the relevant information, but better knowledge of the particulars of the situation can remedy this failure of understanding.

Epistemic empathetic failures can usually be remedied by correcting for the empathizer's emotional proclivities and gaps in knowledge. Ethical empathetic failures are trickier, and their remedies a bit more complicated.

Three examples

Here are a few brief examples of inapt empathy:

1. Animal experimentation and affected ignorance.[8]

   Abby does neurological research on cats. She is interested in grounding some of the behaviorist hypothesis developed in the 1970s in neurophysiology. She regularly

---

[8] This is a term Michel M. Moody-Adams (1994) introduced.

creates lesions in the brains of the cats, causing tremors, seizures, and other behavioral problems. Because anesthetics are likely to have a confounding affect on the brain, when it comes time to study the brains of these cats she removes them without anesthetizing the cats. The cats suffer terribly in this research but Abby doesn't empathize with them. They are just cats.

2. The ideological activist and empathetic bias.

Ben stands before the court after being arrested for burning down Huntington Laboratories and killing one of the researchers that was in the building. Ben does not deny planting a fire bomb but tells the jurors of the horrors that go on in various labs including a Huntington laboratory where Ben's dog Bosco mistakenly ended up, was experimented on, and ultimately killed as a result of malfeasance on the part of the man who died in the fire. Ben holds himself responsible for the trickery that led to Bosco's death at the hands of this experimenter and is bereft. His anger and grief, as well as his ideological commitment to other animals, led him to believe he had no choice but to take action. Two jurors empathize with Ben's situation, the jury is hung, and Ben is released (for now).

3. Atrocity, burnout and empathetic overload.

Jennifer has been an animal activist for many years. Like many activists, when she first came to realize the terrible truths about how we treat animals before we eat them, she thought all she needed to do was let others know what was happening and they would change. She even got a tattoo of the number 286 on her hand – 286 chickens are killed for food every *second*, 24 hours a day, seven days a week. When she wasn't out campaigning she was caring for sick and injured animals. After witnessing so much cruelty, suffering, death, and apathy, her empathy was having a negative impact on her work. She started feeling angry and depressed and finally became completely withdrawn.[9]

In the first case, Abby fails to empathize with the cats and thus violates their most basic interests because she fails to recognize cats as the proper objects of empathetic attention. Sadly, throughout our history there are too many instances of failures of this kind, failures that stem from a cultural blindness to the fact that there are certain others whose well-being has been negatively affected by failures of moral attention. Importantly, this failure is not an epistemic failure. There are social institutions that require this failure for their very existence and thus these institutions have an interest in promoting and naturalizing the failure. Yet, the fact that individuals suffer when whipped, poisoned, tortured, etc., is not mysterious or hard to understand even when dominant social institutions try to convince us otherwise. The failure to see these harms as harms is one of willful or affected ignorance, the phenomena of "choosing not to know what one can and should know" (Moody-Adams, 1994: p. 296). When Abby fails to empathize with the well-being of the cats she is

---

[9] This is loosely based on activist Jasmin Singer, who has the number 267 tattooed on her wrist (the number has actually gone up since she got the tattoo). In reality, Jasmin never burned out because she was aware of the dangers of empathetic overload. Jasmin now co-runs "Our Hen House" – an activist website that helps others find their way to change the world for animals. www.ourhenhouse.org.

choosing to accept misinformation about cats and thus does not allow her empathy to be engaged by the cats' suffering.[10]

The case of Ben and the jurors raises a different kind of ethical failure. Here the two jurors, in empathizing with Ben's plight, apparently condone his actions even though these actions caused death and damage. This is what has been called "here and now bias" and it is fairly widespread.[11] According to social psychologists, there is a tendency for people to "empathize with and help the victim who was at the focus of their attention more than victims who were absent" even when they are reminded that there are a greater number of victims and even when they were told that greater harms could result from empathizing with the individual who has captured their immediate empathetic attention. The failure here is that the jurors are placing unwarranted weight on Ben's distress and failing to adequately empathize with the devastating loss to the family of the man killed and workers whose lives are disrupted by Ben's action.

Jennifer's case is different again. Here, because of her proximity to vast amounts of suffering and the frequency with which she is forced to deal with this suffering, as well as the isolation she experiences by those who cause such suffering, Jennifer is experiencing empathetic saturation. Her failure is in not modulating her empathy. Had she taken time to care for herself and disengaged as a way of coping in the face of so much horror, she would have been able to continue her work and avoid being burned out or overwhelmed. When we empathize with Jennifer, put ourselves in Jennifer's position and understand the circumstances that she is regularly faced with, then we would most likely come to the decision to disengage or distance ourselves too. However, many people choose to ignore the suffering of others, they choose to disengage their empathetic responses to that suffering, because it "feels" like over-load, but their attempts to become disentangled aren't warranted. In a culture in which empathy is discouraged, in which greed and self-promotion (or the promotion of the interests only of those near and dear to one's self) are encouraged, and in which the suffering, humiliation, and distress of others is increasingly becoming a source of entertainment and even pride, this attempt to disentangle should be interrogated.

While some theorists interested in empathy have suggested that empathetic failure is "automatic," it is important to notice the plasticity of empathetic response; even automatic responses can be influenced by social and cultural norms and expectations. I have often been struck by stories about the empathetic response young children have to eating meat when they learn that meat comes from dead animals. In contexts in which concern for animals is encouraged, apparently children resist eating animals and then have to be taught that it is acceptable to do so. This is just one instance of the way that basic empathy can be altered. The failures of empathy, in Abby's case, the case of the

---

[10] Bonnie Steinbock (1978) has admitted to just such a failure when she writes: "I am willing to admit that my horror at the thought of experiments being performed on severely mentally incapacitated human beings in cases in which I would find it justifiable and preferable to perform the same experiments on nonhuman animals (capable of similar suffering) may not be *a moral emotion*." (my emphasis)

[11] Martin L. Hoffman (2000) has an extensive discussion of various empathetic limitations.

jurors judging Ben, and in Jennifer's case, however, represent ethical choices that the individuals in question can make, although they may be initially unaware of these choices as choices. Because they are choices, there is room for altering these choices and remedying these failures.

We can pay more attention to making better choices and promoting the well-being of those with whom we are entangled. Thinking harder not just about the nature of these relations, but also about how to be in *ethical* relations with a range of other beings, is an interesting underexplored project that becomes possible once we stop focusing exclusively on the properties that make us similar.

19 Extending human research protections to non-human animals

Hope Ferdowsian and Chong Choe

The use of non-human animals in experimentation presents opportunities to explore the ethics of how humans use other animals. Human research protections in the United States and elsewhere are guided by a principled approach, as outlined by the Belmont Report and similar documents. This principled approach offers a model in which animal research could be re-evaluated, and opportunities to explore how concepts such as autonomy, vulnerability, beneficence, and justice can be extended to moral considerations about non-human animals.

In our work we have both explored grounds for extending human research principles to non-human animals, in part using emergent evidence of their emotional, cognitive, and moral capacities. One of us (HF) is a physician specialized in internal medicine, preventive medicine, and public health whose clinical, research, and policy work has focused on the prevention and alleviation of suffering in vulnerable human and non-human populations. A consistent thread in the work of the other author (CC), a lawyer and philosopher specialized in moral, political, and legal philosophy, has been to defend the claim that the minimum constraints of justice apply in every social context, whether composed of humans or non-human animals, even against other competing economic and institutional interests.

In the following we describe how concepts such as autonomy, vulnerability, beneficence, and justice could be applied to decisions about non-human animals, similar to how these concepts are used to determine the ethical permissibility of human research. Within human populations, there are significant differences in mental capacity, moral aptitude, and other qualities, which have been used historically to justify differences in the treatment of humans and other animals. However, there is a universal expectation that humans with unequal gradations of these qualities will be treated in accordance with the same basic underlying principles.

The ways in which humans are treated in research and even other areas of society are guided by clear ethical principles. Several of these principles, as highlighted in the Belmont Report (1978),[1] Declaration of Helsinki (1964; see Declaration of Helsinki,

---

[1] The Belmont Report was developed and adopted by the National Commission for the Protection of Human Subjects of Biomedical and Behavioral Research in 1978 and published in 1979 (hereafter referred to as the

The Politics of Species: Reshaping our Relationships with Other Animals, eds R. Corbey and A. Lanjouw. Published by Cambridge University Press. © Cambridge University Press 2013.

Figure 19.1 Concerns about human experimentation, as highlighted in the Nuremberg trials (pictured above), led to the Nuremberg Code (1947). Concerns about unethical human and animal experimentation share a common history of substantive public moral controversies since the late nineteenth century. By contrast, advocates of animal experimentation sought to impress on the public the need for unrestricted access to animals to escape the need for harmful human experimentation. Photo United Nations Archives.

2008), and the Nuremberg Code (1947; see Shuster, 1997), revolve around key concepts including respect for autonomy, beneficence, and justice. Under the Belmont Report, for example, when an individual's autonomy is compromised, either because of being unable to deliberate about personal goals or being prohibited from acting under the direction of such deliberation, she or he is considered vulnerable and receives additional protections.

Currently, laws governing the use of non-human animals in research have resulted from a frequently politicized and fragmented process, rather than through methodical ethical examination. As a result, there are significant inconsistencies in the ways in

"Belmont Report"). The Belmont Report is available at www.hhs.gov/ohrp/humansubjects/guidance/belmont.html [accessed September 14, 2012].

which humans treat other animals, as demonstrated by laws governing research and other areas. For instance, the United States Animal Welfare Act (7 U.S.C. §§ 2131–59, 1966, as amended) arbitrarily excludes birds, rats, and mice bred for research from being considered animals for the purposes of the Act (§ 2132(g)). Animals used for food production are also excluded from the Animal Welfare Act (§ 2132(g)).

Many have justified the use of non-human animals in research by citing their similarities to humans, rather than considering how the common potential for vulnerability and mental and physical suffering influences moral considerations regarding non-human animals. Humans and other animals have certain interests in self-preservation, living free of unnecessary constraints, and meeting their basic needs. Like humans, other animals are susceptible to having their interests neglected. Additionally, characteristics determining quality of life are relevant to the consideration of an individual's interests and qualification for special protections.

The principled approach used in human research offers a model in which animal research could be re-evaluated. It offers opportunities to explore how concepts such as autonomy, vulnerability, beneficence, and justice can be extended toward the protection of non-human animals.

Respect for autonomy

The principle of respect for autonomy incorporates at least two ethical convictions: individuals should be treated as autonomous agents; and individuals with diminished autonomy are entitled to specific protections (Belmont Report).

Autonomy refers to the capacity for self-determination, or the ability of a rational individual to make informed decisions free of undue influence. Respect for autonomy requires that individuals, to the degree they are capable, be provided the opportunity to choose what will or will not happen to them. The principle of autonomy, within the context of biomedical ethics, refers to autonomous choices and, therefore, primarily concerns the relevant attributes of intentionality, understanding, and voluntariness. This principle is the central premise for voluntary informed-consent requirements. Where applicable, three elements are generally required for informed consent. These include disclosure of sufficient information, adequate opportunity for and demonstration of comprehension, and voluntariness. Special provisions are made when comprehension is limited, and these circumstances commonly require the input of third parties who are most likely to understand the individual's situation and act in that individual's best interest. Coercion, which occurs when an individual is threatened with harm in order to guarantee compliance, is prohibited (Belmont Report).

Not all humans are capable of self-determination, and the capacity for self-determination can change over the course of a lifetime. Some individuals are not born with the capacity for self-determination; other individuals achieve the capacity for self-determination through maturation; and others lose the capacity for self-determination as a result of illness, mental disability, or "circumstances that severely restrict liberty" (Belmont Report, Pt. B(1)).

There are many open questions concerning autonomy and how autonomy pertains to other animals. Among these are what attributes are necessary for autonomy and the degrees to which individuals exhibit these attributes across species. It is reasonable to conclude that non-human animals are autonomous to varying degrees, and, even if they fall outside of some accounts of autonomy, they at least have interests that should be protected.

Many non-human animals demonstrate varying degrees of self-determination and choices that protect their best interests. When scientists have examined animals' choices, research has shown that great apes and other animals can demonstrate clear preferences that directly relate to their participation in research procedures. For example, bonobos have demonstrated their preference for choosing and securing their own food, transmitting cultural knowledge to their offspring, seeking solitude, experiencing new places, living free from fear of attack by humans, and maintaining life-long contact with certain individuals (Savage-Rumbaugh et al., 2007).

We may never fully realize the cognitive capacities of non-human animals, or humans, for that matter. In the face of uncertain evidence, it appears most defensible to apply the principle of respect for autonomy to the extent relevant. In practice, this would require respecting the choices of individuals, including non-human animals, to the extent that they are capable of asserting their preferences or to the extent that we are able to discern their preferences from their behavior and interests.

Vulnerability

It has been successfully argued, at least in the context of research on human subjects, that some individuals require extensive protections, even to the point of excluding them from situations that may harm them, as a result of their vulnerability (Belmont Report, Pt. C(3)).

According to the Council for International Organizations of Medical Sciences (2002), vulnerable human subjects of research are those who are relatively or absolutely incapable of protecting their own interests because of insufficient power, education, intelligence, resources, strength, or other needed attributes to protect their own interests.

In practice, research involving vulnerable human subjects cannot involve more than minimal risk, except in specific, well-justified circumstances (see, e.g., 45 C.F.R. §§ 46.204, 46.406). Typically, vulnerable human populations, such as children, cannot be exposed to risks that exceed risks encountered during daily life by normal, average, healthy individuals living in safe environments or during the performance of routine physical or psychological examinations or tests (45 C.F.R. § 46.102(i)). In contrast, non-human animals are commonly used without significant restrictions as a consequence of their vulnerability.

Non-human animals used in research exhibit shared characteristics with vulnerable human populations, in their compromised capacity for self-determination as a result of circumstances that severely restrict liberty. Non-human animals are commonly prohibited from acting under the direction of their own deliberation as a result of their compromised status in society. Moreover, although many animals exhibit intelligence, rationality, and maturity, language barriers prohibit informed consent.

Vulnerability can be considered in several ways: one's ability to provide informed consent; one's ability to protect his or her own interests; and one's chances of being wronged. Each of these accounts of vulnerability can be extended to the case of non-human animals used in invasive research. Further, a combination of these accounts of vulnerability can be considered.

Beneficence

The principle of beneficence generally requires the avoidance of harm and creation of a favorable benefit–risk balance. As described in the Belmont Report, there is also significant emphasis on the correlative principle of non-maleficence. This principle is commonly referred to as "do no harm."

In human research, the obligations of beneficence extend to society at large, in addition to individual investigators. Risk for, or the probability of, harm to the individual subject must be minimized, while potential benefits must be maximized. The nature and scope of physical, psychological, social, and other harms are considered. Regardless of potential benefits to other individuals, subjects' rights must be protected. Inhumane treatment is never justified, and vulnerable populations receive special considerations when weighing risks and potential benefits (Belmont Report).

When considering harms, it is useful to have a conceptual structure. Harms may be normative or non-normative, the latter of which typically involves a setback to interests. Interests typically encompass a welfare condition or advantage and are distinguished from one's preferences or desires (DeGrazia, 1996; Beauchamp, 2011). Interests and needs are nearly synonymous in this context. At the very least, pain, suffering, and death would be considered conditions of harm since they represent setbacks to one's basic interests. Consideration of well-being may broaden the scope of consideration of harms. If core dimensions of well-being were social attachment, security, health, and freedom, then setbacks to those interests would be harmful.

The animal research enterprise involves a high risk of severe harms, including but not limited to mental and physical suffering, and minimal to no benefits to research subjects. Potential benefits to other members of society are difficult to quantify and oftentimes aspirational at best (Horrobin, 2003; Pound et al., 2004; Hackman and Redelmeier, 2006; Ioannidis, 2006; Perel et al., 2006; Matthews, 2008).

It is well established that non-human animals can experience pain, fear, distress, and a range of other emotions and psychological states, including as a result of their use in research (Gregory, 2004; McMillan, 2005a; Institute for Laboratory Animal Research, 2009; Posner, 2009; Sneddon, 2009). In vertebrates, pain is mediated and interpreted by a peripheral and central nervous system. Although the nervous systems of invertebrates are organized differently than those of vertebrates, well-organized nervous systems have also been described in invertebrates. For example, areas of the cephalopod brain have been compared with the cerebral cortex of vertebrates, and it may be more useful

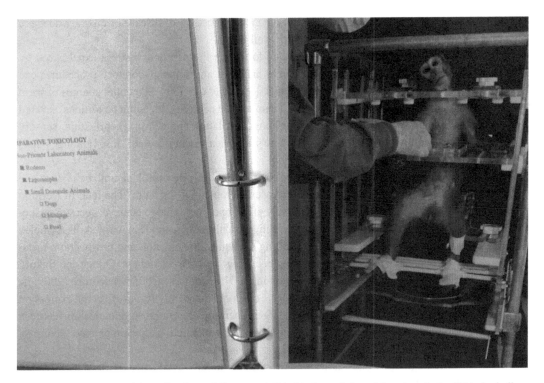

Figure 19.2 A lot of books and files were left behind at a defunct laboratory in the USA, including this vivisection manual. Photographed in 2008 during an investigation by Jo-Anne McArthur of We Animals, with permission to enter.

to examine the convergence of brain functions of invertebrate brains to those of vertebrates, rather than anatomical or physiological details (Mather, 2008).

Animals also experience pain and discomfort associated with disease, which can result in lethargy, depression, anorexia, sleep disturbances, and enhanced sensitivity to pain (Dantzer and Kelley, 2007). This is particularly relevant to the use of animals in research, since diseases are commonly induced in non-human animals to "model" human diseases.

Non-human animals demonstrate many emotional and psychological states similar to humans, increasing the potential for experimental or incidental harms incurred in research. Basic emotional systems such as panic, rage, and fear stem from subcortical operating systems (Panksepp, 2004). As a result, there are homologies across species in terms of mood and anxiety disorders (Brüne, 2008). For example, various forms of depression have been described in animals, including non-human primates, dogs, pigs, cats, birds, and rodents, among others. Complex signs of anxiety have been described in non-human primates, dogs, cats, birds, and other animals (Gregory, 2004; McMillan, 2005a). Post-traumatic stress disorder has been described in chimpanzees and other animals, including those used in laboratory research (see Birkett and McGrew, and King, both this volume; Brüne *et al.*, 2006; Bradshaw *et al.*, 2008; Ferdowsian *et al.*,

2011). However, it is important to note that the absence of disease or disorder does not indicate physical or mental health or well-being.

Although attempts have been made to reduce the pain and distress experienced by animals used in research, even gentle handling and minor laboratory procedures can cause significant distress in animals (Balcombe *et al.*, 2004). Potential sources of harm in animal research are extensive and can include captivity, breeding conditions, severed bonds and neglect, confinement, prolonged isolation and sensory deprivation, sensory over-stimulation, and threats to one's physical integrity and life, among others (Ferdowsian and Merskin, 2012).

It is also important to consider how the inability to fulfill natural behaviors and capacities in laboratory settings increases the potential for harm. The lack of freedom to fulfill these qualities could adversely affect one's potential to achieve a certain quality of life and well-being. For example, many animals demonstrate language skills, complex problem-solving skills, flexible learning strategies, imagination and prospection, meta-cognition, empathy, pleasure-seeking behaviors, and self-awareness (Ortega and Bekoff, 1987; Bekoff, 2004; Emery and Clayton, 2004; Burghardt, 2005; Balcombe, 2009; de Waal, 2009). However, researchers have observed that captivity can affect the expression of certain capacities. For example, a larger repertoire of gestural communication has been observed in chimpanzees living in the wild, compared with chimpanzees living in captivity (Hobaiter and Byrne, 2011).

Although individual subjects of research commonly experience physical and psychological harms, the risk for harm extends beyond harms to the individual. For example, families and social groups are commonly disrupted; ecological niches can be disturbed; and animals can experience distress if they witness harmful procedures being performed on conspecifics.

Justice

The principle of justice involves consideration of which individuals receive the benefits of research and which individuals bear its burdens. This principle partly emerged as a result of concerns about systematic selection of individuals for research because of their easy availability, class, compromised position, or manipulability. Exploitive research such as the Tuskegee syphilis study, hepatitis experiments on mentally challenged children, and other human experiments intensified concerns about the adequacy of human research protections and provided historical context for consideration of the principle of justice (Lederer, 1995; Advisory Committee on Human Radiation Experiments, 1996).

The considerations of justice that may be particularly relevant to non-human animals include the standards applied in determining eligibility in and exclusion from research. Participation in research protocols is usually voluntary and based on consent. In cases involving individuals with impaired autonomy, decisions regarding eligibility or exclusion are based on other criteria, including the degree of impairment, the level of risk involved in the research, and the potential benefits to the individual. The principle of justice requires that individuals, especially those with impaired autonomy, not be subjected to risks disproportionate to the benefits of the research (National Institutes of Health Office of Human Subjects Research, 2006). Moreover, if the individuals are members of groups unlikely to benefit from the research, justice may require that those individuals be excluded from use at all (Belmont Report, Pt. B(3)). The selection of subjects from vulnerable populations raises additional considerations of justice. The application of the principle of justice, as adopted in US federal regulations, requires specific protections for vulnerable populations, including children, prisoners, and pregnant women and fetuses (45 C.F.R. § 46 subparts B,C,D).

Similar to vulnerable human populations, non-human animals represent a convenient research population, since non-human animals are easily available and manipulated, inexpensive (as property), and occupy a dependent status in human society. Like human populations who were historically targeted for use in ethically problematic research, non-human animals also have a "frequently compromised capacity for free consent" (Belmont Report, Pt. B(3)).

It is for the sake of vulnerable populations, such as the mentally ill and those in prison, that the principle of justice became a central feature of our ethical evaluation of research on human subjects. The extension of the principle of justice to non-human animals and inclusion of non-human animals as vulnerable populations would require that their interests be afforded greater protection. That protection may include the consideration of criteria similar to those applied with human subjects, namely, the degree of impaired autonomy, the level of risk involved in the research, and the potential benefits to the individual. The principle of justice, therefore, would demand serious reconsideration of the use of non-human animals, particularly as a vulnerable population sensitive to the risks or burdens of experimentation.

Conclusions

Based on the similarities between humans and non-human animals, including their potential susceptibility as members of vulnerable populations, it is reasonable to conclude that non-human animals could receive similar protections to those offered to vulnerable human subjects of research. Further, their status as a vulnerable population would influence the ways in which other principles, such as beneficence and justice, could be applied to decisions about the use of non-human animals in research. This approach would require greater, rather than fewer, protections for non-human animals than they currently receive. This might include the following scenarios: exclusion from research; research permission only when there is minimal risk to the subject; or use in research when potential benefits to the individual far exceed risks for harm to the individual. However, questions regarding justification will continue to emerge in individuals and populations unable to provide consent. Although these conclusions raise pragmatic issues, these issues should be considered separately.

The applications and conclusions described here may have implications for other areas of society in which animals are used. Opponents to this approach may raise questions about whether animals qualify as legal persons with rights and responsibilities (see Berg, 2007; and Wise, this volume, for possible frameworks responding to these questions). Nevertheless, as has been described previously, the legal order does not necessarily correspond to the natural order; rather, our law reflects a policy determination as to whether legal personality has been extended to various populations (see e.g., *Byrn v. N.Y. City Health & Hosp. Corp.*, 286 N.E.2d 887, 889 (N.Y. 1972); Berg, 2007). Even where personhood is not established, a designation as a vulnerable population offers significant protections.

Acknowledgment

This material is based upon work supported by the National Science Foundation under Grant No. 1058186. Any opinion, findings, and conclusions or recommendations expressed in this material are those of the authors and do not necessarily reflect the views of the National Science Foundation.

20 The capacity of non-human animals for legal personhood and legal rights

Steven Wise

My main concern as a legal scholar and animal protection litigator during three decades now has been and still is the ancient, high, and thick legal wall that presently separates all humans from all non-humans. Humans are "legal persons," non-human animals, "legal things." A legal person has the capacity to possess legal rights; one who possesses a legal right is a legal person (Wise, 1998).

"Legal things" lack the capacity to possess legal rights; they are invisible to civil law and exist solely for legal persons.[1] Only legal persons count in civil court-rooms, for only they exist in law for their own benefits. A court confronted with a plaintiff's claim she possesses a legal right currently needs but to determine her species to see if her claim is possible. If she is human, it is. If she is a non-human animal, it is not.

Not two centuries ago, not all humans were legal persons. James Somerset was a legal thing when he arrived in England in 1769 (Wise, 2005). His 1772 transubstanti-ation from legal thing to legal person at the hands of Lord Mansfield, Chief Justice of the Court of King's Bench, marked the beginning of the end of human slavery (Wise, 1998). A legal transubstantiation for at least one non-human animal from legal thing to legal person is an initial objective of the Nonhuman Rights Project.[2]

The Nonhuman Rights Project will demand that a common law high court declare a non-human animal has the common law capacity to possess at least one legal right, that she is a legal person. It will demand judges engage in what Professor Melvin Aron Eisenberg calls a "transformation" of the common law, whereby judges, using "minim-alist or result-centered approaches... radically reconstruct the precedents and overturn the rule the precedents announce" (Eisenberg, 1988). According to Professor Eisenberg, "what the common law is cannot be determined without consideration of what the common law should be," because there exists a "necessary connection between the content of the common law and [certain] moral norms, policies, and experiential propositions" (Eisenberg, 1988).

---

[1] I refer here to the civil law, as opposed to the criminal law, in common law jurisdictions, not to the civil law inspired by Roman law, as opposed to the common law.

[2] See www.nonhumanrightsproject.org.

The Politics of Species: Reshaping our Relationships with Other Animals, eds R. Corbey and A. Lanjouw. Published by Cambridge University Press. © Cambridge University Press 2013.

The common law should be, and largely is, rooted in social morality, social policy, and human experience. Therefore, the best legal rule to govern any issue is the rule that best reflects these three elements, with appropriate balancing and adjustment when they do not point in exactly the same direction. Although not every common law rule is the best rule at any given moment, over time legal rules tend toward becoming the best.   (Eisenberg, 1990)

This echoes Lord Mansfield who, as Solicitor General, argued "the common law... works itself pure by rules drawn from the fountain of justice," a view held by modern common law judges who do not consider themselves strictly bound by existing common law rules – a group that includes most American state appellate judges (van Barker, 1744). They believe common law structure requires them to bring a common law rule at odds with social morality, public policy, or human experience into harmony with modern understandings.

Once a court recognizes a non-human animal can possess a legal right, its determination of whether she actually has that right she claims will appropriately shift from the current irrational, biased, hyper-formalistic, and overly simplistic question, "What species is the plaintiff?" to the rational, nuanced, value-laden, principled, and policy-enriched question, "What qualities does the plaintiff possess that are relevant to the issue of whether she is entitled to the legal right she claims?"

Recently I proposed a four-tier "legal rights pyramid." Its most basic level is level one: "legal personhood." Level two identifies those "legal rights possessed." Level three asks whether the plaintiff has a private right of action, while level four requires a plaintiff to possess "standing," generally to have been injured by the defendant's wrongful act, which can be redressed by the suit. This chapter will primarily discuss level one legal personhood, though its arguments generally apply equally to a determination of what level two legal rights the plaintiff should possess.

The strongest arguments for level one legal personhood derive from non-comparative liberty and comparative equality. The most critical question for a liberty right is what quality, or qualities, might be sufficient (though not necessary) to generate legal personhood. For humans, the four species of great apes, cetaceans, and individuals of other species with similar fundamental interests, I argue those qualities include bodily integrity and bodily liberty, as they protect fundamental interests (Wise, 2000).

I have never encountered a philosophical or jurisprudential argument that rationally argues that a *Homo sapiens*, human merely in form, is entitled, solely because she is a *Homo sapiens*, to legal personhood (Hoff, 1980). I have encountered bald assertions, but never a jurisprudential or philosophical argument (Fried and Fried, 2010). Therefore imagine a potential rights-bearer as a container that contains no cognitive qualities that could be said rationally to be relevant to creating fundamental interests and generating level one legal personhood. Next imagine filling an imaginary pitcher with qualities objective enough to be proven in court (hence such an unprovable quality as an incorporeal soul would be excluded), and slowly dripping qualities until they are rationally sufficient to generate level one legal personhood.

One important aspect of liberty is autonomy or self-determination. Things don't act autonomously. Persons do. Philosophers may understand autonomy as Kant did two centuries ago. I call Kant's a "full autonomy," whereby fully autonomous beings act

Figure 20.1 Not enough room for belugas: Vancouver aquarium, 2007. One of several species that are likely to be plaintiffs in the first lawsuits intended to pave the way for an American state high court to declare, for the first time, that a non-human animal is a legal person and has the capacity to possess a legal right. Photo Jo-Anne McArthur, We Animals.

completely rationally and should be treated as legal persons. Most moral and legal philosophers and common law judges, however, recognize that lesser autonomies exist and that a being can be autonomous if she has preferences and the ability to act to satisfy them, if she can cope with changed circumstances, make choices, even ones she can't evaluate well, or has desires and beliefs and can make appropriate inferences from them.

I call such lesser autonomies "practical." "Practical autonomy" is not just what most humans have, but is what most judges think is *sufficient* for basic liberty rights. A being has practical autonomy and is entitled to level one legal personhood and level two basic liberty rights if she can desire, intentionally act to try to fulfill her desire, and possesses a sense of self sufficient to allow her to understand, however dimly, that it is she who wants something and is trying to get it. Consciousness, but not necessarily self-conscious, and sentience are implicit.

How do we know one has practical autonomy? The more closely the behavior of a non-human animal resembles ours and the taxonomically closer we are, the more confident we can be that she possesses the practical autonomy we possess. Let us use the example of the common chimpanzee, though we could use other non-human animals, as well. Just a few million years ago, chimpanzees and humans shared a common ancestor. The 2005 draft sequence of the chimpanzee genome reveals that humans and chimpanzees "share more than 98% of our DNA and almost all of our genes." (Gunter and Dhand, 2005). "Overall, human and chimpanzee genes are extremely similar" (Gunter and Dhand, 2005). The divergence of single-nucleotide substitutions between the human and chimpanzee genomes is only about 1.06%, while genomic differences due to insertions and deletions of genetic material, so-called "indels," is about 2.7% (Cheng *et al.*, 2005). Even this small variation may not be significant, as it is nearly impossible to determine whether a human DNA sequence missing in chimpanzees was *added* during the course of human evolution or *lost* to chimpanzees during their evolution (Dennis, 2005). All in all, "given the short time since the human–chimpanzee split, it is likely that a few mutations of large effect are responsible for part of the current physical – phenotypic – differences that separate humans from chimpanzees and other great apes" (Li and Saunders, 2005).

So few genetic differences separate chimpanzees and other primates from human beings that the working group of 22 scientists and philosophers, formed to discuss ethical issues that might arise from human to non-human primate neural stem cell grafting, wrote in 2005 that:

many of the most plausible and widely accepted candidates for determining moral status involve mental capacities such as the ability to feel pleasure and pain, language, rationality, and richness of relationships. To the extent that a [non-human] primate attains those capacities, that creature must be held in correspondingly high moral standing.   (Greene *et al.*, 2005)

Chimpanzees possess these attributes: they demonstrate complex minds, are self-conscious and self-aware, exhibit some or all the elements of a theory of mind (they know what other chimpanzees see or know what other chimpanzees know; see Andrews, Hotto, both this volume), understand abstract symbols, construct complicated societies, transmit culture, use a human language or sophisticated language-like communication system, and engage in such complicated mental operations as deception, pretending, imitation, and insightful solving of difficult problems (Hauser *et al.*, 2002). Humans and chimpanzees are so genetically and evolutionarily close that prominent scientists argue that humans and chimpanzees should be placed in the same subtribe, *Hominina*, and the same genus, *Homo*, to form *Homo sapiens* and *Homo troglodytes* (Wildman and Goodman, 2004). We can be sufficiently confident that such cognitively complex beings possess practical autonomy and the Nonhuman Rights Project will demand that such non-human animals as great apes, cetaceans, and perhaps elephants and African gray parrots are entitled to level one legal personhood and certain level two fundamental legal rights as a matter of liberty.

The comparative right of equality, in its "normative model," as defined in numerous common law jurisdictions, both American and foreign, is an alternate justification for level one legal personhood and fundamental level two legal rights (Yap, 2005: pp. 73–4). In its normative form, equality – the demand that likes be treated alike and unalikes be treated unalike – is determined against the background of a larger social, political, and legal context in which courts identify prohibited judicial ends and determine whether a rational connection exists between the means and ends chosen, or not (Yap, 2005: p. 74).

Prohibited ends may include classifications that burden a plaintiff in a manner that reflects deeply personal social stereotypes regarding features that are biologically "immutable or changeable only at unacceptable personal costs" or that involve morally irrelevant traits (Yap, 2005: pp. 84–6). The Nonhuman Rights Project will argue that such non-human animals as great apes, cetaceans, and perhaps elephants and African gray parrots are entitled to level one legal personhood and certain level two fundamental legal rights as a matter of equality, as well.

Afterword

The relationship between humans and non-human animals figures prominently in most of our lives from the very beginning. For many, the first relationship is with the preferred household pet of one's culture, whether it be a dog, cat, fish, hamster, or boa. Over time, folklore, mythology, agriculture, and food culture converge to give every human, depending where she lives, a particular, complex relationship to other animals. When we consider the range of forms that this relationship takes, it becomes clear how randomly the lines have been drawn between "them" and "us." One culture sees dogs as family members. Another sees them as food. Despite these facts and the reality that our future as humans is inextricably linked to the integrity of a delicate ecosystem that comprises all life, it is rare that a human finds it important to ponder her relationship to non-humans, as have the contributors to this book. Much more predictably, we default to a tribal, narcissistic lens through which we see the rest of the world.

It is rare that we confront the reality that we, too, are animals, more specifically apes. When it does happen, as some have expressed in these pages, there is typically some anomalous event that triggers the process. I was lucky enough to have a transformative experience as a teenager, when I convinced my mother to purchase a pet monkey at a local shopping mall. It didn't take long to see how inhumane it was to attempt keeping the monkey as a pet. I wrote to various zoos until the Lincoln Park Zoo in Chicago accepted the monkey. My experience with Nicky and my subsequent interactions with apes have left me with a permanent sense of commonality with other apes. The similarities between humans and other apes – great apes, in particular – begs the question why we accord ourselves a superior status in nature that we are unable to extend to gorillas, chimpanzees, orangutans, bonobos, and gibbons. Why is it that we find everything that make us apes so critical and defining for ourselves but not for others? Why do we reserve a special place in nature only for ourselves? The fact is that if we do deserve such a place, other apes and animals would be worthy of sharing it with us.

Humans define communities of inclusion, and exclude outsiders, based on various criteria. These can range from race, gender, ethnicity, religion, species, or many other factors. The need to define those included in the in-group as "special," and fundamentally distinct in some way from all others, is inherent in all human cultures. This global

The Politics of Species: Reshaping our Relationships with Other Animals, eds R. Corbey and A. Lanjouw. Published by Cambridge University Press. © Cambridge University Press 2013.

Figure 2 This photograph was taken while a volunteer at Ngamba Island Chimpanzee Sanctuary, Uganda, was grooming a chimpanzee in 2009. Most of the chimpanzees there have been rescued from the bushmeat trade. Photo Jo-Anne McArthur, We Animals.

characteristic of humans has led communities and individuals to "dehumanize," mistreat, abuse, enslave, and otherwise harm those who are excluded. Racism, xenophobia, genocide, and the abuse and torture of other species for our use and consumption are manifestations of our ability to keep an emotional and cognitive distance. We separate ourselves from the need to feel empathy or to take moral and ethical responsibility for the consequences of our actions. Most humans actually see themselves as completely separate from the rest of nature, which is viewed as a resource to be exploited without any consideration for the consequences or implications to these other beings, to our ecosystem, and to our long-term survival. This failure to recognize or even consider our relationship to other animals – and to all of nature – is at our peril and, of course, theirs.

I was fortunate enough to participate in the convening entitled "Humans and Other Apes: Rethinking the Species Interface," that was the impetus for this book. The Arcus Foundation, a private foundation that I founded to advance conservation of the great apes and their habitats and equality for people of all sexual orientations and gender identities, convened a group of academics, writers, and professors from the fields of philosophy, cognitive science and psychology, animal behavior, law, conservation, and animal rights.

During a conversation that lasted three days, participants presented papers and entertained debate among the group. We looked at the broad topic of human/non-human coexistence from various angles too numerous to list in this afterword, but I believe the following questions characterize fairly the scope of our dialogue. Where do humans draw the line in terms of worth and respect relative to non-humans and where should they draw it? Is it possible to come to a consensus on a morality to guide the approach to our relationship to non-human animals? Is the apparent human need to differentiate by species any less dangerous or immoral than human racism? While there was vigorous debate, the group agreed that our behavior toward other species is riddled with inconsistency and even hypocrisy that allows us to justify the ways in which we knowingly inflict suffering on non-human animals, even to the point of extinction.

As the experts presented information and various schools of thought on questions such as how or whether animals grieve, how speciesism undermines social justice, where the lines should be drawn in research, whether eating other animals is moral, and whether non-humans are aware of other animals' thoughts, I was acutely aware of how few people ever have the opportunity to consider these questions in such a forum.

I am confident that the three days we shared were, at times, illuminating, affirming, encouraging, troubling, and maybe even transformative for some. I wish that every human being on Earth could have joined this conversation. Millions more must join if we are to save our planet and the precious life that inhabits it, including ourselves. That's why we were determined to publish this book and bring the conversation to a wider audience.

Whether you are new to the world of conservation, social justice, or ethics, or a seasoned and informed advocate, I hope you have found this book as informative and nourishing as we found our three days and that it informs a thought or an action that helps move all of nature closer to a place of sustainability.

                                                                            Jon Stryker

References

Abraham, J. 2010. Pharmaceuticalisation of society in context: theoretical, empirical and health dimensions. *Sociology* **44**: 603–22.

Acciaioli, G. 1989. Searching for good fortune: The making of a Bugis shore community at Lake Lindu, Central Sulawesi. PhD thesis, Australian National University.

Acciaioli, G. 2000. Kinship and debt: the social organization of Bugis migration and fish marketing at Lake Lindu, Central Sulawesi. *Bijdragen tot de Taal-, Land en Volkenkunde (Journal of the Humanities and Social Sciences of Southeast Asia)* **156**(3): 589–617.

Adams, C. 1993. The feminist traffic in animals. In Gaard, G. (ed.), *Ecofeminism: Women, Animals and Nature*. Philadelphia: Temple University Press.

Adams, C. 1997. 'Mad cow' disease and the animal industrial complex: an ecofeminist analysis. *Organization & Environment* **10**: 26–51.

Advisory Committee on Human Radiation Experiments. 1996. *Final Report: The Human Radiation Experiments*. New York: Oxford University Press.

Aitken, G. 2004. *A New Approach to Conservation: The Importance of the Individual through Wildlife Rehabilitation*. Burlington, VT: Ashgate.

Alaimo, S. and Hekman, S. (eds.). 2008. *Material Feminisms*. Bloomington: Indiana University Press.

Allen, C. Philosophy of cognitive ethology. http://host.uniroma3.it/progetti/kant/field/ceth.htm [Accessed March 26, 2013].

Allen, C. 2004. Animal pain. *Noûs* **38**: 617–43.

Allen, C. and Bekoff, M. 1997. *Species of Mind: The Philosophy and Biology of Cognitive Ethology*. Cambridge, MA: MIT Press.

Allman, J., Watson, K., Tetrault, N. and Hakeem, A. 2005. Intuition and autism: a possible role for Von Economo neurons. *Trends in Cognitive Science*, **9**: 367–73.

Allport, G. 1954. *The Nature of Prejudice*. Cambridge, MA: Addison-Wesley.

Almog, J., Perry, J. and Wettstein, H. (eds.). 1989. *Themes from Kaplan*. Oxford: Oxford University Press.

Altshuler, D. 1999. Novel interactions of non-pollinating ants with pollinators and fruit consumers in a tropical forest. *Oecologia*, **119**: 600–6.

Anderson, J., Gillies, A. and Lock, L. 2010. Pan thanatology. *Current Biology* **20**: R349–51.

Andrews, K. 2003. Knowing mental states: the asymmetry of psychological prediction and explanation. In Jokic, A. and Smith, Q. (eds.), *Consciousness: New Philosophical Perspectives*. Oxford: Oxford University Press, pp. 201–19.

Andrews, K. 2005. Chimpanzee theory of mind: looking in all the wrong places? *Mind and Language* **20**(5): 521–36.

Andrews, K. 2008. It's in your nature: a pluralistic folk psychology. *Synthese* **165**(1): 13–29.

Andrews, K. 2009a. Understanding norms without a theory of mind. *Inquiry* **52**(5): 433–48.

Andrews, K. 2009b. Politics or metaphysics? On attributing psychological properties to animals. *Biology and Philosophy* **24**: 51–63.

Andrews, K. 2012. *Do Apes Read Minds: Toward a New Folk Psychology.* Cambridge, MA: MIT Press.

AP Worldstream. 2004. U.S. zoo allows gorillas to pay last respects to group's leader. December 8, 2004. www.highbeam.com/doc/1P1-103228556.html [Accessed April 10, 2013].

Apperly, I. and Butterfill, S. 2009. Do humans have two systems to track beliefs and belief-like states? *Psychological Review* **116**(4): 953–70.

Apperly, I. and Robinson, E. 2001. Children's difficulties handling dual identity. *Journal of Experimental Child Psychology* **78**(4): 374–97.

Apperly, I. and Robinson, E. 2002. Five-year-olds' handling of reference and description in the domains of language and mental representation. *Journal of Experimental Child Psychology* **83**(1): 53–75.

Apperly, I. and Robinson, E. 2003. When can children handle referential opacity? Evidence for systematic variation in 5- and 6-year-old children's reasoning about beliefs and belief reports. *Journal of Experimental Child Psychology* **85**(4): 297–311.

Appiah, K. 1990. Racisms. In Goldberg, D. (ed.), *Anatomy of Racism*. Minneapolis: University of Minnesota Press, pp. 3–17.

Arhem, K. 1996. The cosmic food web: human–nature relatedness in the northwest Amazon. In Descola, Ph. and Palsson, G. (eds.), *Nature and Society: Anthropological Perspectives*. London: Routledge, pp. 185–204.

Aronson, E. and Patnoe, S. 1997. *The Jigsaw Classroom: Building Cooperation in the Classroom*. New York: Addison Wesley Longman, Inc.

Asquith, P. 1997. Why anthropomorphism is not metaphor: crossing concepts and cultures in animal behavior studies. In Mitchell, R. *et al.* (eds.), *Anthropomorphism, Anecdotes, and Animals*. Albany, NY: State University of New York Press, pp. 22–34.

Asquith, P. 2011. Of bonds and boundaries: what is the modern role of anthropomorphism in primatological studies? *American Journal of Primatology* **73**: 238–44.

Atran, S. 1998. Folk biology and the anthropology of science: cognitive universals and cultural particulars. *Behavioral and Brain Sciences* **21**: 547–609.

Baars, B. 2005. Subjective experience is probably not limited to humans: the evidence from neurobiology and behavior. *Consciousness and Cognition* **14**: 7–21.

Baird, R. 2000. The killer whale: foraging specializations and group hunting. In Mann *et al.* (eds.), *Cetacean Societies: Field Studies of Whales and Dolphins*. Chicago: University of Chicago Press, pp. 127–53.

Bakels, J. 2000. *Het verbond met de tijger. Visies op mensenetende dieren in Indonesië.* Leiden: Research School CNWS.

Bakels, J. 2003. Friend or foe. The perception of the tiger as a wild animal. In Nas, P., Persoon, G. and Jaffe, R. (eds.), *Framing Indonesian Realities. Essays in Symbolic Anthropology in Honour of R. Schefold*. Leiden: KITLV, pp.71–85.

Bakels, J. 2004. Farming the forest edge: perceptions of wildlife among the Kerinci of Sumatra. In: Knight, J. (ed.), *Wildlife in Asia. Cultural Perspectives*. London and New York: RoutledgeCurzon, pp. 147–65.

Bakels, J. 2009. Kerinci's living past: stones, tales and tigers. In Bonatz, D., Miksic, J., Neidel, D. and Tjoa-Bonatz, M. (eds.), *From Distant Tales. Archeology and Ethnohistory in the Highlands of Sumatra*. Cambridge Scholars Publishing, pp. 367–83.

Baker, K. 1997. Straw and forage material ameliorate abnormal behaviors in adult chimpanzees. *Zoo Biology* **16**: 225–36.

Baker, K. and Easley, S. 1996. An analysis of regurgitation and reingestion in captive chimpanzees. *Applied Animal Behavior Science* **49**: 403–15.

Balakrishnan, G. and Schmitt, C. 2000. *The Enemy: An Intellectual Portrait of Carl Schmitt*. New York: Verso.

Balcombe, J. 2009. Animal pleasure and its moral significance. *Applied Animal Behavior Science* **118**(3): 208–16.

Balcombe, J., Barnard, N. and Sandusky, C. 2004. Laboratory routines cause animal stress. *Contemporary Topics in Laboratory Animal Science* **43**(6): 42–51.

Bara, B., Barsalou, L. and Bucciarelli, M. (eds.) 2005. Proceedings of the 27th Annual Conference of the Cognitive Science Society. Mahwah, NJ: Erlbaum.

Barad, K. 2007. *Meeting the Universe Halfway: Quantum Physics and the Entanglement of Matter and Meaning*. Raleigh: Duke University Press.

Barber, T. X. 1993. *The Human Nature of Birds: A Scientific Discovery with Startling Implications*. New York: St. Martin's.

Bard, K., Myowa-Yamakoshi, M., Tomonaga, M. *et al.* 2005. Group differences in the mutual gaze of chimpanzees (*Pan troglodytes*). *Developmental Psychology* **41**: 616–24.

Baron-Cohen, S. 1995. *Mindblindness: an Essay on Autism and Theory of Mind*. Cambridge, MA: MIT Press.

Baron-Cohen, S. (ed.) 1997. Why evolutionary psychopathology? In Baron-Cohen, S. (ed.) *The Maladapted Mind: Classic Readings in Evolutionary Psychopathology*. Hove, Sussex: Psychology Press.

Bateson, D. 1991. *The Altruism Question: Toward a Social Psychological Answer*. Hillsdale: Lawrence Erlbaum.

Bauman, Z. 1989. *Modernity and the Holocaust*. Cambridge: Polity Press.

Beauchamp, T. 2011. Rights theory and animal rights. In Beauchamp, T. and Frey, R. (eds.), *The Oxford Handbook of Animal Ethics*. Oxford and New York: Oxford University Press, pp. 198–227.

Beauchamp, T. and Frey, R. (eds.) 2011. *The Oxford Handbook of Animal Ethics*. Oxford and New York: Oxford University Press.

Bechtel, W. and Graham, G. (eds.) 1998. *A Companion to Cognitive Science*. Oxford: Basil Blackwell.

Becker, H. 1986. *Doing Things Together: Selected Papers*. Evanston, Illinois: Northwestern University Press.

Bekoff, M. 1995. Vigilance, flock size, and flock geometry: Information gathering by Western Evening Grosbeaks (Aves, Fringillidae). *Ethology* **99**: 150–61.

Bekoff, M. 1996. Cognitive ethology, vigilance, information gathering, and representation: who might know what and why? *Behavioural Processes* **35**: 225–37.

Bekoff, M. 1998a. Deep ethology, animal rights, and the Great Ape/animal project: resisting speciesism and expanding the community of equals. *Journal of Agricultural and Environmental Ethics* **10**: 269–96.

Bekoff, M. 1998b. Cognitive ethology: the comparative study of animal minds. In Bechtel, W. and Graham, G. (eds.), *A Companion to Cognitive Science*. Oxford: Basil Blackwell, pp. 371–9.

Bekoff, M. (ed.) 2004. *Encyclopedia of Animal Behavior*. 3 vols. Westport: Greenwood Publishing Group, Inc.

Bekoff, M. 2006. *Animal Passions and Beastly Virtues:Reflections on Redecorating Nature*. Philadelphia: Temple University Press.

Bekoff, M. 2007a. *The Emotional Lives of Animals: A Leading Scientiest Explores Animal Joy, Sorrow and Empathy*. Novato, CA: New World Library.

Bekoff, M. 2007b. Aquatic animals, cognitive ethology, and ethics: questions about sentience and other troubling issues that lurk in turbid water. *Diseases of Aquatic Organisms* **75**, 87–98.

Bekoff, M. 2010. *The Animal Manifesto: Six Reasons for Expanding our Compassion Footprint*. Novato, CA: New World Library.

Bekoff, M. 2011. Cooperation and the evolution of social living: Beyond the constraints and implications of misleading dogmas. In Sussman, R. and Cloninger, R. (eds.), *Origins of Altruism and Cooperation. Developments in Primatology: Progress and Prospects Volume 36*, Part 2, pp. 111–19.

Bekoff, M. 2012. Animal consciousness and science matter. www.psychologytoday.com/blog/animal-emotions/201205/animal-consciousness-and-science-matter. [Accessed March 25, 2012]

Bekoff, M. (ed.) 2013. *Ignoring Nature no More: The Case for Compassionate Conservation*. Chicago: University of Chicago Press.

Bekoff, M. 2014. *Rewilding our Hearts*. Novato, CA: New World Library.

Bekoff, M. and Bexell, S. 2010. Ignoring nature: why we do it, the dire consequences, and the need for a paradigm shift to save animals and habitats and to redeem ourselves. *Human Ecology Review* **17**: 70–4.

Bekoff, M. and Gruen, L. 1993. Animal welfare and individual characteristics: A conversation against speciesism. *Ethics and Behavior* **3**, 163–75.

Bekoff, M. and Pierce, J. 2009. *Wild Justice: The Moral Lives of Animals*. Chicago: University of Chicago Press.

Bekoff, M. and Pierce, J. 2012. Wild justice redux: what we know about social justice in animals and why it matters. *Social Justice Research* **25**: 122–39.

Bekoff, M., Allen, C. and Burghardt, G. (eds.). 2002. *The Cognitive Animal: Empirical and Theoretical Perspectives on Animal Cognition*. Cambridge, MA: MIT Press.

Bennett, E. 2011. Another inconvenient truth: the failure of enforcement systems to save charismatic species. *Oryx* **45**: 476–9.

Bennett, J. 1978. Some remarks about concepts. *Behavioral and Brain Sciences* **1**: 557–60.

Benz-Schwarzburg, J. and Knight, A. 2011. Cognitive relatives yet moral strangers. *Journal of Animal Ethics* **1**: 9–36.

Beran, M. and Smith, J. 2011. Information seeking by rhesus monkeys (*Macaca mulatta*) and capuchin monkeys (*Cebus apella*). *Cognition* **120**: 90–105.

Berg, J. 2007. Of elephants and embryos: a proposed framework for legal personhood. *Hastings Law Journal* **59**: 369–406.

Berkhoffer R. 1979. *The White Man's Indian: Images of the American Indian from Columbus to the Present*. New York: Vintage.

Bermúdez, J. 2009. Mindreading in the Animal Kingdom. In Robert W. Lurz (ed.), *The Philosophy of Animal Minds*. Cambridge: Cambridge University Press, pp. 145–64.

Berry, T. 1999. *The Great Work: Our Way into the Future*. New York: Bell Tower.

Bertolani, P. and Pruetz, J. 2011. Seed reingestion in savannah chimpanzees (*Pan troglodytes verus*) at Fongoli, Senegal. *International Journal of Primatology* **32**: 1123–32.

Best, S. 2009. The rise of critical animal studies: putting theory into action and animal liberation into higher education. *Journal for Critical Animal Studies* **7**: 9–53.

Bikié, H., Collomb, J.-G., Djomo, L. *et al.* 2000. *An Overview of Logging in Cameroon.* A Global Forest Watch Cameroon Report. www.globalforestwatch.org/common/cameroon/english/report.pdf [Accessed March 25, 2013].

Birkett, L. and Newton-Fisher, N. 2011. How abnormal is the behaviour of captive, zoo-living chimpanzees? *PLoS ONE* **6**(6): e20101.

Biro, D., Humle, T., Koops, K. *et al.* 2010. Chimpanzee mothers at Bossou, Guinea carry the mummified remains of their dead infants. *Current Biology* **20**: R351–2.

Blazich, J. 2011. Urban henfare: a model approach to keeping chickens within residential areas. *The Public Servant.* http://governmentandpublicsector.ncbar.org/newsletters/publicservantmarch2011/urbanhenfare.aspx [Accessed 25 March, 2013].

Bleecker, S. E. 1975. Fishes with electric know-how. *Sea Frontiers*, May/June, pp. 142–8.

Blok, A. 2001. *Honour and Violence.* Cambridge: Polity.

Bloomsmith, M. and Lambeth, S. 1995. Effects of predictable versus unpredictable feeding schedules on chimpanzee behavior. *Applied Animal Behavior Science* **44**: 65–74.

Bluff, L., Troscianko, J., Weir, A., Kacelnick, A. and Rutz, C. 2010. Tool use by wild New Caledonian Crows *Corvus moneduloides* at natural foraging sites. *Proceedings of the Royal Society B.* http://rspb.royalsocietypublishing.org/content/early/2010/01/05/rspb.2009.1953.full [Accessed March 25, 2013].

Blüthgen, N., Verhaagh, M., Goitía, W. *et al.* 2000. How plants shape the ant community in the Amazonian rainforest canopy: the key role of extrafloral nectaries and homopteran honeydew. *Oecologia* **125**(2): 229–40.

Boesch, C. 1991. Teaching among wild chimpanzees. *Animal Behavior* **41**: 530–2.

Boesch, C. 1993. Aspects of transmission of tool-use in wild chimpanzees. In Gibson, K. and Ingold, T. (eds.), *Tools, Language and Cognition in Human Evolution.* Cambridge, UK: Cambridge University Press, pp. 171–83.

Boesch, C. 2009. *The Real Chimpanzee: Sex Strategies in the Forest.* Cambridge University Press.

Boesch, C. 2011. Life and death in the forest. In Robbins, M. and Boesch, C. (eds.), *Among African Apes.* Berkeley: University of California Press, pp. 30–42.

Boesch, C. and Boesch-Achermann, H. 2000. *The Chimpanzees of Tai Forest.* Oxford: Oxford University Press.

Bogdan, R. 2009. *Predicative Minds: the Social Ontogeny of Propositional Thinking.* Cambridge, MA: MIT Press.

Bourgeois, S., Vasquez, M. and Brasky, K. 2007. Combination therapy reduces self-injurious behavior in a chimpanzee (*Pan troglodytes troglodytes*): a case report. *Journal of Applied Animal Welfare Science* **10**: 123–40.

Bourke, J. 2011. *What it Means to be Human.* Berkeley, CA: Counterpoint.

Bradshaw, G., Capaldo, T., Lindner, L. and Grow, G. 2008. Building an inner sanctuary: complex PTSD in chimpanzees. *Journal of Trauma and Dissociation* **9**(1): 9–34.

Bradshaw, G., Capaldo, T., Lindner, L. and Grow, G. 2009. Developmental context effects on bicultural post-trauma self repair in chimpanzees. *Developmental Psychology* **45**: 1376–88.

Braithwaite, V. 2010. *Do Fish Feel Pain?* New York: Oxford University Press.

Braithwaite, V. 2011. Researcher explores whether fish feel pain. www.biologynews.net/archives/2010/11/17/researcher_explores_whether_fish_feel_pain.html [Accessed March 25, 2013].

Brakes, P. and Williamson, C. 2007. Dolphin assisted therapy: can you put your faith in DAT? *Whale and Dolphin Conservation Society* **21**.

Brakes, P. and Simmonds, M. (eds.). 2011. *Whales and Dolphins: Cognition, Culture, Conservation and Human Perceptions.* Washington, DC: Earthscan.

Breed, M., Abel, P., Bleuze, T. and Denton, S. 1990. Thievery, home ranges, and nestmate recognition in *Ectatomma ruidum. Oecologia* **84**(1): 117–121.

Breed, M., Snyder, L., Lynn, T. and Morhart, J. 1992. Acquired chemical camouflage in a tropical ant. *Animal Behaviour* **44**: 519–23.

Breed, M., McGlynn, T., Stocker, E. and Klein, A. 1999. Thief workers and variation in nestmate recognition in a ponerine ant *Ectatomma ruidum. Insectes Sociaux* **46**: 327–31.

Brent, L., Koban, T. and Ramirez, S. 2002. Abnormal, abusive, and stress-related behaviors in baboon mothers. *Biological Psychiatry* **52**: 1047–56.

Brewer, J. and Staves, S. (eds.). 1995. *Early Modern Perceptions of Property.* London: Routledge.

Brody, J. E. 1990. Huge study of diet indicts fat and meat. *New York Times*, May 8: C1.

Broom, D. 1998. Welfare, stress and the evolution of feelings. *Advances in the Study of Behavior* **27**: 371–403.

Brown, R. 1984. Blackie was (fin)ished until Big Red swam in. *Weekend Argus* (Cape Town, South Africa), August 18: 15.

Brüne, M. 2008. *Textbook of Evolutionary Psychiatry: The Origins of Psychopathology.* New York/Oxford: Oxford University Press.

Brüne, M., Bruene-Cohrs, U. and McGrew, W. 2004. Psychiatric treatment for great apes? *Science* **306**: 2039.

Brüne, M., Brüne-Cohrs, U., McGrew, W. and Preuschoft, S. 2006. Psychopathology in great apes: concepts, treatment options and possible homologies to human psychiatric disorders. *Neuroscience and Behavioral Reviews* **30**(8): 1246–59.

Bulliet, R. 2005. *Hunters, Herders, and Hamburgers: The Past, Present and Future of Human–Animal Relationships.* New York: Columbia University Press.

Bunting, M. 2011. Occupy London is a nursery for the mind. *The Guardian* (October 30). www.guardian.co.uk/commentisfree/2011/oct/30/occupy-london-nursery-mind [Accessed March 25, 2013].

Burghardt, G. 1997. Review of Cavalieri and Singer. *Society and Animals* **5**: 83–6.

Burghardt, G. 2005. *The Genesis of Animal Play: Testing the Limits.* Cambridge: MIT Press.

Burton, M. 1978. *Just Like an Animal.* New York: Charles Scribner's Sons.

Busch, L. and Juska, A. 1997. Beyond political economy: actor networks and the globalization of agriculture. *Review of International Political Economy* **4**: 688–708.

Buss, D. 2004. *Evolutionary Psychology: The New Science of the Mind.* Boston: Pearson.

Buttelmann, D., Carpenter, M. and Tomasello, M. 2009. Eighteen-month-old infants show false-belief understanding in an active helping paradigm. *Cognition* **112**: 337–42.

Call, J. and Carpenter, M. 2001. Do chimpanzees and children know what they have seen? *Animal Cognition* **4**: 207–20.

Call, J. and Tomasello, M. 1994. The social learning of tool use by orangutans (*Pongo pygmaeus*). *Human Evolution* **9**: 297–313.

Call, J. and Tomasello, M. 1999. A nonverbal false belief task: the performance of children and Great Apes. *Child Development* **70**(2): 381–95.

Call, J. and Tomasello, M. 2008. Does the chimpanzee have a theory of mind? 30 years later. *Trends in Cognitive Science* **12**(5): 187–92.

Call, J., Hare, B., Carpenter, M. and Tomasello, M. 2004. 'Unwilling' versus 'unable': chimpanzees' understanding of human intentional action. *Developmental Science* **7**(4): 488–98.

Cambridge Declaration on Consciousness. 2012. http://fcmconference.org and www.psychology-today.com/blog/animal-emotions/201208/scientists-finally-conclude-nonhuman-animals-are-conscious-beings [Accessed March 25, 2013].

Candea, M. 2010. I fell in love with Carlos the meerkat: engagement and detachment in human–animal relations. *American Ethnologist* **37**(2): 241–58.

Candea, M. 2011. We both wait together: poaching Agustin Fuentes. *Kroeber Anthropological Society Papers* **100**(1), 148–51.

Caro, T. M. 2003. Umbrella species: critique and lessons from East Africa. *Animal Conservation* **6**: 171–81.

Caro, T. M. and Hauser, M. D. 1992. Is there teaching in nonhuman animals. *Quarterly Review of Biology* **67**: 151–74.

Carpendale, J. I. M. and Chandler, M. J. 1996. On the distinction between false belief understanding and subscribing to an interpretive theory of mind. *Child Development* **67**: 1686–706.

Carpenter, N. 2009. *Farm City: The Education of an Urban Farmer*. NY: Penguin.

Carrithers, M. 1996. Nature and culture. In Barnard, A. and Spencer, J. (eds.), *Encyclopedia of Social and Cultural Anthropology*. London and New York: Routledge, pp. 393–6.

Carruthers, P. 1992. *The Animal Issue*. Cambridge: Cambridge University Press.

Carruthers, P. 2000. *Phenomenal Consciousness: A Naturalistic Theory*. Cambridge: Cambridge University Press.

Carruthers, P. 2005. *Consciousness: Essays from a Higher-order Perspective*. Oxford: Oxford University Press.

Carruthers, P. 2006. Higher-order theories of consciousness. In Velmans, M. and Schneider, S. (eds.), *The Blackwell Companion to Consciousness*. Oxford: Blackwell.

Carruthers, P. 2009. Mindreading underlies metacognition. *Behavioral and Brain Sciences*, **32**: 121–82.

Cassidy, R. and Mullin, M. (eds.). 2007. *Where the Wild Things are Now: Domestication Reconsidered*. Oxford: Berg.

Cassil, D., Butler, J., Vinson, S. and Wheeler, D. 2005. Cooperation during prey digestion between workers and larvae in the ant *Pheidole spadonia*. *Insectes Sociaux* **52**: 339–43.

Castree, N. 2002. False antithesis? Marxism, nature and actor-networks. *Antipode* **34**: 111–46.

Cavalieri, P. 2001. *The Animal Question: Why Nonhuman Animals Deserve Human Rights*, 2nd ed. Transl. Catherine Woollard. New York: Oxford University Press.

Cavalieri, P. and Singer, P. 1993. A declaration on Great Apes. In Cavalieri, P. and Singer, P. (eds.), *The Great Ape Project: Equality Beyond Humanity*. London: Fourth Estate, pp. 4–7.

Celli, M., Tomonaga, M., Udono, T., Teramoto, M. and Nagano, K. 2003. Tool use task as environmental enrichment for captive chimpanzees. *Applied Animal Behavior Science* **81**: 171–82.

Chandroo, K., Yue, S. and Moccia, R. 2004. An evaluation of current perspectives on consciousness and pain in fishes. *Fish and Fisheries* **5**: 281–95.

Cheney, D. and Seyfarth, R. 2007. *Baboon Metaphysics: The Evolution of a Social Mind*. Chicago: University of Chicago Press.

Cheng, Z., Ventura, M., She, X. *et al.* 2005. A genome-wide comparison of recent chimpanzee and human segmental duplications. *Nature* **437**: 88–93.

Christman, J. 2009. Autonomy in moral and political philosophy. In Zalta, E. (ed.) *The Stanford Encyclopedia of Philosophy*. http://plato.stanford.edu/entries/autonomy-moral [Accessed 25 March, 2013].

Christopherson, S. and Storper, M. 1986. The city as studio; the world as back lot: the impact of vertical disintegration on the location of the motion picture industry. *Environment and Planning D: Society and Space* **4**: 305–20.

Chrulew, M. 2011. Managing love and death in a zoo: the biopolitics of endangered species preservation. *Australian Humanities Review* **50**: 137–57.

City of Durham. 2013. Zoning districts. www.ci.durham.nc.us/departments/planning/udo/pdf/udo_04.pdf [Accessed 25 March, 2013].

Clark, G. and Willermet, C. M. 1997. *Conceptual Issues in Modern Human Origins Research*. New York: Aldine de Gruyter.

Clarke, A., Juno, C. and Maple, T. 1982. Behavioral effects of a change in the physical environment: a pilot study of captive chimpanzees. *Zoo Biology* **1**: 371–80.

Clayton, S. and Myers, G. 2009. *Conservation Psychology: Understanding and Promoting Human Care for Nature*. New York: Wiley-Blackwell.

Coetzee, J. M. 1999. *The Lives of Animals*. Princeton, NJ: Princeton University Press.

Cohen, C. and Regan, T. 2001. *The Animal Rights Debate*. Lanham, MD: Rowman & Littlefield.

Cole, M. and Morgan, K. 2011. Vegaphobia: derogatory discourses of veganism and the reproduction of speciesism in UK national newspapers. *British Journal of Sociology* **62**: 135–53.

Compassionate Conservation. 2010. http://compassionateconservation.org [Accessed March 25, 2013].

Confer, J., Easton, J., Fleischman, D. *et al.* 2010. Evolutionary psychology: controversies, questions, prospects, and limitations. *American Psychologist* **65**: 110–26.

Connor, R., Wells, R., Mann, J. and Read, A. 2000. The bottlenosed dolphin: social relationships in a fission–fusion society. In Mann, J. *et al. Cetacean Societies: Field Studies of Whales and Dolphins*. Chicago: University of Chicago Press, pp. 91–126.

Cook, O. 1904a. An enemy of the cotton boll weevil. *Science* **19**(492): 862–4.

Cook, O. 1904b. Professor William Morton Wheeler on the Kelep. *Science* **20**(514): 611–12.

Cooney, N. 2011. *Change of Heart: What Psychology can Teach us about Spreading Social Change*. New York: Lantern Books.

Cooper, M. 2008. *Life as Surplus: Biotechnology and Capitalism in the Neoliberal Era*. Washington, DC: University of Washington Press.

Corbey, R. 2005. *The Metaphysics of Apes: Negotiating the Animal–Human Boundary*. Cambridge: Cambridge University Press.

Corbey, R. 2012. *Homo habilis'* humannness: Phillip Tobias as a philosopher. *History and Philosophy of the Life Sciences* **13**: 103–16.

Cormier, L. 2002. Monkey as food, monkey as child: Guaja symbolic cannibalism. In Fuentes, A. and Wolfe, L. (eds.), *Primates Face to Face: The Conservation Implications of Human–Nonhuman Primate Interconnections*. Cambridge: Cambridge University Press, pp. 63–84.

Corning, P. 2011. *The Fair Society: The Science of Human Behavior and the Pursuit of Social Justice*. Chicago: University of Chicago Press.

Cornucopia Institute. 2010. Scrambled eggs: separating factory farm egg production from authentic organic agriculture. www.cornucopia.org/egg-report/scrambledeggs.pdf [Accessed 25 March, 2013].

Council for the International Organizations of Medical Sciences (CIOMS) 2002. *International Guiding Principles for Biomedical Research Involving Animals*. Geneva, Switzerland: CIOMS, WHO.

Crisp, R. and Turner, R. 2009. Can imagined interactions produce positive perceptions? Reducing prejudice through simulated social contact. *American Psychologist* **64**: 231–40.

Csibra, G. 2010. Recognizing communicative intentions in infancy. *Mind and Language* **25**(2): 141–68.

Csibra, G. and Gergely, G. 2006. Social learning and social cognition: the case for pedagogy. In Johnson, M. and Munakata, Y. (eds.), *Processes of Change in Brain and Cognitive Development: Attention and Performance XXI*. Oxford: Oxford University Press, pp. 249–74.

Csibra, G. and Gergely, G. 2011. Natural pedagogy as evolutionary adaptation. *Philosophical Transactions of the Royal Society B* **366**: 1149–57.

Cutler, G. J. 2002. Diseases of the chicken. In Bell, D. D. and Weaver, W. D., Jr. (eds.), *Commercial Chicken Meat and Egg Production*, 5th edn, Norwell, MA: Kluwer Academic, pp. 473–542.

Damasio, A. 2010. *Self Comes to Mind : Constructing the Conscious Brain*. New York: Pantheon Books.

Damerow, G. 2002. *Barnyard in Your Backyard: A Beginner's Guide to Raising Chickens, Ducks, Geese, Rabbits, Goats, Sheep, and Cattle*. Adams, MA: Storey Publishing.

Dantzer, R. and Kelley, K. 2007. Twenty years of research on cytokine-induced sickness behavior. *Brain, Behavior, and Immunity* **21**(2): 153–60.

Darwall, S. 1998. Empathy, sympathy, and care. *Philosophical Studies* **89**: 261–82.

Darwin, C. 1859/1964. *On the Origin of Species*, facsimile of the first edition. Cambridge, MA: Harvard University Press.

Darwin, C. 1871. *The Descent of Man and Selection in Relation to Sex*. London: John Murray.

Daston, L. and Mitman, G. (eds.). 2005. *Thinking with Animals: New Perspectives on Anthropomorphism*. New York: Columbia University Press.

Davis, G. 1976. Parigi: a social history of the Balinese movement to Central Sulawesi, 1907–1974. PhD thesis, Stanford, CA: Stanford University.

Davis, M. 2010. Who will build the ark?. *New Left Review* **61** (January–February) http://new leftreview.org/II/61/mike-davis-who-will-build-the-ark [Accessed March 25, 2013].

Dautenhahn, K. and Nehaniv, C. (eds.). 2002. *Imitation in Animals and Artifacts*. Cambridge, MA: MIT Press.

Dawkins, M. 2012. *Why Animals Matter*. New York: Oxford University Press.

de la Fuente, M. and Marquis, R. 1999. The role of ant-tended extrafloral nectaries in the protection and benefit of a neotropical rainforest tree. *Oecologia* **118**(2): 192–202.

De Merode, E. and Cowlishaw, G. 2006. Species protection, the changing informal economy, and the politics of access to the bushmeat trade in the Democratic Republic of Congo. *Conservation Biology* **20**(4): 1262–71.

de Villiers, J. 2007. The interface of language and theory of mind. *Lingua* **117**(11): 1858–78.

de Waal, F. 1982. *Chimpanzee Politics: Sex and Power Among Apes*. London, UK: Jonathan Cape.

de Waal, F. 1999. Anthropomorphism and anthopodenial: consistency in our thinking about humans and other animals. *Philosophical Topics* **27**(1): 255–80.

de Waal, F. 2009. *The Age of Empathy: Nature's Lessons for a Kinder Society*. New York: Random House, Inc.

Deckha, M. 2012. Toward a postcolonial, posthumanist feminist theory: centralizing race and culture in feminist work on nonhuman animals. *Hypatia* **27**(3): 527–45.

Declaration of Helsinki. 2008. World Medical Association Declaration of Helsinki: Ethical principles for medical research involving human subjects. Seoul, Korea: WMA General Assembly.

DeGrazia, D. 1996. *Taking Animals Seriously*. Cambridge: Cambridge University Press.

DeMaster, D. and Drevenak, J. 1988. Survivorship patterns in three species of captive cetaceans. *Marine Mammal Science* **4**(4): 297–311.

Dennett, D. 1978. Beliefs about beliefs. *Behavioral and Brain Sciences* **4**: 568–70.

Dennett, D. 1991. *Consciousness Explained*. New York: Penguin Books.

Dennis, C. 2005. Branching out. *Nature* **437**, 17–19.

Descola, Ph. and Palsson, G. (eds.). 1996. *Nature and Society: Anthropological Perspectives*, London: Routledge.

Devine, P. and Elliot, A. 1995. Are racial stereotypes really fading? The Princeton Trilogy revisited. *Personality and Social Psychology Bulletin* **21**: 1139–50.

Devisch, R. and Vervaecke, B. 2011. Auto-production, production et reproduction: divination et politique chez les Yaka du Zaïre. *Social Compass* **32**, 111–31.

DeVries, P. J. 1990. Enhancement of symbioses between butterfly caterpillars and ants by vibrational communication. *Science* **248**(4959): 1104–6.

DeVries, P. J. and Baker, I. 1989. Butterfly exploitation of an ant–plant mutualism: adding insult to herbivory. *Journal of the New York Entomological Society* **97**(3), 332–40.

Dinerstein, E., Varma, K., Wikramanayake, E. and Lumpkin, S. 2010.Wildlife Premium Market + REDD: creating a financial incentive for conservation and recovery of endangered species and habitats. www.hcvnetwork.org/resources/folder.2006-09-29.6584228415/Wildlife_Premium-REDD%20Oct%2013%202010%20-2-%20-2.pdf/view [Accessed March 25, 2013].

Donald, P. 2004. Biodiversity impacts of some agricultural commodity production systems. *Conservation Biology* **18**(1): 17–37.

Doris, J. (ed.). 2010. *Moral Psychology Handbook*. Oxford: Oxford University Press.

Douglas, M. 1970a. *Purity and Danger: An Analysis of Concepts of Pollution and Taboo*. Harmondsworth: Penguin.

Douglas, M. 1970b, *Natural Symbols: Explorations in Cosmology*. London: Barrie Cresset.

Duffield, D. and Wells, R. 1991. The combined application of chromosome, protein, and molecular data for investigation of social unit structure and dynamics in *Tursiops truncatus*. In Hoelzel, A. (ed.), *Genetic Ecology of Whales and Dolphins. Reports of the International Whaling Commission, Special Issue* **13** (Cambridge, UK) pp. 155–69.

Dunayer, J. 2001. *Animal Equality: Language and Liberation*. Derwood, MD: Ryce.

Dunbar, R., Gamble, C. and Gowlett, J. 2010. Social brain, distributed mind. *Proceedings of the British Academy* **158**: 269–81.

Dunn, L. C., Dubinin, N. P., Lévi-Strauss, C. *et al.* 1975. *Race, Science and Society*. Paris: The Unesco Press.

Economist, The. 2011. A man made world. May 26.

Edgar, J., Lowe, J., Paul, E. and Nicol, C. 2011. Avian maternal response to chick distress. *Proceedings of the Royal Society B*, published online March 9, 2011.

Ehrlich, P. 1997. *A World of Wounds: Ecologists and the Human Dilemma*. Oldendorf/Luhe: Ecology Institute.

Eiseley, L. 1959. The bird and the machine. In Eiseley, L. (ed.), *The Immense Journey*. New York: Vintage, pp. 179–93.

Eiseley, L. (ed.) 1959. *The Immense Journey*. New York: Vintage.

Eisemann, C. H., Jorgensen, W. K., Merritt, D. J. *et al.* 1984. Do insects feel pain? A biological view. *Experientia (now Cellular and Molecular Life Sciences)* **40**: 164–7.

Eisenberg, M. 1988. *The Nature of the Common Law*. Harvard: Harvard University Press.

Eisenberg, M. 1990. Statement of teaching philosophy. http://teaching.berkeley.edu/dta-recipient/melvin-eisenberg [Accessed March 25, 2013].

Elias, N. 1978. *The Civilizing Process: The History of Manners*. Oxford: Blackwell.

Ellis, C. 2000. An integrated model for conservation: case study on the role of women in the commercial bushmeat trade in Cameroon. Masters thesis, North York, ON: York University.

Emery, N. and Clayton, N. 2004. The mentality of crows: convergent evolution of intelligence in corvids and apes. *Science* **306**: 1903–7.

Engh, A., Beehner, J., Bergman, T. *et al.* 2006. Behavioural and hormonal responses to predation in female chacma baboons (*Papio hamadryas ursinus*). *Proceedings of the Royal Society B* **273**: 707–12.

Ereshefsky, M. 2001. *The Poverty of the Linnaean Hierarchy: A Philosophical Study of Biological Taxonomy.* Cambridge: Cambridge University Press.

Etzkowitz, H. and Leydesdorf, L. (eds.) 1997. *Universities and the Global Knowledge Economy: A Triple Helix of University–Industry–Government Relations.* London: Cassell Academic.

Evans, C. S. and Evans, L. 2007. Representational signalling in birds. *Biology Letters* **3**: 8–11.

Falguères, C., Bahain, J. J., Yokoyama, Y. *et al.* 1999. Earliest humans in Europe: the age of TD6 Gran Dolina, Atapuerca, Spain. *Journal of Human Evolution* **37**: 343–52.

Fashing, P., Nguyen, N., Barry, T. *et al.* 2011. Death among geladas (*Theropithecus gelada*): a broader perspective on mummified infants and primate thanatology. *American Journal of Primatology* **73**: 405–9.

Faucher, L. and Machery, E. 2009. Racism: against Jorge Garcia's moral and psychological monism. *Philosophy of Social Sciences* **39**: 41–62.

Federal Register. 2004. Volume 69, Number 108 (Friday, June 4, 2004). www.gpo.gov/fdsys/pkg/FR-2004-06-04/html/04-12693.htm [Accessed April 1, 2013].

Ferdowsian, H. and Merskin, D. 2012. Parallels in sources of trauma, pain, distress, and suffering in humans and nonhuman animals. *Journal of Trauma and Dissociation* **13**(4): 448–68.

Ferdowsian, H., Durham, D., Kimwele, C. *et al.* 2011. Signs of mood and anxiety disorders in chimpanzees. *PLoS ONE* **6**(6): e19855.

Fichte, J. 2000. *Foundations of Natural Right.* Cambridge: Cambridge University Press.

Fiorito, G. 1986. Is there 'pain' in invertebrates? *Behavioural Processes* **12**: 383–8.

Fitzgerald, A. 2012. Doing time in a slaughterhouse: a critical review of the use of animals and inmates in prison labour programmes. *Journal for Critical Animal Studies*, **10**(2): 12–46.

Flack, J., Girvan, M., de Waal, F. and Krakauer, D. 2006. Policing stabilizes construction of social niches in primates. *Nature* **439**: 426–9.

Flombaum, J. and Santos, L. 2005. Rhesus monkeys attribute perceptions to others. *Current Biology* **15**(5): 447–52.

Fodor, J. 1975. *The Language of Thought.* Cambridge, MA: Harvard University Press.

Fodor, J. 2008. *LOT 2: The Language of Thought Revisited.* Oxford: Oxford University Press.

Fonagy, P. and Luyten, P. 2010. Mentalization: understanding borderline personality disorder. In Fuchs, T. *et al.*, *The Embodied Self: Dimensions, Coherence and Disorders.* Stuggart: Schattauer, pp. 260–77.

Food and Agriculture Organization of the United Nations (2010). *Global Forest Resources Assessment 2010.* Rome: FAO.

Ford, J. 2009. Killer whale. In Perrin, W. *et al.* (eds.), *Encyclopedia of Marine Mammals*, 2nd edn. New York: Academic Press, pp. 650–7.

Ford, J., Ellis, G. and Balcomb, K. 1994. *Killer Whales.* Vancouver: UBC Press, pp. 1–102.

Forth, G. 1989. Animals, witches and wind. Eastern Indonesian variations on the 'Thunder Complex'. *Antropos* **84**: 89–106.

Foucault, M. 1988. Technologies of the self. In Gutman, M. and Hutton, P. (eds.), *Technologies of the Self: A Seminar with Michel Foucault.* Amherst, MA: University of Massachusetts Press, pp. 16–49.

Franklin, S. 2007. *Dolly Mixtures: The Remaking of Genealogy.* Durham, NC: Duke University Press.

Franks, N. and Richardson, T. 2006. Teaching in tandem-running ants. *Nature* **439**: 153.

Freeman, C., Bekoff, M. and Bexell, S. 2011. Giving voice to the 'voiceless': incorporating nonhuman animal perspectives as journalistic sources. *Journalism Studies* **12**(5): 1–18.

Fried, C. and Fried, G. 2010. *Because it is Wrong: Torture, Privacy and Presidential Power in the Age of Terror*. New York: W.W. Norton & Co.

Friedman, S. 1998. *Mappings: Feminism and the Cultural Geographies of Encounter*. Princeton: Princeton University Press.

Fritz, J., Nash, L., Alford, P. and Bowen, J. 1992. Abnormal behaviors, with a special focus on rocking, and reproductive competence in a large sample of captive chimpanzees (*Pan troglodytes*). *American Journal of Primatolology* **27**, 161–76.

Frohoff, T. 2004. Stress in dolphins. In Bekoff, M. (ed.), *Encyclopedia of Animal Behavior*. Westport: Greenwood Publishing Group, Inc., pp. 1158–64.

Fry, D. (ed.). 2012. *War, Peace, and Human Nature*. New York: Oxford University Press.

Frykman, J. and O. Löfgren. 1987. *Culture Builders: A Historical Anthropology of Middle-Class Life*. New Brunswick and London: Rutgers University Press.

Fuchs, T., Sattel, H. and Henningsen, P. 2010. *The Embodied Self: Dimensions, Coherence and Disorders*. Stuggart: Schattauer.

Fuentes, A. 2006. The humanity of animals and the animality of humans: a view from biological anthropology inspired by J.M. Coetzees' Elizabeth Costello. *American Anthropologist* **108**(1): 124–32.

Fuentes, A. 2009a. *Evolution of Human Behavior*. New York: Oxford University Press.

Fuentes, A. 2009b. Re-situating anthropological approaches to the evolution of human behavior. *Anthropology Today* **25**(3):12–17.

Fuentes, A. 2010. Naturecultural encounters in Bali: monkeys, temples, tourists, and ethnoprimatology. *Cultural Anthropology* **25**: 600–24.

Fuentes, A. 2012a. Ethnoprimatology and the anthropology of the human–primate interface. *Annual Review of Anthropology*, **41**: 101–17.

Fuentes, A. 2012b. Pets, property, and partners: macaques as commodities in the human–other primate interface. In Radhakrishna, S. *et al.* (eds.), *The Macaque Connection: Cooperation and Conflict between Humans and Macaques*. Developments in Primatology: Progress and Prospects 43. New York: Springer Science and Business Media, pp. 107–26.

Fuentes, A. 2013. Cooperation, conflict, and niche construction in the genus Homo. In Fry, D. (ed.) *War, Peace, and Human Nature*. Oxford: Oxford University Press, pp. 78–94.

Fuentes, A. and Hockings, K. 2010. The ethnoprimatological approach in primatology. *American Journal of Primatology* **72**: 841–7.

Fuentes, A. and Wolfe, L. (eds.). 2002. *Primates Face to Face: The Conservation Implications of Human–Nonhuman Primate Interconnections*. Cambridge: Cambridge University Press.

Fuentes, A., Wyczalkowski, M. and MacKinnon, K. 2010. Niche construction through cooperation: a nonlinear dynamics contribution to modeling facets of the evolutionary history in the genus *Homo*. *Current Anthropology* **51**(3): 435–44.

Fujimura, J. 1998. Authorizing knowledge in science and anthropology. *American Anthropologist* **100**(2): 347–60.

Gaard, G. ed. 1993. *Ecofeminism: Women, Animals and Nature*. Philadelphia: Temple University Press.

Gallese V. 2005. 'Being like me': self–other identity, mirror neurons and empathy. In Hurley, S. and Chater, N. (eds.), *Perspectives on Imitation: From Neuroscience to Social Science*, Cambridge, MA: MIT Press, pp.101–18.

Gallese, V. 2006. Intentional attunement: a neurophysiological perspective on social cognition and its disruption in autism. *Cognitive Brain Research* **1079**: 15–24.

Gallese, V. 2007. Before and below 'theory of mind': embodied simulation and the neural correlates of social cognition. *Philosophical Transactions of the Royal Society Biological Sciences* **362**: 659–69.

Gallese, V. 2010. Embodied simulation and its role in intersubjectivity. In Fuchs, T., Sattel, H. and Henningsen, P (eds.), *The Embodied Self: Dimensions, Coherence and Disorders*. Stuttgart: Schattauer, pp. 77–91.

Gallese, V. and Goldman, A. 1998. Mirror neurons and the simulation theory of mind-reading. *Trends in Cognitive Sciences* **2**(12): 493–501.

Gallese, V., Fadiga L., Fogassi, L. and Rizzolatti, G. 1996. Action recognition in the premotor cortex. *Brain* **119**: 593–609.

Garcia, J. 1996. The heart of racism. *Journal of Social Philosophy* **27**(1): 5–46.

Garner, J. 2005. Stereotypies and other abnormal repetitive behaviors: potential impact on validity, reliability, and replicability of scientific outcomes. *ILAR Journal* **46**: 106–17.

Garner, J., Meehan, C. and Mench, J. 2003. Stereotypies in caged parrots, schizophrenia and autism: evidence for a common mechanism. *Behavioral & Brain Research* **145**: 125–34.

Garner, J., Weisker, S., Dufour, B. and Mench, J. 2004. Barbering (fur and whisker trimming) by laboratory mice as a model of human trichotillomania and obsessive–compulsive spectrum disorders. *Comparative Medicine* **54**: 216–24.

Garrick, L. D. and Lang, J. W. 1977. Social signals and behaviors of adult alligators and crocodiles. *American Zoologist* **17**: 225–39.

Gaskin, D. 1982. *The Ecology of Whales and Dolphins*. London: Heinemann.

Gaulin, S. and McBurney, D. 2004. *Evolutionary Psychology*. Upper Saddle River, NJ: Prentice-Hall.

Gazda, S., Connor, R., Edgar, R. and Cox, F. 2005. A division of labor with role specialization in group-hunting bottlenosed dolphins *(Tursiops truncatus)* off Cedar Key, Florida. *Proceedings of the Royal Society of London Series, B* **272**: 135–40.

Geertz, C. 1973. Deep play: notes on the Balinese cockfight. In Geertz, C. (ed.), *The Interpretation of Cultures*. New York: Basic Books, pp. 412–53.

Geertz, C. (ed.). 1973. *The Interpretation of Cultures*. New York: Basic Books.

Gelman, S. 2003. *The Essential Child: Origins of Essentialism in Everyday Thought*. New York: Oxford University Press.

Gergely, G., Nadasdy, Z., Csibra, G. and Bíró, S. 1995. Taking the intentional stance at 12 months of age. *Cognition* **56**(2): 165–93.

Gerth, H. and Wright Mills, C. (eds.). 1946. *From Max Weber: Essays in Sociology*. New York: Oxford University Press.

Ghiselin, M. 1974. A radical solution to the species problem. *Systematic Zoology* **23**: 536–44.

Gilbert, P. 1992. *Depression: The Evolution of Powerlessness*. Hove, Sussex: Lawrence Erlbaum.

Gil-White, F. 2001. Are ethnic groups biological 'species' to the human brain? Essentialism in our cognition of some social categories. *Current Anthropology* **42**: 515–54.

Gingerich, P. and Uhen, M. 1998. Likelihood estimation of the time of origin of cetaceans and the time of divergence of cetacean and Artiodactyla. *Paleo-electronica* **2**: 1–47.

Giurfa, M., Zhang, S., Jenett, A., Menzel, R. and Srinivasan, M. V. 2001. The concepts of 'sameness' and 'difference' in an insect. *Nature* **410**: 930–3.

Gogtay, N., Giedd, J. N., Lusk, L. *et al.* 2004. Dynamic mapping of human cortical development during childhood through early adulthood. *Proceedings of the National Academy of Sciences* **101**(21): 8174–9.

Goldberg, D. (ed.). 1990. *Anatomy of Racism*. Minneapolis: University of Minnesota Press.

Goodall, J. 1986. *The Chimpanzees of Gombe: Patterns of Behavior*. Cambridge, MA: Harvard University Press.

Goodall, J. 1990. *Through a Window*. New York: Mariner Books.

Goodall, J. 1999. *Reason for Hope: A Spiritual Journey*. New York: Warner Books.

Goodall, J., Maynard, T. and Hudson, G. 2009. *Hope for Animals and their World: How Endangered Species are being Rescued from the Brink*. New York: Grand Central Publishing.

Goodman, D. and Watts, M. 1997. *Globalising Food: Agrarian Questions and Global Restructuring*. London: Routledge.

Gould, E. and Bress, M. 1986. Regurgitation and reingestion in captive gorillas: description and intervention. *Zoo Biology* **5**: 241–50.

Greene, M., Schill, K., Takahashi, S. *et al.* 2005. Moral issues of human: nonhuman primate neural grafting. *Science* **309**: 385.

Gregory, N. 2004. *Physiology and Behavior of Animal Suffering*. Oxford: Blackwell Science.

Griffin, D. 1976. *The Question of Animal Awareness: Evolutionary Continuity of Mental Experience*. New York: Rockefeller University Press.

Griffin, D. 1984. *Animal Thinking*. Cambridge, MA: Harvard University Press.

Griffin, D. 1992. *Animal Minds*. Chicago: University of Chicago Press.

Griffin, D. and Speck, G. 2003. New evidence of animal consciousness. *Animal Cognition* **7**: 5–18.

Griffiths, P. 2004. Emotions as natural and normative kinds. *Philosophy of Science* **71**: 901–11.

Gruen, L. 2004. Empathy and vegetarian commitments. In Sapontzis, S. (ed.), *Food for Thought: The Debate Over Eating Meat*. New York: Prometheus Press, pp. 284–94.

Gruen, L. 2009. Attending to nature. *Ethics and the Environment* **14**(2): 23–38.

Gruen, L. 2011. *Ethics and Animals: An Introduction*. Cambridge: Cambridge University Press.

Gruen, L. 2012. Navigating difference (again): animal ethics and entangled empathy. In Zucker, G. (ed.), *Strangers to Nature: Animal Lives and Human Ethics*. New York: Lexington Books, pp. 213–33.

Gunter, C. and Dhand, R. 2005. The chimpanzee genome. *Nature* **437**: 47.

Gutman, M. and Hutton, P. (eds.). 1988. *Technologies of the Self: A Seminar with Michel Foucault*. Amherst, MA: University of Massachusetts Press.

Gutmann, A. (ed.). 1994. *Multiculturalism: Examining the Politics of Recognition*. Princeton: Princeton University Press.

Gutmann, A. 1999. Introduction. In Gutmann, A. (ed.), *The Lives of Animals*. Princeton: Princeton University Press, pp. 107–20.

Hackman, D. and Redelmeier, D. 2006. Translation of research evidence from animals to humans. *Journal of the American Medical Association* **296**: 1731–2.

Haidt, J. 2001. The emotional dog and its rational tail: a social intuitionist approach to moral judgement. *Psychological Review* **108**(4): 814–34.

Hakeem, A., Sherwood, C., Bonar, C. *et al.* 2009. Von Economo neurons in the elephant brain. *The Anatomical Record* **292**: 242–8.

Ham, R. 1998. *Nationwide Chimpanzee Survey and Large Mammal Survey, Republic of Guinea*. Unpublished report for the European Communion, Conakry, Guinea.

Hamilton, D. (ed.). 1981. *Cognitive Processes in Stereotyping and Intergroup Behavior*. Hillsdale, NJ: Erlbaum.

Hamman, K., Warneken, F., Greenberg, J. and Tomasello, M. 2011. Collaboration encourages equal sharing in children but not in chimpanzees. *Nature* **476**: 328–31.

Hammond, P., Mizroch, S. and Donovan, G. (eds.). 1990. Individual recognition of cetaceans: use of photo-identification and other techniques to estimate population parameters. *Reports of the International Whaling Commission, Special Issue* **12**.

Haralson, J. V., Groff, C. I. and Haralson, S. J. 1975. Classical conditioning in the sea anemone, *Cribrina xanthogrammica*. *Physiology and Behavior* **15**: 455–60.

Haraway, D. 1988. Remodeling the human way of life. In Stocking, G. W. (ed.), *Bones, Bodies, Behavior: Essays in Biological Anthropology*. Madison, WI: University of Wisconsin Press, pp.206–59.

Haraway, D. 2003. *The Companion Species Manifesto: Dogs, People, and Significant Otherness*. Chicago: Prickly Paradigm Press.

Haraway, D. 2007. Speculative fabulations for technoculture's generations: taking care of unexpected country. In Piccinini, P. (ed.), *(Tiernas) Criatura: Catalogue of the Artium Exhibition of Patricia Piccinini's Art*. Vitoria: Egileak, pp. 100–7.

Haraway, D. 2008. *When Species Meet*. Minneapolis: University of Minnesota Press.

Hardin, G. 1968 The tragedy of the commons. *Science* **162**(3859): 1243–8.

Hardt, M. and Negri, A. 2004. *Multitude: War and Democracy in the Age of Empire*. New York: The Penguin Press.

Hare, B., Call, J., Agnetta, B. and Tomasello, M. 2000. Chimpanzees know what conspecifics do and do not see. *Animal Behaviour* **59**(4): 771–85.

Hare, B., Call, J. and Tomasello, M. 2001. Do chimpanzees know what conspecifics know? *Animal Behaviour* **61**(1): 139–51.

Hare, B., Addessi, E., Call, J., Tomasello, M. and Visalberghi, E. 2003. Do capuchin monkeys, *Cebus paella*, know what conspecifics do and do not see? *Animal Behaviour* **65**: 131–42.

Hare, B., Call, J. and Tomasello, M. 2006. Chimpanzees deceive a human competitor by hiding. *Cognition* **101**(3): 495–514.

Hare, B., Melis, A., Woods, V., Hastings, S. and Wrangham, R. 2007. Tolerance allows bonobos to outperform chimpanzees on a cooperative task. *Current Biology* **17**: 619–23.

Harman, G. 1978. Studying the chimpanzees' theory of mind. *Behavioral and Brain Sciences* **1**: 576–7.

Harper, A. B. (ed.) 2010a. *Sistah Vegan: Black Female Vegans Speak on Food, Identity, Health, and Society*. New York: Lantern Books.

Harper, A. B. 2010b. Race as a 'feeble matter' in veganism: interrogating whiteness, geopolitical privilege, and consumption philosophy of 'cruelty-free' products. *Journal for Critical Animal Studies*, **8**(3): 5–27.

Harris, P., Johnson, C. N., Hutton, D., Andrews, G. and Cooke, T. 1989. Young children's theory of mind and emotion. *Cognition and Emotion* **3**: 379–400.

Harrison, L. M., Kastin, A. J., Weber, J. T. *et al.* 1994. The opiate system in invertebrates. *Peptides* **15**: 1309–29.

Hartman, M. 2012. In Detroit, half the street lights could go dark. *Marketplace* (May 24), American Public Media.

Hauser, M., Chomsky, N. and Fitch, T. 2002. The faculty of language: what is it, who has it, and how did it evolve? *Science* **298**: 1569–79.

Hayward, E. 2010. Fingeryeyes: impressions of cup corals. *Cultural Anthropology* **25**(4): 577–99.

Heidegger, M. 2010. *The Fundamental Concepts of Metaphysics*. Bloomington: Indiana University Press.

Heinsohn, G. 2000. What makes the holocaust a uniquely unique genocide? *Journal of Genocide Research* **2**: 411–30.

Hendry, R. 2006. Elements, compounds and other chemical kinds. *Philosophy of Science* **73**: 864–75.

Henson, P. 2002. Invading arcadia: women scientists in the field in Latin America, 1900–1950. *The Americas* **58**(4): 577–600.

Herman, L. 2002. Vocal, social, and self-imitation by bottlenosed dolphins. In Dautenhahn, K. and Nehaniv, C. (eds.), *Imitation in Animals and Artifacts*. Cambridge, MA: MIT Press, pp. 63–108.

Herman, L., Matus, D., Herman, E., Ivancic, M. and Pack, A. 2001. The bottlenosed dolphin's *(Tursiops truncatus)* understanding of gestures as symbolic representations of its body parts. *Animal Learning & Behaviour* **29**: 250–64.

Herrmann, E., Call, J., Hernandez-Lloreda, M., Hare, B. and Tomasello, M. 2007. Humans have evolved specialized skills of social cognition: the cultural intelligence hypothesis. *Science* **317**: 1360–6.

Herzog, H. 2010. *Some We Love, Some We Hate, Some We Eat*. New York: HarperCollins Publishers.

Higgins, R. 1994. Race, pollution, and the mastery of nature. *Environmental Ethics* **16**: 251–64.

Hill, S. 2009. Do gorillas regurgitate potentially injurious stomach acid during regurgitation and reingestion? *Animal Welfare* **18**: 123–7.

Himmler, H. 1942. *Der Untermensch*. Berlin: DHM, Do 56/685.

Hirschfeld, L. 1998. *Race in the Making: Cognition, Culture, and the Child's Construction of Human Kinds*. New York: Bradford Books.

Hobaiter, C. and Byrne, R. 2011. The gestural repertoire of the wild chimpanzee. *Animal Cognition* **15**(5): 745–67.

Hockings, K., Yamakoshi, G., Kabasawa, A. and Matsuzawa, T. 2010. Attacks on local persons by chimpanzees in Bossou, Republic of Guinea: long-term perspectives. *American Journal of Primatology* **72**: 887–96.

Hodder, I. 1990. *The Domestication of Europe: Structure and Contingency in Neolithic Societies*. Birmingham: Blackwell Publishing.

Hodson, G. 2011. Do ideologically intolerant people benefit from intergroup contact? *Current Directions in Psychological Science* **20**: 154–9.

Hoelzel, A. (ed.). 1991. *Genetic Ecology of Whales and Dolphins. Reports of the International Whaling Commission, Special Issue* **13** (Cambridge, UK).

Hof, P. and Van der Gucht, E. 2007. The structure of the cerebral cortex of the humpback whale, *Megaptera novaeangliae* (Cetacea, Mysticeti, Balaenopteridae). *The Anatomical Record* **290**: 1–31.

Hof, P., Chanis, R. and Marino, L. 2005. Cortical complexity in cetacean brains. *The Anatomical Record* **287**: 1142–52.

Hoff, C. 1980. Immoral and moral uses of animals. *New England Journal of Medicine* **302**(115): 115.

Hoffman, M. 2000. *Empathy and Moral Development*. Cambridge: Cambridge University Press.

Hölldobler, B. and Wilson, E. 1990. *The Ants*. Cambridge: Harvard University Press.

Holloway, L. and Morris, C. 2007. Exploring biopower in the regulation of farm animal bodies: genetic policy interventions in UK livestock. *Genomics, Society and Policy* **3**: 82–98.

Homsy, J. 1999. *Ape Tourism and Human Diseases*. Report of a consultancy for the International Gorilla Conservation Programme.

Hook, M., Lambeth, S., Periman, J. *et al.* 2002. Inter-group variation in abnormal behavior in chimpanzees (*Pan troglodytes*) and rhesus monkeys (*Macaca mulatta*). *Applied Animal Behaviour Science* **76**: 165–76.

Horrobin, D. 2003. Modern biomedical research: an internally self-consistent universe with little contact with medical reality? *Nature Reviews Drug Discovery* **2**(2): 151–4.

Hosey, G. 2005. How does the zoo environment affect the behaviour of captive primates? *Applied Animal Behaviour Science* **90**: 107–29.

Hosey, G. and Skyner, L. 2007. Self-injurious behavior in zoo primates. *International Journal of Primatology* **28**: 1431–7.

Hrdy, S. 2009. *Mothers and Others: the Evolutionary Origins of Mutual Understanding*. Cambridge, MA: The Belknap Press of Harvard University Press.

Hughes, A. and Reimer, S. (eds.) 2004. *Geographies of Commodity Chains*. London: Routledge.

Hull, D. 1978. A matter of individuality. *Philosophy of Science* **45**: 335–60.

Humanimalifesto. 2009. Editorial, *Humanimalia*. www.depauw.edu/humanimalia/humanimalifesto.html [Accessed March 26, 2013].

Hume, D. 1975 (1777). *An Enquiry Concerning the Principles of Morals*. Nidditch, P. H. (ed.). Oxford: Oxford University Press.

Humle, T., Colin, C., Laurans, M. and Raballand, E. 2011. Group release of sanctuary chimpanzees (*Pan troglodytes*) in the Haut Niger National Park, Guinea, West Africa: ranging patterns and lessons so far. *International Journal of Primatology* **32**: 456–73.

Humphries, T. 2003. Effectiveness of dolphin-assisted therapy as a behavioral intervention for young children with disabilities. *Bridges: Practical-based Research Syntheses, Research and Training Centre of Early Childhood Development* **1**(6): 1–9.

Hutto D. 2008. *Folk Psychological Narratives: The Socio-cultural Basis of Understanding Reasons*. Cambridge, MA: MIT Press.

Hutto D. 2009. Mental representation and consciousness. In Banks, W. (ed.). *Encyclopedia of Consciousness*. Vol. 2. Elsevier, pp. 19–32.

Hutto D. and Myin, E. 2013. *Radicalizing Enactivism: Basic Minds without Content*. Cambridge, MA: MIT Press.

Hutto, D., Herschbach, M. and Southgate, V. 2011. Social cognition: mindreading and alternatives. *Review of Philosophy and Psychology*. **2**(3): 375–95.

Hyslop, A. 2010. Other minds. Stanford Encyclopedia of Philosophy (Fall 2010 edition), Edward, Z. (ed.), http://plato.stanford.edu/archives/fall2010/entries/other-minds [Accessed March 26, 2013].

Institute for Laboratory Animal Research. 2009. *Guide for the Care and Use of Laboratory Animals*, 8th edn. Washington, DC: National Academies Press. http://oacu.od.nih.gov/regs/guide/guide.pdf [Accessed April 10, 2013].

Ioannidis, J. 2006. Evolution and translation of research findings: from bench to where? *PLoS Clinical Trials* **1**(7): e36.

IUCN and UNEP-WCMC. 2012. *The World Database on Protected Areas (WDPA)*. Cambridge, UK: UNEP-WCMC.

Jacobs, M., McFarland, W. and Morgane, P. 1979. The anatomy of the brain of the bottlenosed dolphin (*Tursiops truncatus*). Rhinic lobe (rhinencephalon): the archicortex. *Brain Research Bulletin* **4**(Suppl 1): 1–108.

Jerison, H. 1973. *Evolution of the Brain and Intelligence*. New York: Academic Press.

John Paul II, Pope 1996. Truth cannot contradict truth: address to the Pontifical Academy of Sciences.(October 22), *L'Osservatore Romano* (English edition), October 30.

Johnson, A. 1991. *Factory Farming*. Oxford: Blackwell.

Johnson, M. and Munakata, Y. (eds.). 2006. *Processes of Change in Brain and Cognitive Development: Attention and Performance XXI*. Oxford: Oxford University Press.

Jokic, A. and Smith, Q. (eds.) 2003. *Consciousness: New Philosophical Perspectives*. Oxford: Oxford University Press.

Juma, C. 2011 Africa's new engine. *Finance & Development* 6–11. www.imf.org/external/pubs/ft/fandd/2011/12/pdf/juma.pdf [Accessed March 26, 2013].

Kaminski, J., Call, J. and Tomasello, M. 2006. Goats' behaviour in a competitive food paradigm: evidence for perspective taking? *Behaviour* **143**: 1341–56.

Kaminski, J., Call, J. and Tomasello, M. 2008. Chimpanzees know what others know, but not what they believe. *Cognition* **109**: 224–34.

Kant, I. 1974. *Anthropology from a Pragmatic Point of View*. Trans. Gregor, M. Berlin: Springer Verlag.

Kant, I. 1993. *Groundwork of the Metaphysics of Morals*. Trans. Ellington, J. Indianapolis, IN: Hackett.

Kaplan, D. 1989. Demonstratives. In Almog, J. *et al.* (eds.) *Themes from Kaplan*. Oxford: Oxford University Press, pp. 481–563.

Katz, D. 1992. Roger and me revisited. *Detroit News*. Reprinted on http://dogeatdog.michael moore.com/films/detroitnews.html [Accessed March 26, 2013].

Kavaliers, M. 1988. Evolutionary and comparative aspects of nociception. *Brain Research Bulletin* **21**: 923–31.

Keeley, B. 2004. Anthropomorphism, primatomorphism, mammalomorphism: understanding cross-species comparisons. *Biology and Philosophy* **19**(4): 521–40.

Kelley, J. L. and Magurran, A. E. 2003. Learned predator recognition and antipredator responses in fishes. *Fish and Fisheries* **4**: 216–26.

Kelley, P., Kowalewski, M. and Hansen, T. (eds.). 2003. *Predator–Prey Interactions in the Fossil Record*. New York: Springer.

Kelly, D., Faucher, L. and Machery, E. 2010a. Getting rid of racism: assessing three proposals in light of empirical evidence. *Journal of Social Philosophy* **41**: 293–322.

Kelly, D., Machery, E. and Mallon, R. 2010b. Racial cognition and normative theory. In Doris, J. (ed.), *Moral Psychology Handbook*. Oxford: Oxford University Press, pp. 433–72.

Keltner, D. 2009. *Born to be Good: The Science of a Meaningful Life*. New York: W. W. Norton & Company.

Kennair, L. 2001. Review of subordination and defeat: an evolutionary approach to mood disorders and their therapy. *Human Ethology Bulletin* **16**: 16–18.

Kennair, L. 2003. Evolutionary psychology and psychopathology. *Current Opinion in Psychiatry* **16**: 691–9.

Kenner, R. (ed.). 2009. *Food, Inc.* Movie, produced by Magnolia Home Entertainment.

Kenway, J., Bullen, E., Fahey, J. and Robb, S. 2006. *Haunting the Knowledge Economy*. London: Routledge.

Kesarev, V. 1971. The inferior brain of the dolphin. *Soviet Science Review* **2**: 52–8.

Kim, C. 2007. Multiculturalism goes imperial: immigrants, animals, and the suppression of moral dialogue. *Du Bois Review* **4**(1): 1–17.

King, B. 1998. Death notices. *Anthropology News* **49**: 38–40.

King, B. 2011. Retirement home or research lab? Report weighs fate of U.S. chimpanzees. www.npr.org/blogs/13.7/2011/12/15/143735486/retirement-home-or-research-lab-report-weighs-fate-of-u-s-chimpanzees [Accessed March 26, 2013].

King, B. 2013. *How Animals Grieve*. Chicago: University of Chicago Press.

King, J and Figueredo, A. 1997. The five-factor model plus dominance in chimpanzee personality. *Journal of Research in Personality* **31**(2): 257–71.

Kinnzey, W. (ed.). 1997. *New World Primates: Ecology, Evolution, and Behavior.* New York: Aldine Gruyter.

Kirby, A. 2004. Parrot's oratory stuns scientists. *BBC News Online*, January 26.

Kirksey, E. 2012. *Freedom in Entangled Worlds: West Papua and the Architecture of Global Power.* Durham, NC: Duke University Press.

Kirksey, E. and Helmreich, S. 2010. The emergence of multispecies ethnography. *Cultural Anthropology* **25**(4): 545–76.

Kitcher, Ph. 1984. Species. *Philosophy of Science* **51**: 308–33.

Kockelman, P. 2011. A Mayan ontology of poultry: selfhood, affect, animals, and ethnography. *Language in Society* **40**(2011): 427–54.

Koh, L. P., Dunn, R. R., Sodhi, N. S. *et al.* 2004. Species coextinctions and the biodiversity crisis. *Science* **305**: 1632–4.

Kohlberg, L. 1981. *The Philosophy of Moral Development* (Vol. 1). New York: Harper and Row.

Koontz, C. 2003. *The Nazi Conscience.* Cambridge, MA: Harvard University Press.

Kornblith, H. 1993. *Inductive Inference and its Natural Ground.* Cambridge, MA: MIT Press.

Korsgaard, C. 2005. Fellow creatures: Kantian ethics and our duties to animals. In Peterson, G. (ed.), *The Tanner Lectures on Human Values; Volume 25.* Salt Lake City: University of Utah Press, pp. 19–112.

Korsgaard, C. 2006. Morality and the distinctiveness of human action. In Macedo, S. and Ober, J. (eds.), *Primates and Philosophers: How Morality Evolved.* Princeton: Princeton University Press, pp. 98–119.

Korsgaard, C. ms. A Kantian case for animal rights.

Kosek, J. 2010. Ecologies of empire: on the new uses of the honeybee. *Cultural Anthropology* **25**(4): 650–78.

Koslicki, K. 2008. Natural kinds or natural kind terms. *Philosophy Compass* **3**: 789–802.

Krachun, C., Carpenter M., Call, J. and Tomasello, M. 2009. A competitive nonverbal false belief task for children and apes. *Developmental Science* **12**(4): 521–35.

Krützen M., Mann, J., Heithaus, M. *et al.* 2005. Cultural transmission of tool use in bottlenosed dolphins. *Proceedings of the National Academy of Sciences USA* **102**: 8939–43.

Kumar, S. and Blair Hedges, S. 1998. A molecular timescale for vertebrate evolution. *Nature*, **392**: 917–20.

Kunda, Z. 1990. The case for motivated reasoning. *Psychological Bulletin* **108**: 480–98.

Kuroshima, H., Fujita, F., Fuyuki, A. and Masuda, T. 2002. Understanding of the relationship between seeing and knowing by tufted capuchin monkeys (*Cebus apella*). *Animal Cognition* **5**: 41–8.

Kuroshima, H., Fujita, K., Adachi, I., Iwata, K. and Fuyuki, A. 2003. A capuchin monkey (*Cebus paella*) recognizes when people do and do not know the location of food. *Animal Cognition* **6**: 283–91.

Laden, G. and Wrangham, R. 2005. The rise of the hominids as an adaptive shift in fallback foods. *Journal of Human Evolution* **49**: 482–98.

Lambert, J. E. and Garber, P. A. 1998. Evolutionary and ecological implications of primate seed dispersal. *American Journal of Primatology* **45**: 9–28.

Landenberger, D. E. 1966. Learning in the Pacific starfish *Pisaster giganteus. Animal Behaviour* **14**: 414–18.

Langford, D. 2006. Social modulation of pain as evidence for empathy in mice. *Science* **312**(5782): 1967–70.

Latour, B. 1987. *Science in Action*. Cambridge, MA: Harvard University Press.

Latour, B. 2004. *Politics of Nature*. Cambridge, MA: Harvard University Press.

Latour, B. and Weibel, P. (eds.). 2005. *Making Things Public: Atmospheres of Democracy*. Cambridge, MA: MIT Press.

Latta, M. 2011. *Kant's Problem Regarding Others*. Unpublished manuscript.

Leakey, R. and Lewin, R. 1993. *Origins Reconsidered: In Search of What Makes us Human*. New York: Anchor.

Lebrecht, S., Pierce, L., Tarr, M. and Tanaka, J. 2009. Perceptual other-race training reduces implicit racial bias. *Plos ONE* **4**: e4215.

Lederer, S. 1995. *Subjected to Science*. Baltimore: Johns Hopkins University Press.

Lee, P. 2010. Sharing space: can ethnoprimatology contribute to the survival of nonhuman primates in human-dominated globalized landscapes? *American Journal of Primatology* **72**: 925–31.

Leslie, A. M. and Polizzi, P. 1998. Inhibitory processing in the false belief task: two conjectures. *Developmental Science* **1**: 247–53.

Leslie, A. M., German, T. and Polizzi. P. 2005. Belief–desire reasoning as a process of selection. *Cognitive Psychology* **50**: 45–85.

Lewin, R. and Foley, R. 2004. *Principles of Human Evolution*. New York: Blackwell.

Li, W. and Saunders, W. 2005. The chimpanzee and us. *Nature* **437**: 50–1.

Lihoreau, M., Lars C. and Nigel, R. 2010. Travel optimization by foraging bumblebees through readjustments of traplines after discovery of new feeding locations. *American Naturalist* **176**(6): 744–57.

Lindsay-Poland, J. 2003. *Emperors in the Jungle: The Hidden History of the U.S. in Panama*. Durham, NC: Duke University Press.

Litt, R. and Litt, H. 2011. *A Chicken in Every Yard: The Urban Farm Store's Guide to Chicken Keeping*. Berkeley: Ten Speed Press.

Livingstone Smith, D. 2011. *Less than Human: Why we Demean, Enslave and Exterminate Others*. London: MacMillan.

Lock, A. (ed.). 1978. *Action, Gesture, and Symbol: The Emergence of Language*. New York: Academic Press.

Lowe, C. 2010. Viral clouds: becoming H5N1 in Indonesia. *Cultural Anthropology* **25**(4): 625–49.

Lurz, R. 2011. *Mindreading Animals*. Cambridge, MA: MIT Press.

Lusseau, D. 2006. Why do dolphins jump? Interpreting the behavioural repertoire of bottlenosed dolphins (*Tursiops* sp.) in Doubtful Sound, New Zealand. *Behavioural Processes* **73**: 257–65.

Lusseau, D. 2007. Evidence for social role in a dolphin social network. *Evolutionary Ecology* **2**: 357–66.

Lutz, C., Well, A. and Novak, M. 2003. Stereotypic and self-injurious behavior in rhesus macaques: a survey and retrospective analysis of environment and early experience. *American Journal of Primatology* **60**: 1–15.

Lycan W. 1996. *Consciousness and Experience*. Cambridge, MA: MIT Press.

Macedo, S. and Ober, J. (eds.). 2006. *Primates and Philosophers: How Morality Evolved*. Princeton: Princeton University Press, pp. 98–119.

Macfie, E. and Williamson, E. 2010. *Best Practice Guidelines for Great Ape Tourism*. Gland, Switzerland: IUCN/SSC Primate Specialist Group (PSG).

Machery, E. and Faucher, L. 2005. Social construction and the concept of race. *Philosophy of Science* **72**, 1208–19.

Machin, K. L. 2001. Fish, amphibian, and reptile analgesia. *Veterinary Clinics of North America: Exotic Animal Practice* **4**: 19–33.

MacKinnon, K. and Fuentes, A. 2011. Primates, niche construction, and social complexity: the roles of social cooperation and altruism. In Sussman, R. and Cloninger, R. (eds.) *Origins of Altruism and Cooperation. Developments in Primatology: Progress and Prospects Volume 36*, Part 2, pp. 121–43.

Malone, N., Fuentes, A. and White, F. 2012. Variation in the social systems of extant Hominoids: comparative insight into the social behavior of early Hominins. *International Journal of Primatology* doi:10.10007/s10764–012–9617–0.

Mann, J., Connor, R., Tyack, P. and Whitehead, H. (eds.). 2000. *Cetacean Societies: Field Studies of Whales and Dolphins*. Chicago: University of Chicago Press.

Marcus, E. 2001. *Vegan: The New Ethics of Eating*, revised edn. Ithaca, NY: McBooks.

Marino, L. 1998. A comparison of encephalization between odontocete cetaceans and anthropoid primates. *Brain, Behaviour and Evolution* **51**: 230–8.

Marino, L. 2006. Absolute brain size: have we thrown the baby out with the bathwater? Invited commentary in *Proceedings of the National Academy of Sciences USA*, **103**(37): 13,563–4.

Marino, L. 2009. Brain size evolution. In Perrin, W. *et al.* (eds.), *Encyclopedia of Marine Mammals*, 2nd edn. New York: Academic Press, pp.149–52.

Marino, L. 2011. Dolphin assisted therapy: from ancient myth to modern snake oil. *Phi Kappa Phi Forum Magazine*: 4–6.

Marino, L. and Frohoff, T. 2011. Towards a new paradigm of non-captive research on cetacean cognition. *PLoS ONE*, **6**(9): e24121. doi:10.1371/journal.pone.0024121.

Marino, L. and Lilienfeld, S. 1998. Dolphin-assisted therapy: flawed data, flawed conclusions. *Anthrozoos*, **11**(4): 194–9.

Marino, L. and Lilienfeld, S. 2007. Dolphin assisted therapy: more flawed data, more flawed conclusions. *Anthrozoos*, **20**: 239–49.

Marino, L., Butti, C., Connor, R. *et al.* 2008. A claim in search of evidence: reply to Manger's thermogenesis hypothesis of cetacean brain structure. *Biological Reviews of the Cambridge Philosophical Society* **83**: 417–40.

Marino, L., Lilienfeld, S., Malamud, R., Nobis, N. and Broglio, R. 2010. Do zoos and aquariums promote attitude change in visitors? A critical evaluation of the American Zoo and Aquarium Study. *Society and Animals* **18**: 126–38.

Martin, J. 2002. Early life experiences: activity levels and abnormal behaviors in resocialised chimpanzees. *Animal Welfare* **11**: 419–36.

Mason, G. and Rushen, J. (eds.) 2006. *Stereotypic Animal Behaviour: Fundamentals and Applications to Welfare*, 2nd edn. Wallingford: CABI.

Mather, J. 2008. Cephalopod consciousness: behavioral evidence. *Conscious Cognition* **17**(1): 37–48.

Matsuzawa, T., Humle, T. and Sugiyama, Y. (eds.) 2011. *The Chimpanzees of Bossou and Nimba*. Tokyo: Springer.

Matthews, R. 2008. Medical progress depends on animal models – doesn't it? *Journal of the Royal Society of Medicine* **101**(2): 95–8.

Mauss, M. 1990. *The Gift: The Form and Reason for Exchange in Archaic Societies*, transl. by W. D. Halls. London: Routledge.

Mayr, E. 1988. *Toward a New Philosophy of Biology*. Cambridge, MA: Harvard University Press.

McComb, S. 2009. Hair-plucking and hair-loss in captive chimpanzees *(Pan troglodytes)*. Unpublished BA dissertation, University of Cambridge.

McCracken, G. 1988. *Culture and Consumption: New Approaches to the Symbolic Character of Consumer Goods and Activities*. Bloomington: Indiana University Press.

McGrew, W. C. 1981. Social and cognitive capabilities of non-human primates: lessons from the wild to captivity. *International Journal for the Study of Animal Problems* **2**: 138–49.

McGrew, W. C. 1992. Tool-use by free-ranging chimpanzees: the extent of diversity. *Journal of Zoology* **228**: 689–94.

McGrew, W. C. 2004. *The Cultured Chimpanzee: Reflections on Cultural Primatology*. Cambridge: Cambridge University Press.

McGrew, W. C., Tutin, C. E. G. and Baldwin, P. J. 1979. Chimpanzees, tools and termites: cross-cultural comparisons of Senegal, Tanzania and Rio Muni. *Man* **14**(2): 185–214.

McMillan, F. D. (ed.) 2005a. *Mental Health and Well-being in Animals*. Oxford: Blackwell Publishing Professional.

McMillan, F. D. 2005b. The concept of quality of life in animals. In McMillan, F. D. (ed.), *Mental Health and Well-being in Animals*. Oxford: Blackwell Publishing Professional, pp. 183–200.

Medin, D. 1989. Concepts and conceptual structure. *American Psychologist* **44**: 1469–81.

Meijaard, E. 2012. Not by science alone: why orangutan conservationists must think outside the box. *Annals New York Academy of Sciences* **1249**: 29–44. doi: 10.1111/j.1749–6632.2011.06288.x, pp 29–44.

Melis, A., Call, J. and Tomasello, M. 2006. Chimpanzees conceal visual and auditory information from others. *Journal of Comparative Psychology* **120**: 154–62.

Melis, A., Schneider, A. and Tomasello, M. 2011. Chimpanzees, *Pan troglodytes*, share food in the same way after collaborative and individual food acquisition. *Animal Behaviour* **82**: 485–93.

Mercado III, E., Murray, S., Uyeyama, R., Pack, A. and Herman, L. 1998. Memory for recent actions in the bottlenosed dolphin *(Tursiops truncatus)*: repetition of arbitrary behaviours using an abstract rule. *Animal Learning and Behaviour* **26**: 210–18.

Mercado III, E., Uyeyama, R., Pack, A. and Herman, L. 1999. Memory for action events in the bottlenosed dolphin. *Animal Cognition* **2**: 17–25.

Meyers, D. (forthcoming). *Victims' Stories and the Advancement of Human Rights*. New York: Oxford University Press.

Mikkelsen, T. 2004. What makes us human? *Genome Biology* **5**: 238.

Millikan, M. 1936. Pareto's sociology. *Econometrica* **4**(4): 324–37.

Millikan, R. 2005. *Language: A Biological Model*. Oxford: Oxford University Press.

Mitchell, R., Thompson, N. and Miles, H. (eds.). 1997. *Anthropomorphism, Anecdotes, and Animals*. Albany, NY: State University of New York Press.

Mizelle, B. 2012. The visibility and invisibility of pigs, part two: the disappearing slaughterhouse. *Humane Research Council* www.humanespot.org/content/disappearing-slaughterhouse [Accessed March 26, 2013].

Montgomery, S. 2011. Deep intellect: inside the mind of an octopus. *Orion* (November/December). www.orionmagazine.org/index.php/articles/article/6474 [Accessed March 26, 2013].

Moody-Adams, M. 1994. Culture, responsibility, and affected ignorance. *Ethics* **104**(2): 291–309.

Mooney, C. 2011. The science of why we don't believe science. *Mother Jones* http://motherjones.com/politics/2011/03/denial-science-chris-mooney [Accessed March 26, 2013].

Moore, L. J. and Kosut, M. 2013. *Buzz: Urban Beekeeping and the Power of the Bee*. New York: New York University Press.

Moore, M. (Dir.). 1989. *Roger & Me*. Movie, produced by Warner Bros.

Morris, A. 2000. Development of logical reasoning: children's ability to verbally explain the nature of the distinction between logical and nonlogical forms of argument. *Developmental Psychology* **36**: 741–58.

Morris, B. 2004. *Insects and Human Life*. New York: Berg.

Moshman, D. 2004. From inference to reasoning: the construction of reality. *Thinking and Reasoning* **10**: 221–39.

Moshman, D. and Franks, B. 1986. Development of the concept of inferential validity. *Child Development* **57**: 153–65.

Mountain, M. 2010. What SeaWorld doesn't want you to know. www.zoenature.org/2010/09/what-seaworld-doesnt-want-you-to-know [Accessed March 26, 2013].

Mullin, M. 1999. Mirrors and windows: sociocultural studies of human–animal relationships. *Annual Reviews in Anthropology* **28**: 201–24.

Mullin, M. 2002. Animals and anthropology. *Society and Animals* **10**(4): 387–93.

Mullin, M. 2007. Feeding the animals. In Cassidy, R. and Mullin, M. (eds.), *Where the Wild Things are Now: Domestication Reconsidered*. Oxford: Berg, pp. 277–303.

Murtoff, J. 2012. Chicken retirement. *Home to Roost: Helping City Folks Raise Chickens*. http://urbanchickenconsultant.wordpress.com/2012/05/01/chicken-retirement [Accessed March 26, 2013].

Nagel, T. 1974. What is it like to be a bat? *Philosophical Review* **83**(4): 435–50.

Nagel, T. 1999. *Other Minds: Critical Essays 1969–1994*. New York: Oxford University Press.

Nash, L., Fritz, J., Alford, P. and Brent, L. 1999. Variables influencing the origins of diverse abnormal behaviors in a large sample of captive chimpanzees (*Pan troglodytes*). *American Journal of Primatology* **48**: 15–29.

National Commission for the Protection of Human Subjects of Biomedical and Behavioral Research. 1979. *The Belmont Report: Ethical Principles and Guidelines for the Protection of Human Subjects of Research*. Washington, DC: US Department of Health, Education, and Welfare.

National Institutes of Health Office of Human Subjects Research. 2006. *Sheet 7: Research Involving Cognitively Impaired Subjects: A Review of Some Ethical Considerations*. http://ohsr.od.nih.gov/info/sheet7.html.

National Research Council (US) Committee on Recognition and Alleviation of Pain in Laboratory Animals. 2009. *Recognition and Alleviation of Pain and Distress in Laboratory Animals*. Washington, DC: National Academy Press.

Nesse, R. 2000. Is depression an adaptation? *Archives of General Psychiatry* **57**: 14–20.

Nesse, R. and Williams, G. 1994. *Why we Get Sick: the New Science of Darwinian Medicine*. New York: Vintage.

Newton, L. 2007. What makes us human? *Bioscience Reports* **27**: 185–7.

Nibert, D. 2002. *Animal Rights/Human Rights: Entanglements of Oppression and Liberation*. Lanham, MD: Rowman and Littlefield.

Nichols, S. and Stich, S. 2003. *Mindreading: An Integrated Account of Pretence, Self-Awareness and Understanding of Other Minds*. Oxford: Oxford University Press.

Nielsen, M., Collier-Baker, E., Davis, J. and Suddendorf, T. 2005. Imitation recognition in a captive chimpanzee (*Pan troglodytes*). *Animal Cognition* **8**: 31–6.

Nishida, T., Zamma, K., Matsusaka, T., Inaba, A. and McGrew, W. 2010. *Chimpanzee Behavior in the Wild: An Audiovisual Encyclopedia*. Tokyo: Springer.

Norgaard, K. 2011. *Living in Denial: Climate Change, Emotions, and Everyday Life*. Cambridge, MA: MIT Press.

Nosek, B., Hawkins, C. and Frazier, R. 2011. Implicit social cognition: from measures to mechanisms. *Trends in Cognitive Sciences* **15**: 152–9.

Noske, B. 1989. *Humans and Other Animals: Beyond the Boundaries of Anthropology*. London: Pluto Press.

Noss, R. F. 1990. Indicators for monitoring biodiversity: a hierarchical approach. *Conservation Biology* **4**: 355–64.

Noss, R., Dobson, A.P., Bladwin, R. *et al.* 2012. Bolder thinking for conservation. *Conservation Biology* **26**: 1–4.

Nowak, M. and Highfield, R. 2011. *Supercooperators: Altruism, Evolution, and Why We Need Each Other to Succeed*. New York: Free Press.

Odling-Smee, J., Laland, K. and Feldman, M. 2003. *Niche Construction: The Neglected Process in Evolution*. Princeton: Princeton University Press.

Oelschlager, H. and Oelschlager, J. 2009. Brains. In Perrin, W. *et al.* (eds.), *Encyclopedia of Marine Mammals*, 2nd edn. New York: Academic Press, pp. 133–58.

Olesiuk, P. F., Bigg, M. and Ellis, G. 1990. Life history and population dynamics of resident killer whales (*Orcinus orca*) in the coastal waters of British Columbia and Washington State. *Reports of the International Whaling Commission Special Issue*, **12**: 209–44.

Olmert, M. 2009. *Made for Each Other: The Biology of the Human–Animal Bond*. Philadelphia: Da Capo Press.

Onishi, K. and Baillargeon, R. 2005. Do 15-month-old infants understand false beliefs? *Science* **308**: 255–8.

Ortega, J. and Bekoff, M. 1987. Avian play: comparative evolutionary and developmental trends. *The Auk* **104**: 338–41.

Owen, D. and Strong, T. B., 2004. *Max Weber, the Vocation Lectures*. Indianapolis: Hacket Company.

Panksepp, J. 2004. *Affective Neuroscience: The Foundations of Human and Animal Emotions*. New York: Oxford University Press.

Passic, F. 2001. US census in Albion. *Morning Star*, p 6. www.albionmich.com/history/histor_notebook/010408.shtml [Accessed March 26, 2013].

Patrick-Goudreau, C. 2013. www.compassionatecooks.com/prayer.htm [Accessed March 26, 2013].

Patterson, F. and Gordon, W. 1993. The case for the personhood of gorillas. In Cavalieri, P. and Singer, P. (eds.), *The Great Ape Project: Equality Beyond Humanity*. London: Fourth Estate, pp. 58–77.

Pearce, F. 2012. Turning point: what future for forest peoples and resources in the emerging world order? www.redd-monitor.org/2012/02/02/turning-point-what-future-for-forest-peoples-and-resources-in-the-emerging-world-order [Accessed March 26, 2013].

Pearsall, J. and Hank, P. (eds.) 1998. *The New Oxford Dictionary of English*. Oxford: Clarendon Press.

Pedersen, H. and Stănescu, V. 2012. What is 'critical' about animal studies? From the animal 'question' to the animal 'condition'.Series editors' introduction to Socha, K. *Women, Destruction and the Avant-Garde: A Paradigm for Animal Liberation*. New York: Rodopi.

Pepperberg, I. M. 1988. Comprehension of 'absence' by an African grey parrot: learning with respect to questions of same/different. *Journal of the Experimental Analysis of Behavior* **50**: 553–64.

Pepperberg, I. M. 1994. Numerical competence in an African gray parrot (*Psittacus erithacus*). *Journal of Comparative Psychology* **108**: 36–44.

Pepperburg, I. M. 2002. *The Alex Studies: Cognitive and Communicative Abilities of Grey Parrots*. Cambridge, MA: Harvard University Press.

Perel, P., Roberts, I., Sena, E. *et al.* 2006. Comparison of treatment effects between animal experiments and clinical trials: systematic review. *BMJ* **334**: 197.

Perfecto, I. and Vandermeer, J. 1993. Cleptobiosis in the ant *Ectatomma ruidum* in Nicaragua. *Insectes Sociaux* **40**: 295–9.

Perner, J. 2010. Who took the cog out of cognitive science? Mentalism in an era of anti-cognitivism. *Perception, Attention, and Action: International Perspectives on Psychological Science (Volume 1)*. New York: Psychology Press.

Perrin, W., Wursig, B. and Thewissen, J. (eds.). 2009. *Encyclopedia of Marine Mammals*, 2nd edn. New York: Academic Press.

Peterson, G. (ed.). 2005. *The Tanner Lectures on Human Values, Volume 25*. Salt Lake City: University of Utah Press.

Pettigrew, T. 1998. Intergroup contact theory. *Annual Review of Psychology* **49**: 65–85.

Pettigrew, T. and Tropp, L. 2006. A meta-analytic test of intergroup contact theory. *Journal of Personality and Social Psychology* **90**: 751–83.

Piccinini, P. (ed.). 2007. *(Tiernas) Criatura: Catalogue of the Artium Exhibition of Patricia Piccinini's Art*. Vitoria: Egileak.

Pierce, J. and Bekoff, M. 2012. Wild justice redux: what do we know about the moral lives of animals and why it matters. *Social Justice Research* **25**: 122–139.

Pillow, B. 1999. Children's understanding of inferential knowledge. *Journal of Genetic Psychology* **160**: 419–28.

Plooij, F. 1978. Some basic traits of language in wild chimpanzees. In Lock, A. (ed.), *Action, Gesture, and Symbol: The Emergence of Language*. New York: Academic Press, pp. 111–31.

Plumptre, A. J. 1995. The effects of trampling damage by herbivores on the vegetation of the Parc National des Volcans, Rwanda. *African Journal of Ecology* **32**: 115–29.

Plumwood, V. 1993. *Feminism and the Mastery of Nature*. London: Routledge.

Polimeni, J. and Reiss, J. 2003. Evolutionary perspectives on schizophrenia. *Canadian Journal of Psychiatry* **48**: 34–9.

Pollan, M. 2007. *The Omnivore's Dilemma: A Natural History of Four Meals*. New York: Harper.

Pollard, K. 2009. What makes us human? *Scientific American* **300**: 44–9.

Posner, L. 2009. Introduction: pain and distress in fish. A review of the evidence. *Institute for Laboratory Animal Research Journal* **50**(4): 327–8.

Potter, W. 2011. *Green is the New Red: The Journey from Activist to Eco-terrorist*. San Francisco: City Lights Books.

Potts, A. 2012. *Chicken*. London: Reaktion Books.

Potts, R. 2003. Early human predation. In Kelley, P. *et al.* (eds), *Predator–Prey Interactions in the Fossil Record*. New York: Springer, pp. 359–76.

Poulton, R. and Menzies, R. 2002. Non-associative fear acquisition: a review of the evidence from retrospective and longitudinal research. *Behavioral Research and Therapy* **40**: 127–49.

Pound, P., Ebrahim, S., Sandercock, P., Bracken, M. and Roberts, I. 2004. Where is the evidence that animal research benefits humans? *BMJ* **328**: 514–17.

Povinelli, D. J. and Vonk, J. 2004. We don't need a microscope to explore the chimpanzee's mind. *Mind and Language* **19**: 1–28.

Pratt, S. 1989. Recruitment and other communication behavior in the ponerine ant *Ectatomma ruidum*. *Ethology* **81**: 313–31.

Premack, D. and Woodruff, G. 1978. Does the chimpanzee have a theory of mind? *Behavioral and Brain Sciences* **1**(4): 515–26.

Preston, S. and de Waal, F. 2002. Empathy: its ultimate and proximate bases. *Behavioral and Brain Sciences* **25**(1): 1–71.

Pruetz, J. and McGrew, W. 2001. What does a chimpanzee need? Using natural behavior to guide the care of captive populations. In Brent, L. (ed.), *The Care and Management of Captive Chimpanzees*. American Society of Primatologists, pp. 16–37.

Rachels, J. 1990. *Created from Animals: The Moral Implications of Darwinism*. Oxford: Oxford University Press.

Rachman, S. 2002. Fears born and bred: non-associative fear acquisition? *Behavioral Research and Therapy* **40**: 121–6.

Radhakrishna, S., Huffman, M. and Sinha, A. (eds.). 2012. *The Macaque Connection: Cooperation and Conflict between Humans and Macaques*. Developments in Primatology: Progress and Prospects 43. New York: Springer Science and Business Media.

Raffles, H. 2010. *The Illustrated Insectopedia: Insect Love from A–Z*. New York: Pantheon/ Vintage.

Raha, R. 2006. Animal liberation: an interview with Professor Peter Singer. *The Vegan*, Autumn, 18–19.

Raihani, N. and Ridley, A. 2008. Experimental evidence for teaching in wild pied babblers. *Animal Behaviour* **75**: 3–11.

Raihani, N., Grutter, A. and Bshary, R. 2010. Punishers benefit from third-party punishment in fish. *Science* **327**(5962): 171.

Rakoczy, H., Warneken, F. and Tomasello, M. 2008. The sources of normativity: young children's awareness of the normative structure of games. *Developmental Psychology* **44**: 875–81.

Ramachandran, V. 2011. *The Tell-tale Brain: A Neuroscientist's Quest for What Makes us Human*. New York: WW Norton & Company.

Ramsey W. 2007. *Representation Reconsidered*. Cambridge: Cambridge University Press.

Reeves, R., Smith, B., Crespo, E. and Notarbartolo di Sciara, G. 2003. Dolphins, whales and porpoises: 2002–2010. *Conservation Action Plan for the World's Cetaceans*. IUCN, Gland, Switzerland and Cambridge, UK: IUCN/SSC Cetacean Specialist Group.

Regan, T. 2001. The case for animal rights. In Cohen, C. and Regan, T. (eds.), *The Animal Rights Debate*. Lanham, MD: Rowman & Littlefield, pp. 125–222.

Reiss, D. and Marino, L. 2001. Self-recognition in the bottlenose dolphin: a case of cognitive convergence. *Proceedings of the National Academy of Sciences USA* **98**: 5937–42.

Reznikova, Z. 2007. *Animal Intelligence: From Individual to Social Cognition*. Cambridge: Cambridge University Press.

Rhoads, D. and Goldsworthy, R. 1979. The effects of zoo environments on public attitudes toward endangered wildlife. *International Journal of Environmental Studies* **13**: 238–87.

Rifkin, J. 2009. *The Empathic Generation: The Race to Global Consciousness in a World in Crisis*. New York: Jeremy P. Tarcher/Penguin.

Riley, E. 2005. Ethnoprimatology of *Macaca tonkeana*: the interface of primate ecology, human ecology, and conservation in Lore Lindu National Park, Sulawesi, Indonesia. PhD dissertation, University of Georgia.

Riley, E. 2006. Ethnoprimatology: toward reconciliation between biological and cultural anthropology. *Ecological and Environmental Anthropology* **2**(2): 75–86.

Riley, E. 2007. The human–macaque interface: conservation implications of current and future overlap and conflict in Lore Lindu National park, Sulawesi, Indonesia. *American Anthropologist* **109**(3): 473–84.

Riley, E. 2010. The importance of human–macaque folklore for conservation in Lore Lindu National Park, Sulawesi, Indonesia. *Oryx* **44**(2): 235–40.

Riley, E. and Fuentes, A. 2011. Conserving social-ecological systems in Indonesia: human–nonhuman primate interconnections in Bali and Sulawesi. *American Journal of Primatology* **73**: 62–74.

Ritter, F. 2007. Behavioral responses of rough-toothed dolphins to a dead newborn calf. *Marine Mammal Science* **23**: 429–33.

Ritvo, H. 1995. Possessing mother nature: genetic capital in eighteenth-century Britain. In Brewer, J. and Staves, S. (eds.), *Early Modern Perceptions of Property*. London: Routledge, pp. 413–26.

Rizzolatti, G. and Sinigaglia, C. 2006. *Mirrors in the Brain: How our Minds Share Actions and Emotions*. Oxford, Oxford University Press.

Rizzolatti, G. and Sinigaglia, C. 2010. The functional role of the parieto-frontal mirror circuit: interpretations and misinterpretations. *Nature Reviews Neuroscience* **11**(16): 264–74.

Robbins, J. 1987. *Diet for a New America*. Walpole, NH: Stillpoint.

Robbins, M. and Boesch, C. (eds.), 2011. *Among African Apes*. Berkeley: University of California Press.

Roberts, P. 2009. Spoiled: organic and local is so 2008. *Mother Jones* (March/April) www.motherjones.com/environment/2009/02/spoiled-organic-and-local-so-2008?page=1 [Accessed March 26, 2013].

Rodman, P. 1999. Whither primatology? The place of primates in contemporary anthropology. *Annual Review of Anthropology* **28**: 311–39.

Rogers, M. E., Voysey, B. C., McDonald, K. E., Parnell, R. J. and. Tutin C. E. G. 1998. Lowland gorillas and seed dispersal: the importance of nest sites. *American Journal of Primatology* **45**: 45–68.

Rollin, B. E 1989. *The Unheeded Cry: Animal Consciousness, Animal Pain, and Science*. New York: Oxford University Press.

Rollin, B. E. 1992. *Animal Rights and Human Morality*, 2nd edn. Buffalo, NY: Prometheus.

Rollin, B. E 2006. *Science and Ethics*. Cambridge: Cambridge University Press.

Rorty, R. 1991. *Objectivity, Relativism, and Truth: Philosophical Papers, Volume 1*. Cambridge: Cambridge University Press, p. 13.

Rose, A. 2011 Bonding, biophilia, biosynergy, and the future of primates in the wild. *American Journal of Primatology* **73**: 245–52.

Rose, D. 2009. Introduction: writing in the Anthropocene. *Australian Humanities Review* **49**: 87.

Rose, D. and van Dooren, T. 2011. Unloved others: death of the disregarded in the time of extinctions. *Australian Humanities Review* **50**.

Rosenthal, D. 2005. *Consciousness and Mind*. New York: Oxford University Press.

Ross, S., Vreeman, V. and Lonsdorf, E. 2011. Specific image characteristics influence attitudes about chimpanzee conservation and use as pets. *PLoS ONE* **6**: e22050.

Rozin, P. 1999. The process of moralization. *Psychological Science* **10**(3): 218–21.

Rozin, P., Markwith, M. and Stoess, C. 1997. Moralization and becoming a vegetarian: the transformation of preferences into values and the recruitment of disgust. *Psychological Science* **8**: 67–73.

Ruffman T. and Keenan, T. R. 1996. The belief-based emotion of surprise: the case for a lag in understanding relative to false belief. *Developmental Psychology* **32**: 40–9.

Ryder, R. 1992. An autobiography. *Between the Species* (Summer): 168–73.

Sahley, C. L. and Ready, D. F. 1988. Associative learning modifies two behaviors in the leech, *Hirudo medicinalis*. *Journal of Neuroscience* **8**: 4612–20.

Sahril. 2009. *Terdakwa mengaku tak tahu batas TNLL* (The defendants did not know the National Park boundaries). *Media Alkhairaat*.

Sampson, W. A. 1986. Desegregation and racial tolerance in academia. *Journal of Negro Education* **55**(2): 171–84.

Sanbonmatsu, J. (ed.) 2011. *Critical Theory and Animal Liberation*. Lanham: Rowman & Littlefield.

Sapontzis, S. (ed.) 2004. *Food for Thought: The Debate Over Eating Meat*. New York: Prometheus Press.

Savage-Rumbaugh, S., Wamba, K., Wamba, P., and Wamba, N. 2007. Welfare of apes in captive environments: comments on, and by, a specific group of apes. *Journal of Applied Animal Welfare Science* **10**(1): 7–19.

Saxe, R. 2009a. The neural evidence for simulation is weaker than I think you think it is. *Philosophical Studies* **144**(3): 447–56.

Saxe, R. 2009b. Theory of mind (neural basis). In Banks, W. P. (ed.) *Encyclopedia of Consciousness*. Amsterdam: Elsevier and Academic Press, pp. 401–9.

Saxe, R. and Wexler, A. 2005. Making sense of another mind: the role of the right temporoparietal junction. *Neuropsychologia* **43**(10): 1391–9.

Saxe, R., Whitfield-Gabrieli, S., Scholz, J. and Pelphrey, K. 2009. Brain regions for perceiving and reasoning about other people in school-aged children. *Child Development* **80**(4): 1197–209.

Schefold, R. 1988. *Lia: Das grosse Ritual auf den Mentawai-Inseln (Indonesien)*. Berlin: Dietrich Reimer.

Schefold, R. 2002. Visions of the wilderness on Siberut in a comparative Southeast Asian perspective. In Benjamin, G. and Chou, C. (eds), *Tribal Communities in the Malay World: Historical, Cultural and Social Perspectives*. Singapore: International Institute for Asian Studies, pp. 422–38.

Schefold, R. 2012. *Wees goed voor je ziel. Mijn jaren bij de Sakuddei*. Amsterdam: Nieuw Amsterdam.

Schlosser, E. 1998. The prison-industrial complex. *The Atlantic Monthly*, www.theatlantic.com/magazine/archive/1998/12/the-prison-industrial-complex/4669 [Accessed March 26, 2013].

Schlosser, E. 2005. *Fast Food Nation: The Dark Side of the All-American Meal*. New York: Harper.

Schmiegelow, F., Cumming, S., Harrison, S. *et al.* 2006. *Conservation Beyond Crisis Management: A Conservation-Matrix Model*. Canadian BEACONs project discussion paper No. 1. Edmonton: University of Alberta.

Schmitt, D. 2003. Insights into the evolution of human bipedalism from experimental studies of humans and other primates. *Journal of Experimental Biology* **206**: 1437–48.

Schultz, W. 2011. Conservation means behavior. *Conservation Biology* **25**: 1080–3.

Schweithelm, J., Wirawan, N., Elliott, J. and Asmeen, K. 1992. Sulawesi Parks program land use and socio-economic survey Lore Lindu National Park and Morowali Nature Reserve. Report for Sulawesi Parks program, a colloborative program between the directorate general of forest protection and nature conservation (PHPA), ministry of forestry RI and the Nature Conservancy (TNC).

Scott, J. 1988. Deconstructing equality-versus-difference: or, the uses of poststructuralist theory for feminism. *Feminist Studies* **14**(1): 33–50.

Searle, J. 1983. *Intentionality: An Essay in the Philosophy of Mind*. Cambridge: Cambridge University Press.

SeaWorld. 2011. Beluga whales: http://seaworld.org/animal-info/info-books/beluga/index.htm; Bottlenose dolphins: http://seaworld.org/animal-info/info-books/bottlenose/index.htm; Killer whales: http://seaworld.org/animal-info/info-books/killer-whale/index.htm [Accessed March 26, 2013].

Seyfarth, R. and Cheney, D. 2012. The evolutionary origin of friendship. *Annual Review of Psychology* **63**: 153–77.

Shapin, S. and Schaffer, S. 1985. *Leviathan and the Air-pump: Hobbes, Boyle, and the Experimental Life*. Princeton: Princeton University Press.

Sharot, T. 2011. *The Optimism Bias*. Pantheon Books, New York. www.time.com/time/health/article/0,8599,2074067,00.html [Accessed March 26, 2013].

Shermer, M. 2011. *The Believing Brain*. New York: Times Books, Henry Holt and Company.

Shir-Vertesh, D. 2012. Flexible personhood: loving animals as family members in Israel. *American Anthropologist* **114**(3): 420–32.

Shove, E., Pantzar, M. and Watson, M. 2012. *The Dynamics of Social Practice: Everyday Life and How it Changes*. London: Sage.

Shukin, N. 2009. *Animal Capital: Rendering Life in Biopolitical Times*. Minneapolis: University of Minnesota Press.

Shumaker, R. W., Walkup, K. R. and Beck, B. B. 2011. *Animal Tool Behavior: The Use and Manufacture of Tools by Animals*, revised edn. Baltimore: Johns Hopkins University Press.

Shuster, E. 1997. Fifty years later: the significance of the Nuremberg code. *New England Journal of Medicine* **337**: 1436–40.

Siddique, A. 2011. Radio collar can save Bengal tiger. *Daily Sun* (10 April). www.daily-sun.com/details_yes_10-04-2011_Radio-collar-can-save–Bengal-Tigers:-Experts_185_1_2_1_7.html [Accessed March 26, 2013].

Sierra, R. 1999. Traditional resource-use systems and tropical deforestation in a multi-ethnic region in North-west Ecuador. *Environmental Conservation* **26**(2): 136–45.

Silk, J. 2007. Social component of fitness in primate groups. *Science* **317**: 1347–51.

Singer, P. 1975. *Animal Liberation*. New York: Random House.

Singer, P. 1990. *Animal Liberation*, 2nd edn. New York: New York Review of Books.

Singer, P. 1993. *Practical Ethics*, 2nd edn. Cambridge: Cambridge University Press.

Singer, P. 1997. The drowning child and the expanding circle. *New Internationalist*, April.

Singer, P. 2001–2002. *Animal Equality: Language and Liberation* by Joan Dunayer (book review). *Vegan Voice* (December–February): 36.

Singer, P. 2003. Animal liberation at 30. *New York Review of Books*, May 15, pp. 23–6.

Singer, P. and Mason, J. 2006. *The Way We Eat: Why Our Food Choices Matter*. Emmaus, PA: Rodale.

Singleton, I., Wich, S. A., Husson, S. *et al.* 2004. *Orangutan Population and Habitat Viability Assessment: Final Report*. Apple Valley, MN, USA: IUCN / SSC Conservation Breeding Specialist Group.

Sinigaglia, C. 2009. Mirror in action. *Journal of Consciousness Studies* **16**(6–8): 309–34.

Skinner, K. 2010. *The City Chicken*. http://thecitychicken.com/hhotm.html [Accessed March 26, 2013].

Sleigh, C. 2007. *Six Legs Better: A Cultural History of Myrmecology*. Baltimore: Johns Hopkins University Press.

Small, D. and Loewenstein, G. 2003. Helping a victim or helping the victim: altruism and identifiability. *Journal of Risk and Uncertainty* **26**: 5–16.

Small, D., Loewenstein, G. and Slovic, P. 2007. Sympathy and callousness: the impact of deliberative thought on donations to identifiable and statistical victims. *Organizational Behavior and Human Decision Processes* **102**: 143–53.

Small, R. and DeMaster, D. 1995. Survival of five species of captive marine mammals. *Marine Mammal Science* **11**(2): 209–26.

Smith, C. M. 2008. Foucauldian technologies of the self, participatory cultural production and new media. *NMEDIAC: Journal of New Media and Culture* **5**(1) www.ibiblio.org/nmediac/summer2008/techself.html [Accessed April 1, 2013].

Smith, D. 2011. *Less than Human: Why we Demean, Enslave, and Exterminate Others*. New York: St. Martin's Press.

Smith, J., Schull, J., Strote, J. *et al.* 1995. The uncertain response in the bottlenosed dolphin (*Tursiops truncatus*). *Journal of Experimental Psychology: General* **124**: 391–408.

Smith, J., Shields, W., Schull, J. and Washburn, D. 1997. The uncertain response in humans and animals. *Cognition* **62**: 75–97.

Smith, L. 1949. *Killers of the Dream*. New York: W.W. Norton & Co.

Smulewicz-Zucker, G. (ed.), *Strangers to Nature: Animal Lives and Human Ethics*. Lanham: Lexington Press.

Smuts, B. 1999. Reflections. In Gutmann, A. (ed.), *The Lives of Animals*. Princeton: Princeton University Press, pp. 107–20.

Sneddon, L. U. 2003. The evidence for pain in fish: the use of morphine as an analgesic. *Applied Animal Behaviour Science* **83**: 153–62.

Sneddon, L. U. 2009. Pain perception in fish: indicators and endpoints. *Institute for Laboratory Animal Research Journal* **50**(4): 338–42.

Socha, K. 2012. *Women, Destruction and the Avant-Garde: A Paradigm for Animal Liberation*. New York: Rodopi.

Southgate, V., Senju, A. and Csibra, G. 2007. Action anticipation through attribution of false belief by 2-year-olds. *Psychological Science*, **18**: 587–92.

Specter, M. 2009. *Denialism: How Irrational Thinking Hinders Scientific Progress, Harms the Planet, and Threatens our Lives*. New York: Penguin Press.

Sponsel, L. 1997. The human niche in Amazonia: explorations in ethnoprimatology. In Kinnzey, W. (ed.), *New World Primates: Ecology, Evolution, and Behavior*. New York: Aldine Gruyter, pp. 143–65.

Squier, S. 2011. *Poultry Science, Chicken Culture: A Partial Alphabet*. New Brunswick, NJ: Rutgers University Press.

Stamos, D. 2003. *The Species Problem: Biological Species, Ontology, and the Metaphysics of Biology*. Lanham, MD: Lexington Books.

Steichen, E. 1955. *The Family of Man: The Greatest Photographic Exhibition of All Time, 503 Pictures from 68 Countries*, New York: Museum of Modern Art.

Steinbock, B. 1978. Speciesism and the idea of equality. *Philosophy* **53**.

Stengers, I. 2005. The cosmopolitical proposal. In Latour, B. and Weibel, P. (eds.), *Making Things Public: Atmospheres of Democracy*. Cambridge, MA: MIT Press, pp. 994–1003.

Stengers, I. 2011. *Cosmopolitics II*. Minneapolis: University of Minnesota Press.

Sterelny, K. 2012. *The Evolved Apprentice: How Evolution Made Humans Unique*. Cambridge: MIT Press.

Stewart, K. and Marino, L. 2009. Dolphin–human interaction programs: policies, problems, and practical alternatives. *Policy Paper for Animals and Society Institute*, **39**.

Stocking, G. W. (ed.). 1988. *Bones, Bodies, Behavior: Essays in Biological Anthropology.* Madison, WI: University of Wisconsin Press.

Stoczkowski, W. 2009. UNESCO's doctrine of human diversity. *Anthropology Today* **25**: 7–11.

Stolzenburg, W. 2011. *Rat Island: Predators in Paradise– And the World's Greatest Wildlife Rescue.* New York: Bloomsbury.

Strain, E. 1996/1997. Stereoscopic visions: touring the Panama Canal. *Visual Anthropology Review* **12**(2): 44–58.

Strathern, M. 1996. Cutting the network. *The Journal of the Royal Anthropological Institute* **2**(3): 517–35.

Subiaul, F., Vonk, J., Okamoto-Barth, O. and Barth, J. 2008. Chimpanzees learn the reputation of strangers by observation. *Animal Cognition* **11**(4): 611–23.

Suddendorf, T. and Whiten, A. 2003. Reinterpreting the mentality of apes. In Sterelny, K. and Fitness, J. (eds.), *From Mating to Mentality: Evaluating Evolutionary Psychology.* New York: Psychology Press, pp. 173–96.

Sugiyama, Y., Kurita, H., Matsu, T., Kimoto, S. and Shimomura, T. 2009. Carrying of dead infants by Japanese macaque (*Macaca fuscata*) mothers. *Anthropological Science* **117**: 113–19.

Sumner, W. 2002. *Folkways: The Study of Mores, Manners, Customs and Morals.* Mineola, NY: Dover.

Sundarbans Tiger Project. 2011. http://sundarbantigerproject.info/news.php?readmore=77 [Accessed March 26, 2013].

Supriatna, J., Froehlich, J., Erwin, J. and Southwick, C. 1992. Population, habitat and conservation status of *Macaca maurus, Macaca tonkeana* and their putative hybrids. *Tropical Biodiversity* **1**(1): 31–48.

Surian, L., Caldi, S. and Sperber, D. 2007. Attribution of beliefs by 13-month-old infants. *Psychological Science* **18**: 580–6.

Sussman, R. and Cloninger, R. (eds.). 2011. *Origins of Altruism and Cooperation. Developments in Primatology: Progress and Prospects Volume 36* Part 2. Springer.

Swaisgood, R. and Sheppard, J. 2011. Reconnecting people to nature is a prerequisite for the future conservation agenda. *BioScience* **61**(2): 94–5.

Swanagan, J. 2000. Factors influencing zoo visitors' conservation attitudes and behavior. *Journal of Environmental Education* **31**: 26–31.

Taylor, Ch. 1994. The politics of recognition. In Gutmann, A. (ed.), *Multiculturalism: Examining the Politics of Recognition.* Princeton: Princeton University Press, pp. 25–73.

TEEB. 2010. The economics of ecosystems and biodiversity: mainstreaming the economics of nature – synthesis of the approach, conclusions and recommendations of TEEB. www.teebweb.org [Accessed March 26, 2013].

Teleki, G. 1973. Group response to the accidental death of a chimpanzee in Gombe National Park, Tanzania. *Folia Primatologica*, **20**: 81–94.

Thomas, C. D., Cameron, A., Green, R. E. *et al.* (2004). Biodiversity conservation: uncertainty in predictions of extinction risk/effects of changes in climate and land use/climate change and extinction risk (reply). *Nature*, **430**(6995) (July 1, 2004). doi:10.1038/nature02719.

Thompson, E. 2007. *Mind in Life: Biology, Phenomenology, and the Sciences of Mind.* Cambridge, MA: Harvard University Press.

Thomson, J. 1971. A defense of abortion. *Philosophy and Public Affairs* **1**: 47–66.

Thornton, A. and McAuliffe, K. 2006. Teaching in wild meerkats. *Science* **313**: 227–9.

Tinbergen, N. 1963. On aims and methods of ethology. *Zeitschrift fuer Tierpsychologie* **20**: 410–433.

Tomasello, M. 2008. *Origins of Human Communication*. Cambridge: MIT Press.

Tomasello, M. and Carpenter, M. 2005. The emergence of social cognition in three young chimpanzees. *Monographs of the Society for Research in Child Development* **70**(1): 1–131.

Tomasello, M. and Herrmann, E. 2010. Ape and human cognition: what's the difference? *Current Directions in Psychological Science* **19**: 3–8.

Tomasello, M., Davis-Dasilva, M., Camak, L. and Bard, K. 1987. Observational learning of tool-use by young chimpanzees. *Human Evolution* **2**(2): 175–83.

Tomasello, M., Call, J. and Hare, B. 2003. Chimpanzees understand psychological states: the question is which ones and to what extent. *Trends in Cognitive Sciences* **7**(4): 153–6.

Tomasello, M., Carpenter, M., Call, J., Behne, T. and Moll, H. 2005. Understanding and sharing intentions: the origins of cultural cognition. *Behavioral and Brain Sciences* **28**: 675–91.

Tomlinson, I. 2011. Doubling food production to feed the 9 billion: a critical perspective on a key discourse of food security in the UK. *Journal of Rural Studies* **29**: 81–90.

Tooley, M. 1972. Abortion and infanticide. *Philosophy and Public Affairs* **2**: 37–65.

Treaty of Lisbon. n.d. http://ec.europa.eu/food/animal/welfare/policy/index_en.htm [Accessed March 16, 2013].

Turner, W. R., Brandon, K., Brooks, T. M. *et al.* 2011. Global biodiversity conservation and the alleviation of poverty. *BioScience* **62**(1): 85–92.

Twine, R. 2007. Searching for the win-win? Animals, genomics and welfare. *International Journal of Sociology of Agriculture and Food* **16**: 1–18.

Twine, R. 2010a. *Animals as Biotechnology: Ethics, Sustainability and Critical Animal Studies*. London: Earthscan.

Twine, R. 2010b. Intersectional disgust? Animals and (eco)feminism. *Feminism and Psychology* **20**: 397–406.

Uller, C. 2004. Disposition to recognize goals in infant chimpanzees. *Animal Cognition* **7**(3): 154–61.

UNEP/CBD/COP. 2010.The strategic plan for biodiversity 2011–2020 and the Aichi biodiversity targets.

UNESCO. 1950. *Statement on Race*. Paris: United Nations Educational, Scientific and Cultural Organisation.

United Nations. 1952. *Universal Declaration of Human Rights: Final Authorized Text*. New York: United Nations.

United Nations Development Programme. 2011. *Human Development Report 2011. Sustainability and Equity: a Better Future for All*. United Nations.

United Poultry Concerns. 2009. Coalition of animal sanctuaries' draft statement on urban chicken-keeping. www.upc-online.org/backyard/backyard_poultry.html [Accessed March 26, 2013].

United States Census Bureau. 2010. Quick facts: Albion, Michigan. http://quickfacts.census.gov/qfd/states/26/2600980.html [Accessed March 26, 2013].

U.S. Marine Mammal Inventory Report. 2010. *National Marine Fisheries Service*. Office of Protected Resources.

Van der Voo, L. 2012. New homes beckon for old chickens, outside the pot. *New York Times* (April 25), www.nytimes.com/2012/04/26/us/new-homes-beckon-for-city-chickens-in-retirement.html?_r=1 [Accessed March 26, 2013].

Van Schaik, C. 2004. *Among Orangutans: Red Apes and the Rise of Human Culture*. Belknap Press.

Varela, F., Thompson, E. and Rosch, E. 1991. *The Embodied Mind: Cognitive Science and Human Experience.* Cambridge, MA: MIT Press.

Vaughn, A. 1995. *Roots of AMERICAN Racism: Essays on the Colonial Experiment.* Oxford: Oxford University Press.

Velleman, J. 2000. *The Possibility of Practical Reason.* Oxford: Oxford University Press.

Verbeek, P. 2008. Peace ethology. *Behaviour* **145**: 1497–524.

Verheijen, F. and Flight, W. 1997. Decapitation and brining: experimental tests show that after these commercial methods for slaughtering eel, *Anguilla anguilla* (L.), death is not instantaneous. *Aquaculture Research* **28**: 361–6.

Vining, J. 2003. The connection to other animals and caring for nature. *Human Ecology Review* **10**: 87–99.

Voci, A. and Hewstone, M. 2003. Intergroup contact and prejudice toward immigrants in Italy: the mediational role of anxiety and the moderational role of group salience. *Group Processes & Intergroup Relations* **6**: 37–54.

Voysey, B. C., McDonald, K. E., Rogers, M. E., Tutin, C. E. G. and Parnell, R. J. 1999. Gorillas and seed dispersal in the Lopé Reserve, Gabon. *Journal of Tropical Ecology* **15**: 39–60.

Waddell, S. and Quinn, W. G. 2001. Flies, genes, and learning. *Annual Review of Neuroscience* **24**: 1283–309.

Walker, A. 2011. *The Chicken Chronicles.* New York: New Press.

Walker, A., Painemilla, K., Rylands, A. B., Woofter, A. and Hughes, C. (eds.). 2010. *Indigenous Peoples and Conservation: From Rights to Resource Management.* Arlington, VA: Conservation International.

Walsh, S., Bramblett, C. and Alford, P. 1982. A vocabulary of abnormal behaviors in restrictively reared chimpanzees. *American Journal of Primatology* **3**: 315–19.

Waples, K. and Gales, N. 2002. Evaluating and minimizing social stress in the care of captive bottlenose dolphins (*Tursiops aduncus*). *Zoo Biology* **21**: 5–26.

Warneken, F. and Tomasello, M. 2006. Altruistic helping in infants and young chimpanzees. *Science* **311**(5765): 1301–3.

Wasser, S. (ed.). 2004. *Evolutionary Theory and Processes. Modern Horizons: Papers in Honour of Eviatar Nevo.* Dordrecht: Kluwer.

Watson, M. 2011. *Towards a Subalternist Cosmopolitics.* ms. presented at the 110th annual meeting of the American Anthropological Association, Montreal, November 17.

Watts, D., Muller, M., Amsler, S., Mbabazi, J. and Mitani, J. 2006. Lethal intergroup aggression by chimpanzees in Kibale National Park, Uganda. *American Journal of Primatology* **68**: 161–80.

Watts, M. 2004. Are hogs like chickens? Enclosure and mechanization in two 'white meat' filières. In Hughes, A. and Reimer, S. (eds.), *Geographies of Commodity Chains.* London: Routledge, pp. 39–42.

Weber, M. 1919. Politics as vocation. http://socialpolicy.ucc.ie/Weber_Politics_as_Vocation.htm [Accessed March 26, 2013].

Weber, M. 1946. Politics as a vocation. In Gerth, H. and Wright Mills, C. (eds.), *From Max Weber: Essays in Sociology.* New York: Oxford University Press, pp. 77–128.

Weis, T. 2007. *The Global Food Economy: The Battle for the Future of Farming.* London: Zed Books.

Weiss, A., King, J. and Perkins, L. 2006. Personality and subjective well-being in orangutans (*Pongo pygmaeus* and *Pongo abelii*). *Journal of Personality and Social Psychology* **90**(3): 501–11.

Weiss, B. 2012. Configuring the authentic value of real food: farm-to-fork, snout-to-tail, and local food movements. *American Ethnologist* **39**(3): 614–26.

Wellman, H., Cross, D. and Watson, J. 2001. Meta-analysis of theory-of-mind development: the truth about false belief. *Child Development* **72**(3): 655–84.

Wells, R. and Scott, M. 1990. Estimating bottlenose dolphin population parameters from individual identification and capture-release techniques. In Hammond, P. *et al.* (eds.), *Reports of the International Whaling Commission, Special Issue* **12**, pp. 407–15.

Whatmore, S. and Thorne, L. 1997. Nourishing networks: alternative geographies of food. In Goodman, D. and Watts, M. (eds.), *Globalising Food: Agrarian Questions and Global Restructuring*. London: Routledge, pp. 287–304.

Whatmore, S. and Thorne, L. 2000. Elephants on the move: spatial formations of wildlife exchange. *Environment and Planning D: Society and Space* **18**: 185–203.

Wheeler, M. E. and Fiske, S. T. 2005. Controlling racial prejudice: social-cognitive goals affect amygdala and stereotype activation. *Psychological Science* **16**: 56–63.

White, L. 1967. The historical roots of our ecologic crisis. *Science* **155**: 1203–12.

White, T. 2007. *In Defense of Dolphins: The New Moral Frontier*. Oxford: Blackwell.

White, T. 2011. What is it like to be a dolphin? In Brakes, P. and Simmonds, M. (eds.), *Whales and Dolphins: Cognition, Culture, Conservation and Human Perceptions*. Washington, DC: Earthscan, pp. 188–206.

Whitehead, H. 2011. The culture of whales and dolphins. In Brakes, P. and Simmonds, M. (eds.), *Whales and Dolphins: Cognition, Culture, Conservation and Human Perceptions*. Washington, DC: Earthscan, pp. 149–65.

Whiten, A. 2000. Primate culture and social learning. *Cognitive Science: A Multidisciplinary Journal* **24**: 477–508.

Wich, S. A., Riswan , Jenson, J., Refisch, J. and Nelleman, C. 2011. *Orangutans and the Economics of Sustainable Forest Management in Sumatra*. UNEP/GRASP/ PanEco/YEL/ ICRAF/GRID-Arendal.

Wich, S., Gaveau, D., Abram, N. *et al.* 2012. Understanding the impacts of land-use policies on a threatened species: is there a future for the Bornean orang-utan?', *PLoS ONE* **7**(11): e49142.

Wicks, D. 2011. Silence and denial in everyday life: the case of animal suffering. *Animals* **1**: 186–99. www.mdpi.com/2076-2615/1/1/186 [Accessed March 26, 2013].

Wigglesworth, V. B. 1980. Do insects feel pain? *Antenna* **4**: 8–9.

Wilcox, S. and Jackson, R. 2002. Jumping spider tricksters. In Bekoff, M. *et al.*, *The Cognitive Animal: Empirical and Theoretical Perspectives on Animal Cognition*. Cambridge, MA: MIT Press, pp. 27–34.

Wilder, D. 1981. Perceiving persons as a group: categorization and intergroup relations. In Hamilton, D. (ed.), *Cognitive Processes in Stereotyping and Intergroup Behavior*. Hillsdale, NJ: Erlbaum, pp. 213–57.

Wildman, D. and Goodman, M. 2004. Humankind's place in a phylogenetic classification of living primates. In Wasser, S. (ed.), *Evolutionary Theory and Processes. Modern Horizons: Papers in Honour of Eviatar Nevo*. Dordrecht: Kluwer, pp. 293–311.

Williams, V. 2001. Captive orcas. Dying to entertain you. A Report for Whale and Dolphin Conservation Society. Chippenham, UK: Whale and Dolphin Conservation Society.

Wilson, E. O. www.brainyquote.com/quotes/keywords/vanish.html [Accessed March 26, 2013].

Wimmer, H. and Perner, J. 1983. Beliefs about beliefs: representation and constraining function of wrong beliefs in young children's understanding of deception. *Cognition* **13**(1): 103–28.

Wise, S. 1998. Hardly a revolution: the eligibility of nonhuman animals for dignity rights in a liberal democracy. *Vermont Law Review* **793**: 823–4.

Wise, S. 2000. *Rattling the Cage: Toward Legal Rights for Animals*. New York: Perseus Books.

Wise, S. 2002. *Drawing the Line: Science and the Case for Animal Rights*. New York: Perseus Books.

Wise, S. 2005. *Though the Heavens may Fall: The Landmark Trial that led to the End of Human Slavery*. Cambridge: Da Capo Press.

Woodley, T., Hannah, J. and Lavigne, D. 1997. A comparison of survival rates for captive and free-ranging killer whales (*Orcinus orca*). *International Marine Mammal Association Inc. Draft Technical Report* no 93–01.

Woolloff, A. 2010. Hair loss in captive chimpanzees: experimental environmental enrichment effects on stress-related behaviours. Unpublished BA dissertation, University of Cambridge.

Wolpert, L. 2001. *Malignant Sadness: The Anatomy of Depression*. London: Faber & Faber.

Wrangham, R. W, Chapman, C. A. and Chapman, L. J. 1994. Seed dispersal by forest chimpanzees in Uganda. *Journal of Tropical Ecology* **10**: 355–68.

Wright, A. A., Cook, R. G. and Rivera, J. J. 1988. Concept learning by pigeons: matching-to-sample with trial-unique video picture stimuli. *Animal Learning and Behavior* **16**: 436–44.

Würbel, H., Stauffacher, M. and von Holst, D. 1996. Stereotypies in laboratory mice: quantitative and qualitative description of the ontogeny of 'wire-gnawing' and 'jumping' in Zur:ICR and Zur:ICR nu. *Ethology* **102**: 371–85.

Wyer, R. and Frey, D. 1983. The effects of feedback about self and others on the recall and judgments of feedback-relevant information. *Journal of Experimental Social Psychology* **19**: 540–59.

Wynne, C. 2004. The perils of anthropomorphism. *Nature* **428**: 606.

Xu, F. and Rhemtulla, M. 2005. In defense of psychological essentialism. In Bara, B. *et al.*, Proceedings of the 27th Annual Conference of the Cognitive Science Society. Mahwah, NJ: Erlbaum. pp. 2377–80.

Yap, P. 2005. Four models of equality. *International and Comparative Law Review* **27**(63): 73–4.

Young, I. 1990. *Justice and the Politics of Difference*. Princeton: Princeton University Press.

Zack, N. 1998. *Thinking about Race*. Belmont, CA: Wadsworth.

Zammit-Lucia, J. 2011. Conservation is not about nature. IUCN. www.iucn.org/involved/opinion/ ?8195/Conservation-is-not-about-nature [Accessed March 26, 2013].

Zerner, Ch. 2011. Stealth nature: biomimesis and the weaponization of life. In Feldman, I and Tictin, M. (eds.), *In the Name of Humanity: the Government of Threat and Care*. Durham, NC: Duke University Press.

Zucker, G. (ed.) 2012. *Strangers to Nature: Animal Lives and Human Ethics*. New York: Lexington Books.

Index